Impact *of* Hazardous Waste *on* Human Health

Hazard, Health Effects, Equity, and Communications Issues

Barry L. Johnson

Agency for Toxic Substances and Disease Registry,
Atlanta, GA

LEWIS PUBLISHERS

Boca Raton London New York Washington, D.C.

Acquiring Editor:	Skip DeWall
Project Editor:	Sylvia Wood
Marketing Managers:	Barbara Glunn / Jane Stark
Cover design:	Jonathan Pennell
Manufacturing Manager:	Carol Slatter

Library of Congress Cataloging-in-Publication Data

Johnson, Barry. (Barry Lee), 1938–
    Impact of hazardous waste on human health : hazard health effects, equity and
communication issues / Barry L. Johnson
        p.   cm.
Includes bibliographical references and index.
ISBN 0-8493-4107-8
    1. Hazardous wastes–Health aspects. 2. Hazardous waste sites—Health aspects
3. Environmental toxicology 4. Hazardouse wastes–risk assessment. I. Title.
    RA579.J64 1999
615.9'02—dc21   98-29720
                                                                                CIP

Disclaimer

The opinions contained in this book represent those of the author and do not represent any stated or implied official position of the Agency for Toxic Substances and Disease Registry, the Public Health Service, or the U.S. Department of Health and Human Services.

Acknowledgments

The author is indebted to many colleagues at the Agency for Toxic Substances and Disease Registry (ATSDR) and the U.S. Environmental Protection Agency (EPA) who provided valuable data, insight, and discussions that helped shape the content of this book. In particular, the author is indebted to Charles Xintaras, Sc.D., for his careful editing and technical corrections to the book's manuscript. The author also gratefully acknowledges the contributions of John Andrews, Jr., M.D., Mark Bashor, Ph.D., Christopher DeRosa, Ph.D., Maureen Lichtveld, M.D., Steve Luftig, Jeffrey Lybarger, M.D., Molla Sarros, Sandra Susten, Ph.D., Timothy Tinker, Dr.P.H., and Robert Williams, D.E.E.

Four persons gave indispensable advice as readers of the draft manuscript. I express my appreciation to Dr. Thomas A. Burke, Co-Director, Risk Sciences and Public Policy Institute, School of Hygiene and Public Health, Johns Hopkins University, Baltimore, Maryland; Ms. Doris Cellarius, Member, Sierra Club Community Health Committee, Olympia, Washington; Ms. Marie Flickinger, Publisher, *South Belt-Ellington Leader*, Community Leader, Brio Superfund Site, Houston, Texas; and Dr. Philip J. Landrigan, Ethel H. Wise Professor of Community Medicine, Mount Sinai School of Medicine, New York, New York.

I am indebted to Jeanne Bucsela, ATSDR, who kindly and patiently edited most of this manuscript. Her editorial contributions substantially improved my work. I also thank Cindy Allen and Laura Northen for preparing the artwork in this book.

In particular, I express my appreciation to Drs. James O. Mason, William Roper, and David Satcher, all of whom are leaders in protecting the health of all Americans, and who served as Administrators of ATSDR.

I acknowledge the direct assistance of many persons in ATSDR, EPA, and state/local health departments who helped me prepare this book. However, any errors of fact or misunderstanding of data are the sole responsibility of the author.

About the Author

Barry L. Johnson is the Assistant Administrator of the Agency for Toxic Substances and Disease Registry (ATSDR), Atlanta, Georgia. ATSDR is one of eight agencies that constitute the Public Health Service. The ATSDR was established by Congress to act on the public health consequences of hazardous waste and uncontrolled releases of toxic substances into the environment. As Assistant Administrator, Dr. Johnson has administrative responsibility for the agency's programs and direction. He holds the rank of Assistant Surgeon General in the Commissioned Corps of the Public Health Service. Prior to assuming his current position in 1986, he was Director of the Division of Biomedical and Behavioral Science at the National Institute for Occupational Safety and Health (NIOSH). His NIOSH responsibilities included administrative responsibility for research programs in the areas of toxicology, ergonomics, physical agents, and occupational stress.

Dr. Johnson's professional contributions have been in the areas of neurotoxicology, ergonomics, environmental toxicology, and environmental policy. He is, or has served as, a member of the editorial boards of *Archives of Environmental Health, NeuroToxicology, Toxicology and Industrial Health, Contaminated Soils, Risk Analysis, International Journal of Occupational and Environmental Health,* and *Journal of Clean Technology and Environmental Sciences.* He is a consultant to the World Health Organization in the areas of neurotoxicology and environmental science. He received his Ph.D. in biomedical engineering from Iowa State University.

He is the senior or coeditor of the following books: *Behavioral Toxicology; The Prevention of Neurotoxic Illnesses in Working Populations; Advances in Neurobehavioral Methods in Environmental and Occupational Health; Minority Health: Focus on Environmental Contamination; Hazardous Waste and Public Health; Hazardous Waste: Impacts on Human and Ecological Health; International Symposium on Neurobehavioral Methods and Effects in Occupational and Environmental Health;* and *Hazardous Waste: Toxicology and Health Effects.*

Preface

With apology to the work of Charles Dickens, the Comprehensive Environmental Response, Compensation, and Liability Act (CERCLA) is the worst of laws; it is the best of laws. Within these two extremes lies a predictably vast span of opinion about CERCLA's importance. Much less debatable, however, is the proposition that CERCLA is the single most controversial federal environmental statute. Why is this? And where are the "worst" and "best" of this much maligned statute, also known as the Superfund law?

As a brief background, the U.S. Congress enacted CERCLA in 1980, and President Carter signed it into law. Predicate events that shaped the statute included widespread media attention about abandoned hazardous waste dumps leaking into homes in Love Canal, New York, and the discovery of thousands of leaking, buried drums of chemicals in Kentucky, the so-called "Valley of the Drums." The public expressed great concern that chemicals released from such sites would cause human health problems. By enacting CERCLA, Congress acted on this concern to protect human health and prevent further damage to natural resources and environmental quality.

The news media quickly named the new statute "Superfund" because the law established a multibillion dollar trust fund. The trust fund was created by a tax on segments of private industry. Other revenue was to be generated by granting broad authority to the federal government to identify parties responsible for the waste in a particular waste site and to require these "potentially responsible parties" to pay for cleanup costs. The issue of who pays and in what amounts quickly forged an outpouring of litigation from parties great and small. Costs of litigation and the statute's mandate to assess blame further fueled discontent and confrontation between government and the private sector. All this, in turn, has led to numerous criticisms of the statute and occasionally some praise.

An observer looking for the "worst" of CERCLA has ready access to a mountain of material. Consider only the congressional debates that attend periodic attempts to reauthorize CERCLA. Criticisms abound from disparate sources. For example, some private industry sources assert that costs for remediating uncontrolled hazardous waste sites[1] (i.e., CERCLA sites) are too great, are unfairly parsed among responsible parties, and are incommensurate with the health risks posed by the sites. On the other hand, some grassroots community groups voice concerns that sites take too long to be remediated, economic impacts of diminished property values are unaddressed, and adverse health effects in CERCLA communities are inadequately examined. Media crit-

---

[1] Uncontrolled hazardous waste sites are distinguished in federal law from operating, permitted, hazardous waste management facilities. CERCLA covers the former; the Resource Conservation and Recovery Act covers the latter.

ics also abound, noting that, in their reckoning, too few CERCLA sites have been struck from the National Priorities List. Many of these criticisms have merit, regardless of their source, but as with any criticism, there is usually much more to the story.

Many would aver that an observer looking for the "best" in CERCLA will find significant outcomes and benefits from the statute and its attendant programs. Several hundred hazardous waste sites have been remediated, emergency responding has been enhanced through community preparedness and response actions, a large number of public health actions have occurred in communities near waste sites, and toxicologic and other scientific databases have been advanced. National environmental organizations have been active supporters of CERCLA because it fits into their agenda for reducing the impact of environmental hazards, in general, and the effects of toxic substances, in particular. And some grassroots environmental organizations have actively supported CERCLA because it is the only supply of resources and authorities that deal directly with uncontrolled hazardous waste sites. Also, over time, some private industry organizations and state and municipal governments have voiced support of CERCLA's intent of environmental remediation and protection of human and ecologic health.

As a person involved in the public health aspects of CERCLA since 1986, I believe much good has been accomplished, although not without errors and poor judgments in both the law's content and its administration. Clearly CERCLA has brought the problem of hazardous waste generation and management onto the agenda of environmental and public health agencies. Absent CERCLA, very little attention would be paid to the problem of abandoned hazardous waste in the environment and what to do about it. This is because many other environmental priorities are perceived by some as having higher priority. Furthermore, CERCLA has made substantial contributions to a better understanding of the effects hazardous substances have on humans and ecosystems. Similarly, knowledge about how to clean up hazardous wastes would still be marginal without the statute.

Given the divergence of opinion about the worth of CERCLA, it would be impossible in one work to address all the points of contention. It is possible, however, in a single volume to address one point of contention: To what extent do hazardous wastes deleteriously impact human health?

What then is the impact of hazardous waste and waste sites on human health? Again opinion is divergent. Statements from some members of Congress, buttressed by opinions and data from some risk assessors, conclude that CERCLA sites have a small impact. On the other end of the scale, other members of Congress and some grassroots environmentalists assert that major health problems are attributable to CERCLA sites. Where lies the truth? Moreover, is the question of human health impact really relevant? Can't sites simply be cleaned up as resources permit and won't that be enough to assuage health concerns?

This work will attempt to provide a data-based response to the foregoing questions. Many of the health data described in this book have been collected or commissioned by the Agency for Toxic Substances and Disease Registry (ATSDR), the federal public health agency established in 1980 by the CERCLA statute. Also, a substantial amount of environmental contamination data is presented, largely collected by the U.S. Environmental Protection Agency (EPA). This book is intended as a successor to earlier scholarly works by persons concerned about the human health implications of mismanaged hazardous waste. In particular, I refer to the following works as key antecedents to this book: J.B. Andelman and D.W. Underhill, *Health Effects from Hazardous Waste Sites*; J.

Grisham, Ed., *Health Aspects of the Disposal of Waste Chemicals*; and National Research Council (NRC), *Environmental Epidemiology: Public Health and Hazardous Wastes, Vol. 1.*

In many ways it is unfortunate that the CERCLA statute, because of its controversial nature, has become a program focused almost exclusively on site cleanups and litigation. What gets lost in this kind of calculus is the considerable body of improved scientific data, improved risk assessment methods, and positive impacts on community health that have resulted over the years of the CERCLA statute. When funds are redirected away from science and human health programs into litigation and other transactional costs, the public will ultimately be the poorer for this *de facto* de-emphasis of health issues. Stagnant environmental and toxicologic science and less clear impacts on human health will be the result.

The chapters that follow will provide up-to-date depictions of the human health impacts of hazardous waste and attendant public health responses. I hope this work will be useful to policy makers, environmentalists, toxicologists, public health officials, academicians, health care providers, and community-based organizations. That was my goal.

The Book's Structure and Themes

This book describes the association between hazardous waste and human health and the role of public health programs in addressing this association. As stated in the Preface, this is intended for use by policy makers, environmentalists, toxicologists, public health officials, academic personnel, and health care providers. Because this is a broad and diverse audience it is important to make clear certain key definitions of terms used throughout the book. One key definition needed is "public health," a term most Americans have a sense of, but probably cannot define. Indeed, it may be surprising that no single, operative definition is used by all public health agencies. The following are cited as illustrations of the diverse definitions of public health.

Definition of Public Health

The Institute of Medicine is a component of the National Academy of Sciences. The U.S. Congress created the academy in the middle of the nineteenth century to advise federal government agencies on matters of science. The Institute of Medicine (IOM) focuses on medical science and practice. In 1988 the IOM constituted a Committee for the Study of the Future of Public Health, which reviewed the current status and future of public health in the United States and produced the report, *The Future of Public Health* (IOM, 1988). As part of their work, the IOM's committee developed a definition of public health, dividing it into three parts: mission, substance, and organizational framework:

> The committee defines the *mission* of public health as: the fulfillment of society's interest in assuring the conditions in which people can be healthy.
> The committee defines the *substance* of public health as: organized community efforts aimed at the prevention of disease and promotion of health. It links many disciplines and rests upon the scientific core of epidemiology.
> In summary, the committee defines the *organizational framework* of public health to encompass both activities undertaken within the formal structure of government and the associated efforts of private and voluntary organizations and individuals.

The IOM's committee therefore chose to define public health in terms of its perceived components. This was purposeful for the committee's work, but seems to beg the question of defining public health more prescriptively.

Gordon and McFarlane (1996) define public health as: "The art and science of preventing disease and disability, prolonging life, promoting the health and efficiency of populations, and ensuring a healthful environment through organized community efforts." This definition is more coherent than the IOM statement, and brings together the concepts of science, quality of life, health promotion, and community effort.

However, the definition by Detels and Breslow (1991) is preferred for this book. They state, **"Public health is the process of mobilizing local, state, national, and international resources to solve the major health problems affecting communities."** This definition is appealing because it promotes the ideas of partnership, communities, and action—all of which are themes in this book.

Several points should be noted in the Detels and Breslow definition. Public health is a process; that is, a system of understood ways of addressing problems. Public health mobilizes resources, bring together various agencies or groups in a common cause to prevent or eradicate a community health problem. Public health begins and ends at the local level. Disease outbreaks occur locally and communities expect local authorities to solve *their* health problems. Public health involves cooperative interaction and sharing of resources across lines that extend from local to international agencies. This intersectoral approach to public health has become increasingly important because of international trade and travel. The AIDS epidemic is an example. Depletion of the earth's protective ozone layer is another example. For the most part, public health solves major health problems. Trifling and low-significance problems receive little or no public health response. This book takes the view that hazardous waste problems are major hazards to the public's health.

Public health is generally not the means for providing an individual's personal health care, at least not in the United States. Most clinical medical care is provided by community physicians and other health care providers, local hospitals, and private health insurance providers. Some local public health agencies, however, do provide some health care for indigent populations; for example, community health centers provide prenatal care for pregnant women.

Public health in the United States is organized along levels of government. Federal public health programs are found in the Department of Health and Human Services (DHHS). State programs are lodged in state health departments, and local public health services are located in county or city health departments. Regarding hazardous waste problems, each level of government can, and often does, get involved. Local health departments are frequently the first place where community residents express their concerns about an abandoned waste site or a toxic substance. The response to chemical spills always involves hazardous materials (HazMat) teams. These teams are most often found in fire or police departments but can involve local health departments as well (Chapter 6). State health departments get involved with hazardous waste issues through interaction with federal and state Superfund programs. Health departments of some states work with local health departments to give local health care providers training and education on matters of hazardous substances and waste issues. Federal health agencies provide technical assistance (e.g., epidemiologists) and financial and personnel resources to state and some local health departments, in response to communities' health concerns about hazardous substances.

Themes in the Book

This book is about one specific environmental hazard, hazardous waste, and its impact on human health. Several themes will characterize the material in subsequent chapters as a coherent body of knowledge. These themes are the product of both the author's experience as a public health officer and the science and findings to be described throughout the book. There are six primary themes in this book, as follows:

1. The large number of uncontrolled hazardous waste sites present in the United States is a major human health hazard. Although the precise nature and extent of the impact of waste sites on human health is difficult to quantify, the CERCLA program on human health effects has produced sufficient epidemiologic and toxicologic findings to warrant serious concern for the effects of substances released from uncontrolled waste sites on the public's health.

2. It is not only possible, but essential, to set priorities for risk management actions (e.g., site remediation) at uncontrolled waste sites on the basis of their hazard to human health. Indeed, prioritization is a key concept throughout this book. It is possible to rank sites on the basis of health hazard, to prioritize substances released from sites on the basis of toxicity, and to rank putative health effects caused by human exposure to substances released from sites.

3. Risk management actions at CERCLA sites must include more than simply interdicting any ongoing human exposure to substances released from sites. The trend has been to consider the CERCLA statute and its resources as merely a site remediation program. While interdicting releases of hazardous substances from sites is surely the first step in any effort to prevent further hazard to human health, other actions must also be taken when human exposures have been documented. These actions include educating physicians and other health care providers in communities impacted by releases of substances from sites, communicating health risks in ways that empower communities, characterizing exposure of persons at a health risk, and conducting health surveillance and health promotion (including medical monitoring when warranted) to identify any early signs of illness or disease.

4. Prevention should be the operative word and action in any program intended to reduce the toll of hazards on human health and ecologic systems. The public health literature abundantly documents that preventing a disease is less costly than treating it. This adage applies equally well to environmental hazards. Preventing the production of hazardous waste through industrial recycling or preventing contact with any generated hazardous waste will obviously lessen any adverse effects on human health or the ecosystem. Even when hazardous waste has been produced and released into the environment, early interdiction of contact with the released substances can help minimize or prevent subsequent human health effects.

5. Partnerships are required if the legacy of uncontrolled hazardous waste is to be effectively counteracted. Government, community, business and industry, and academia are all needed if cost-effective, responsible means for protecting the public's health and the environment are to be achieved.

6. The best science must be applied to the environmental health and remediation challenges presented by uncontrolled waste sites and similar sources of environmental contaminants. The combination of hazardous substances released from waste sites or through unintended release (e.g., chemical spills) and their conditions of release present severe challenges to scientists. The toxicology of many substances commonly released from waste sites is often incomplete, and the toxicology of mixtures of substances is even more dimly understood. Yet costly remediation actions at sites and well-intended health responses for communities exposed to substances must be predicated on the best current scientific knowledge and should not be forestalled because of inadequate science. Indeed, nothing should prevent the allocating of financial and personnel resources to conduct the science necessary to improve site risk assessments and population-based health actions.

Table of Contents

Chapter One

Nature and Extent of Hazardous Waste

How many uncontrolled hazardous waste sites are there in the United States and what will be the cost of their remediation (i.e., cleanup)? What are the broad characteristics of these sites, and where does the public rank the health hazard of hazardous waste sites? Moreover, what is meant by *hazardous waste* and *public health* in the context of this book? This chapter addresses these questions and introduces CERCLA and RCRA, the two key federal statutes on waste management.

Industrial development and modern agricultural practices have both contributed mightily to human advancement and well-being. The former serves as the engine for modern national economies; the latter is essential for producing food and fibers. It can be argued that advances in industrialization and agriculture have improved the quality of life, including the public's health. For instance, post–World War II industrialization in the United States produced the pharmaceutical industry, which, in turn, has produced an abundance of drugs essential for treatment or prevention of human disease. Moreover, industrialization has led to development and production of consumer products that make work easier and safer. Similarly, agricultural improvements have increased food production in many parts of the world, with obvious benefits for human health.

However, both industrial development and modern agriculture have undeniably harmed environmental quality and ecosystem integrity. One significant adverse effect is the generation of waste. Waste is defined as a by-product that is deemed useless or worthless by the discarder. By this definition, chemicals discarded into the environment are a form of waste. Some effects of environmental contaminants on human health and ecosystems are relatively well understood; others are still being investigated and debated. For example, the association between vehicle exhaust and ambient air pollution is well-known, including the deleterious release of lead from leaded gasoline into the environment and its effect on human health as lead accumulates in the body.[1] Less well understood is how chlorofluorocarbons contribute to the reduction of ozone in the atmosphere, leading ultimately to a predicted increase in the incidence of skin cancer (NRC, 1991a; Rom, 1992).

---

[1] Lead is banned as a gasoline additive in the United States but not in some other countries.

Some agricultural practices have also contributed to environmental degradation and damage to ecologic systems. For example, excessive reliance on chemical fertilizers increased levels of chemical contaminants in groundwater and surface water in some areas of the United States. Another example is the effects on ecologic systems of DDT from past agricultural, consumer, and public health uses (e.g., mosquito control). DDT reduced the numbers of some wild birds through eggshell thinning. A contemporary controversy concerns the effects on the global environment as rain forests are destroyed for conversion into cropland. These examples, and others that could be cited, show that agricultural practices, if not adequately evaluated for their ecologic effects, can seriously damage the environment and ecologic systems.

Modern industrial societies produce a multitude of products intended for consumption. Indeed, consumerism fostered by free enterprise undergirds national economies in many countries. The production and distribution of consumer goods has brought many benefits to individual consumers; for example, products such as video equipment used for education and entertainment can improve the quality of life. But these positive contributions have come with some deleterious effects. One such effect is the environmental impact of increased waste generation. Discarded consumer products soon become waste that requires disposal. In 1988, each person in the United States produced about four to five pounds of waste per day (PHS, 1990; Plotzman, 1992). Household waste adds to the municipal waste stream. More than 50,000 tons of waste are produced each day in Los Angeles County, California. Figures available for several industrial countries show a pattern similar to that in the United States—large amounts of waste produced per person per year (Plotzman, 1992).

Studies conducted by the University of Arizona have produced a comprehensive database on residential waste. Detailed characterization of household hazardous waste showed that 11 million hazardous waste items were generated by the approximately 120,000 households in Tucson, Arizona (Wilson and Rathje, 1989). They also collected curbside household waste in Tucson and Phoenix, Arizona; New Orleans, Louisiana; and Marin County, California. The household hazardous waste fraction ranged between 0.42 and 0.61% of the refuse weight. The investigators noted that residential waste sent to landfills may contain higher concentrations of hazardous materials than does commercial waste.

The kinds of hazardous materials discarded in household waste and the householders' knowledge about hazardous waste management were studied in two counties in Arizona (Wolf et al., 1997). Telephone and face-to-face interviews with residents about their hazardous waste practices were conducted in Pima and Maricopa counties, Arizona. The results indicated residents were improperly disposing of significant amounts of household hazardous waste. Of note was that although Pima County had a household hazardous waste program and public awareness support, only 5% of the population participated. The investigators grouped hazardous materials in household waste for 1995–1996 into five categories, shown in Figure 1.1.[2] The category *automotive* includes batteries and antifreeze. Two categories, *automobile* and *paint/paint products*, represented more than 80% of household hazardous waste. The investigators concluded that household hazardous waste programs must be supplemented by strong educational programs, including how the waste can cause adverse effects on human health and the environment.

---

[2] Reprinted with permission. Copyright 1997, *Journal of Environmental Health*, Denver, Colorado.

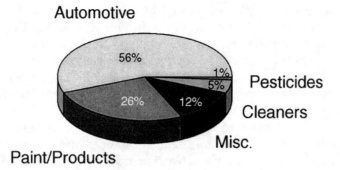

Figure 1.1. Household hazardous waste by weight in Pima County, Arizona, 1995–1996 (Wolf et al., 1997).

Improper disposal of common household waste can cause health problems by, for example, creating noxious smoke from burning trash or providing breeding grounds for vermin. In addition, hazardous substances in municipal waste released from landfills or other methods of waste disposal can contribute to air, water, and food-chain pollution. Releases of hazardous substances from hazardous waste sites into groundwater can significantly threaten drinking water supplies. To reduce the amount of household waste produced per household will require greater efforts in recycling, biodegradable packaging for consumer products, and changes in personal lifestyles.

Modern economies are built on both industrialization and agriculture, with benefits that accrue from both. But environmental degradation, with consequent implications for human health and ecologic systems, has accompanied this development. In some sense, the old cliche "Haste makes waste" applies here. Hazardous waste left expediently in the environment because it was often inexpensive to do so has now become a major concern. What is the basis for this concern and what does it mean for human health and its impact on ecologic systems? Some background about the two major U.S. statutes on hazardous waste management will be helpful.

CERCLA AND RCRA

In the 1970s, discoveries of toxic substances in communities at Love Canal, New York and the "Valley of the Drums" in Kentucky, and a chemical plant fire in Elizabeth City, New Jersey, among others, riveted the attention of the U.S. public and galvanized legislative action. Hazardous waste and its presence in residential environments became a concern of environmentalists, legislators, and public health officials alike. The legislative outcome was the Comprehensive Environmental Response, Compensation, and Liability Act of 1980 (CERCLA), quickly christened the Superfund Act, after the multibillion-dollar Hazardous Substance Superfund created by the act.

The Hazardous Substance Superfund[3] is a trust fund held by the U.S. Treasury and funded by taxes on select parts of private industry: petroleum excise taxes, a chemical

---

[3] The Hazardous Substance Superfund was authorized by CERCLA, section 9507. In 1995, Congress allowed the authority to impose the Hazardous Substances Superfund tax to expire. CERCLA programs have been financed after 1995 on unexpended funds in the Superfund.

feedstock tax, a corporate environmental tax, general revenues, and other sources (Probst, 1992). Ultimate responsibility for the fund rests with the U.S. Congress. Appropriations from the trust fund are made annually to the U.S. Environmental Protection Agency (EPA) for administration. This budget appropriation to the EPA funds CERCLA's directives of identifying uncontrolled hazardous waste sites and sites where emergency spills have occurred, protecting the public's health and environmental media from the effects of releases from identified sites, and remediating (i.e., cleaning up) those sites that merit such action. That same budget appropriation to the EPA contains funds for CERCLA mandates to the Agency for Toxic Substances and Disease Registry (ATSDR) and a program of basic research and worker training for site remediation workers administered by the National Institute of Environmental Health Sciences.

The word *Superfund* is sometimes used to refer to both the CERCLA statute and the Hazardous Substance Superfund. However, in the context of this book, CERCLA will be used exclusively to describe the statute and programs of the same name.

The CERCLA statute of 1980 contains what is called a "sunset clause," which is a legislative device that sets a time limit on a specific law. The sun sets on the statute after a prescribed passage of time, unless Congress continues the law through reauthorization. For CERCLA, this time limit is five years. Reauthorization means that Congress takes action to extend a law's authorities and mandates. This process usually involves congressional hearings, debate, and ultimately, a vote to extend or terminate the law under consideration. In 1986, Congress reauthorized CERCLA, but with substantive changes through various amendments to the 1980 statute. In 1991, Congress voted simply to extend CERCLA, as amended in 1986, for another five years. As of this writing, CERCLA has not been reauthorized again, but the EPA has made several changes in how it administers various parts of the statute.

At the heart of CERCLA is the philosophy, evidenced by enforcement powers given to the EPA, to hold accountable those parties responsible for the consequences of hazardous substances released into the environment from uncontrolled hazardous waste sites. As noted in the Preface, CERCLA is a federal environmental law without parallel in its controversy. Much of the controversy derives from the "polluter pays" philosophy, because parties can be held liable for costs they may deem inappropriate or excessive.

Opinion is widely divided on CERCLA's importance for protecting human health and remediating environmental contamination. Because CERCLA was enacted by Congress primarily from concern that substances released from uncontrolled waste sites were causing cancer and other dire health problems, some of the controversy attending CERCLA gets voiced as issues of human health. The 1986 congressional debate over CERCLA's reauthorization illustrates the polar extremes of view on issues of health:

> On this point we should be absolutely clear. Uncontrolled releases of hazardous substances present a very real threat to the public health. Superfund is the way to clean up the contaminated water and soil so that our children do not become ill from toxic chemicals (Senator George Mitchell, 1986).
> It is time to move faster, to rid our environment of the toxics that are poisoning our land and water, and threatening our citizens (Senator Frank Lautenberg, 1985).
> As a matter of fact, at this moment I honestly do not know if there are any victims of diseases caused by exposure to releases from Superfund sites (Senator James Abdnor, 1985).

The opinions quoted probably reflect similar polar differences of view between segments of private industry and environmental organizations about hazardous waste as an environmental health problem. (Environmental health is used here to denote the area of public health that is concerned with effects on human health of socioeconomic conditions and physical factors in the environment.) These differences will narrow as scientific information and experience accrue. It is hoped that the information in this book will contribute to a better base of technical information on the costs and benefits of the CERCLA statute.

Distinct from CERCLA, which covers uncontrolled hazardous waste sites, is the Resource Conservation and Recovery Act (RCRA), which pertains to the permitted, controlled management of hazardous and solid wastes. RCRA is the federal statute that regulates waste generators, waste transporters, and waste management facilities. RCRA began as an amendment to the Solid Waste Disposal Act in 1965, was enacted into law in 1976, and was amended in 1980 and 1984. Sites covered by RCRA include landfills, waste piles, surface impoundments, land treatment units, tanks, containment areas, and satellite accumulation areas (Ruttenberg et al., 1996). The EPA is the leading federal agency administering the terms of RCRA under four broad programs: hazardous solid waste, nonhazardous solid waste, medical waste, and underground storage.

Remediation of RCRA sites is handled differently from that of CERCLA sites. No funding source like the Hazardous Substance Superfund is available for RCRA sites requiring remediation (Ruttenberg et al., 1996). Instead, corrective actions issued by the EPA under RCRA are funded by site owners or operators. Such actions can involve site cleanups. Releases from solid waste management units (SWMUs) at treatment, storage, and disposal facilities also are covered by the RCRA corrective action program.

PUBLIC PERCEPTION OF HAZARDOUS WASTE

According to opinion polls, the American public continues to place a high priority on the need to reduce hazardous waste and to repair existing environmental damage. This concern has mobilized many community groups and national environmental organizations to oppose the generation, transportation, and disposal of hazardous waste. This opposition has given rise to the term NIMBY, which stands for Not In My BackYard. Community groups, in particular, often express NIMBY concerns through local opposition to the placement of landfills, incinerators, and other forms of waste storage or destruction facilities within their locales. An example is the vocal opposition of communities to construction of incinerators because of concern that dioxins will be released as incinerator emissions.

NIMBY actions are usually rooted in concerns about human health effects of hazardous waste and economic consequences to homeowners (such as reduced property values), although NIMBY actions also are being based on environmental inequity (Chapter 9). According to Szasz (1994), local community groups gradually enlarge their sphere of contact beyond local agencies and officials and engage national grassroots organizations. The result is a broadening of environmental awareness and involvement. This kind of grassroots networking is a consequential political force that has had significant impact on local and federal government policies on waste reduction, recycling, waste site cleanups, and waste management policies.

In 1996 the Superfund Reform Coalition, an alliance of business organizations interested in reforming CERCLA, surveyed public opinion of the major federal environmental laws and priorities (Silverman, 1996). More than 1,000 registered voters between the ages of 18 and 65 from the two major U.S. political parties were surveyed. Although nearly 60% of those surveyed expressed support for less government regulation in general, the results changed with respect to environmental regulation. According to the survey, 36% of respondents indicated there was not enough environmental regulation; 21% said there was too much; and 21% indicated the amount of environmental regulation was about right.

The Coalition's survey found hazardous waste cleanup and safe disposal of hazardous waste were top priorities for 27% of respondents, a figure that exceeded priorities for air and water pollution. When surveyors presented the respondents with "historical data on the performance of Superfund," 49% of respondents said the program had been unsuccessful and needed an immediate overhaul (Silverman, 1996). The extent of the public's dissatisfaction with Superfund cannot be ascertained from this survey because of possible biasing due to how the surveyors presented the Superfund program's performance data to respondents.

The Coalition's survey and results of other public opinion polls show the public is highly concerned about hazardous waste hazards but also believes the CERCLA statute and its implementation need improvement. The high priority ascribed by the public is at odds with the priority given by scientific experts. Chapter 8 discusses how the public perceives risk, and Chapter 10 describes the results of comparative risk assessments of environmental hazards, including hazardous waste.

DEFINITIONS OF HAZARDOUS WASTE AND HAZARDOUS SUBSTANCES

Before proceeding further, *hazardous waste* and *hazardous substance* must be defined. In a general sense, *hazardous waste* is, as the name implies, discarded material that has the potential to do harm. However, laws and international directives on hazardous waste require more precise definitions.

As defined in RCRA, "Hazardous waste means a solid waste, or combination of solid wastes, that because of its quantity, concentration, or physical, chemical, or infectious characteristics, may (A) cause or significantly contribute to an increase in mortality or an increase in serious, irreversible, or incapacitating reversible illness; or (B) pose a substantial present or potential hazard to human health or the environment when improperly treated, stored, transported, or disposed of, or otherwise managed" (ELI, 1989).

The European Community (EC)[4] has taken a more operational approach to defining *hazardous waste* (ELI, 1992). By their definition, hazardous waste means wastes that have one or more of the following 15 properties: explosive, oxidizing, highly flammable, flammable, irritant, harmful, toxic, carcinogenic, corrosive, infectious, teratogenic, mutagenic, ecotoxic; substances and preparations that release toxic or very toxic gases in contact with water, air, or an acid; substances and preparations capable by any means, after disposal, of yielding another substance, for example, a leachate, that possesses any of the other 14 properties. Using these criteria, lists of hazardous wastes are developed by the EC and used for administrative purposes such as regulating transport and disposal of individual hazardous wastes.

---

[4] The European Community is now called the European Union.

The term *hazardous substance* is defined by CERCLA in operational terms as any substance designated by CERCLA itself and other federal statutes. At its root, *hazardous substance* is a legal term but in application refers to a specific pollutant or contaminant. CERCLA defines *pollutant or contaminant* to "...include, but not be limited to, any element, substance, compound, or mixture, including disease-causing agents, which after release into the environment and upon exposure, ingestion, inhalation, or assimilation into any organism, either directly from the environment or indirectly by ingestion through food chains, will or may reasonably be anticipated to cause death, disease, behavioral abnormalities, cancer, genetic mutation, physiologic malfunctions (including malfunctions in reproduction) or physical deformations, in such organisms or their offspring; except that the term 'pollutant or contaminant' shall not include petroleum ...and shall not include natural gas, liquefied natural gas, or synthetic gas..." (ELI, 1989). This definition was obviously written by attorneys, for attorneys!

Hazardous waste and *hazardous substance* will be used in this book to denote waste and substances which possess properties that can cause harm to human health and ecologic systems. However, the reader is advised that these terms take on specific meanings under various legal and regulatory acts, such as CERCLA. Also included as hazardous waste for the purposes of this book are substances released during chemical spills and similar uncontrolled releases of hazardous substances into the environment.

EXTENT OF THE HAZARDOUS WASTE PROBLEM

Industrialized countries produce huge quantities of hazardous waste. This often results from inefficient or uneconomical production methods and use of energy and materials. In the United States, the "solution" frequently chosen in the past was to dispose of the waste in landfills or waterways (rivers, lakes), or to incinerate it. The end result is a toxic legacy in the environment. Too often an attitude of "The solution to pollution is dilution" has prevailed. That practice doesn't work, especially for pollutants that are persistent in the environment and that accumulate in ecologic systems. An attitude of "The solution to pollution is diminution" should characterize current and future waste management practices. From the perspective of human health and ecologic systems, prevention of environmental degradation leads to prevention of adverse health effects because exposure to toxicants does not occur. In public health parlance, this is *primary prevention* at work; that is, elimination of the hazard.

The exact volume of hazardous waste produced globally is unknown. International comparisons are almost impossible because of differences in definition and classification of hazardous wastes from country to country. Batstone et al. (1989) estimate that, for several western European countries, hazardous waste production is about 5,000 tons per billion U.S. dollars of gross domestic products (GDP). On this basis, the United States is estimated to produce 75,000 tons per annum (tpa); and Canada, 10,000 tpa. Waste production per GDP was estimated to be 10,000 tons in the former Soviet Union; that in other countries with mature industry, 5,000 tons; in newly industrialized countries, 2,000 tons; and in developing countries, 1,000 tons.

Other estimates of the annual generation of hazardous waste in the United States were developed by the Office of Technology Assessment (OTA), the Congressional Budget Office (CBO), and the EPA. The OTA estimates that 255–275 million tonnes[5] of

---

[5] 1 tonne = 1,000 kilograms.

hazardous waste are produced (OTA, 1983). The EPA estimated in 1991 that 275 million tons of hazardous waste was produced (Skinner, 1991). Moreover, the EPA cites estimates from the Congressional Budget Office (CBO) on the major generators of hazardous waste, showing that chemical and allied products account for about half of all hazardous waste generated (Figure 1.2).

By whatever yardstick, the quantities of hazardous wastes generated each year are enormous, and without proper management of its storage or disposal, have the potential for compromising the public's health and the well-being of ecosystems.

Number of CERCLA Hazardous Waste Sites

Hazardous waste has often been discarded into the environment in landfills and on industrial properties. In turn, these properties have themselves been discarded. At the end of December 1996, the EPA's inventory of uncontrolled hazardous waste sites, called the Comprehensive Environmental Response and Liability Information System, listed 41,266 sites (EPA, 1996a). However, in 1995 the EPA archived approximately 30,000 of these sites because they posed little or no threat to health or the environment; no further federal remedial action is planned for them (Browner, 1997). As of December 31, 1996, 1,296 sites were listed on or proposed for the EPA's National Priorities List (NPL), which are the sites posing the greatest threat to the public's health and the environment (EPA, 1996b,c). NPL sites also become the subject of federal funding and enforcement efforts. More specifically, NPL sites are identified by the EPA for remediation, using CERCLA authorities for recovering remediation costs from potentially responsible parties (PRPs).

Uncontrolled hazardous waste sites associated with U.S. federal government facilities are of particular importance because of their potentially high remediation costs. These facilities include sites operated by the Department of Defense (DOD, military bases) and the Department of Energy (DOE, weapons complexes). According to the National Research Council, 17,482 contaminated sites were located at 1,855 DOD installations as of September 1990 and 3,700 sites at 500 DOE facilities (NRC, 1994). Some DOE sites cover large geographic areas and are toxicologically very complex in terms of both the radioactive and chemical wastes released into the environment.

As of April 1995, federal agencies had placed 2,070 facilities on the federal facility docket, the EPA's listing of the facilities awaiting evaluation for possible cleanup (GAO, 1996). The EPA has placed 154 federal facilities on the NPL, and, as of February 1996, had proposed another five facilities for NPL listing (GAO, 1996). Of the 154 federal facilities on the NPL, the largest number are DOD facilities (127), followed by the Departments of Energy (20), Interior (2), Agriculture (2), Transportation (1), and other departments (2).

Other Hazardous Waste Sites

Remedial actions are needed at other hazardous waste sites in addition to CERCLA sites. These include RCRA sites and those under the control of states and private parties. According to Ruttenberg et al. (1996), the number of treatment, storage, and disposal facilities covered under RCRA is in the range of 4,700–5,100, with 1,500–3,500 requiring some kind of corrective action (Ruttenberg et al., 1996). Corrective actions were under way or completed at 247 facilities, about 3,500 facilities had undergone RCRA facility

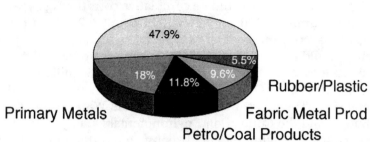

Figure 1.2. Top five generators of hazardous waste by percentage of total hazardous waste generated in the United States (Skinner, 1991).

assessments, and 614 were undergoing RCRA facilities investigations. Of these RCRA sites, about half use off-site disposal remedies and half use innovative treatment technologies.

Furthermore, there are an estimated 21,575 large-quantity waste generators; 190,431 small-quantity waste generators; and 2,389 treatment, storage, and disposal facilities acting as waste generators are in existence (Ruttenberg et al., 1996). The corrective actions and remediation efforts at these sites are unknown.

In addition to the federal Superfund program, states and U.S. territories administer major programs of removal and remedial actions at non-NPL sites that parallel those of the federal CERCLA. The U.S. Office of Technology Assessment (OTA) estimated that as many as 439,000 state sites might need to be evaluated for remediation (OTA, 1989). According to the Association of State and Territorial Solid Waste Management Officials (ASTSWMO), which acquired data from 39 states and one territory, as of December 31, 1992, 21,905 hazardous waste sites had been identified (ASTSWMO, 1994).

Anecdotal information indicates that many hazardous waste sites have been voluntarily remediated by private parties. However, no source is available to the public for keeping track of the number of sites voluntarily remediated, the method of remediation, the costs, or other pertinent data. The numbers of such sites and the effects of the remediation are therefore unknown.

Costs of Uncontrolled Hazardous Waste Sites

In 1991, the EPA estimated that cleaning up the nonfederal NPL sites could cost more than $30 billion. Reisch et al. (1996) estimated the average cost for cleaning up nonfederal sites is approximately $30 million; the EPA (1998a) estimates a figure of $18–20 million. In addition to NPL sites, the EPA estimated that about 4,000 facilities, representing 64,000 solid waste management units covered under RCRA, may require cleanup, but no cost estimates were provided (Habicht, 1991).

Russell et al. (1991) projected that cumulative cleanup costs for all sites from the year 1990 through 2020, using current remediation practices, will be approximately $750 billion, with a plausible lower bound at something less than $500 billion and an upper bound at approximately $1 trillion. Their analysis included both federal and nonfederal sites and covered CERCLA NPL sites, state and private-sector waste remediation programs, underground storage tanks, and RCRA sites requiring corrective action. Specific to CERCLA NPL sites, Russell et al. estimate cumulative costs through year 2020 to range from $106 billion (assuming 2,100 nonfederal sites) to $302 billion (assuming 6,000 nonfederal sites). Less stringent cleanup would, of course, lower these projected costs; greater stringency would increase them. Their "best guess" of cumulative remediation costs for federal facilities is $240 billion for DOE sites and $30 billion for DOD sites.

The Federal Facilities Policy Group (FFPG) conducted an analysis in 1995 of costs to remediate federal sites. The group was convened by the White House's Office of Management and Budget and the Council on Environmental Quality to review the current status and future course of environmental response and restoration of federal facilities. In the course of the review, estimates were derived of the numbers and costs of federal facilities and sites contaminated with hazardous substances. These estimates are listed in Table 1.1, which profiles the federal departments with the largest number of contaminated sites. Included are DOE, DOD, Department of Interior (DOI), Department of Agriculture (USDA), and the National Aeronautics and Space Administration (NASA).

The federal departments identified in Table 1.1 have the greatest number of facilities and contaminated sites that require remediation. Cleanup costs are given in 1994 dollars. The DOI has the largest number of active sites requiring review and potential remediation. The DOE, because of the complex nature of their facilities and contaminated sites, represents the single largest cost for site remediation. According to the Federal Facilities Policy Group (FFPG), site remediation and attendant costs for DOE, DOD, and NASA will come from federal funds, that is, from the federal treasury. Sites that are the responsibility of the DOI and the USDA will derive their cleanup costs from a mix of federal, private party, and local government funds (FFPG, 1995). The FFPG estimates the total cost to remediate federal sites at approximately $400 billion in 1994 dollars.

The CBO (1994) released in 1994 a set of cost estimates for nonfederal sites, analyzing three scenarios for CERCLA costs after 1992. The CBO's estimates were based on assumptions related to the number of nonfederal NPL sites. The base-case estimate (4,000 nonfederal NPL sites) is $74 billion in discounted, present-worth dollars; the low-case estimate (2,300 sites) is $42 billion; and the high-case estimate (7,000 sites) is $120 billion. Annual undiscounted costs in the base-case peak are $9.1 billion in the year 2003; they average $2.9 billion per year through the year 2070 (CBO, 1994).

The CBO's estimates of future CERCLA costs are different from earlier estimates developed by the EPA (Habicht, 1991) and Russell et al. (1991). The main factors that explain the differences are CBO's broader coverage of costs and use of discounted dollars, different average cleanup costs per site, and different numbers of sites on the NPL. Specifically, the CBO figure includes all future public and private CERCLA expenditures; the Russell et al. estimates cover public and private costs for study and cleanup at NPL sites, but omit administrative and legal expenses and the costs of screening and removals at non-NPL sites. Furthermore, the CBO (1994) estimate is in present-worth dollars; Russell et al.'s (1991) figures are expressed in undiscounted dollars.

In addition to the costs of site remediation, other site-specific costs add to the overall burden of protecting the public and ecologic systems against the legacy of uncontrolled releases of hazardous substances. In particular, operations and maintenance (O&M) activities will be necessary at many sites. These activities are meant to ensure that the remedy implemented at a site continues to operate effectively. The GAO reviewed information from 275 NPL sites where the cleanup remedy had been implemented. Of

Table 1.1. Profile of Key Federal Departments' Facilities and Sites (FFPG, 1995).

	DOE	DOD	DOI	USDA	NASA
Nature of contamination	Radioactive, hazardous, and mixed waste and fissile material	Fuels and solvents, industrial waste, and unexploded ordnance	Mining, municipal, and industrial wastes	Hazardous, mining, and chemical waste	Fuels, solvents, and industrial waste
Number of potentially contaminated sites (and site types)	10,000 sites (former weapons-production facilities)	21,425 sites (underground tanks, landfills, spill areas, storage areas)	26,000 sites (abandoned mines, oil & gas production, landfills)	3,000 sites (abandoned mines, landfills)	730 sites (underground storage tanks, spill areas)
Number of potentially contaminated facilities	137	1,769	Not applicable	Not applicable	17
Number of active sites	10,000	11,785	26,000	3,000	575
Current estimate to complete cleanup	$200–$350 billion	$26.2 billion	$3.9–$8.2 billion	$2.5 billion	$1.5–$2 billion

this number, 173 sites were found to require O&M activities by the federal government, states, and parties responsible for cleanup costs.

In addition to the federal CERCLA program, states and U.S. territories have major programs under way for removal and remedial actions at non-NPL sites. According to ASTSWMO (1994), 3,527 sites had at least one completed state removal, 4,834 sites were in a state's version of the remedial investigation/feasibility study phase, 2,689 sites have completed construction through state remedial processes, and 11,000 sites were described as still active in some part of their state remedial process. The costs associated with 3,395 sites amounted to approximately $1.2 billion. This program represents a substantial commitment on the part of states and territories to removing toxic wastes from the environment.

The costs of O&M activities vary according to the kind of remediation used at a site. For example, the GAO found that when the remedial action uses a technology to contain surface waste, the ensuing O&M activities that follow the containment system construction could typically cost $5 million per site over 30 years. When the cleanup remedy included treating contaminated groundwater, O&M activities could cost $17 million per site over 30 years. The GAO concluded, "For clean-up remedies that EPA or the responsible parties have already undertaken or will undertake from now to fiscal year 2005, we estimate that about $32 billion will be needed for operations and maintenance costs nationwide through fiscal year 2040" (GAO, 1995a). Of this amount, $18 billion would come from private sector parties, $8 billion from states, and $5 billion would be paid by the federal government.

By any measure, the cost in money and human resources to remediate the legacy of hazardous wastes left in the American environment will be huge if current policies on site identification, prioritization, and remediation are maintained. Because of the extensive resources committed to remediating hazardous waste sites and related environmental hazards, knowing the magnitude of the human health hazard is important to permit balancing costs and benefits. This book is intended to provide data relevant to the cost/benefit calculus.

Numbers of Sites Remediated

The ultimate environmental impact of the CERCLA statute is the remediation of uncontrolled hazardous waste sites. This action removes environmental contamination that can potentially cause adverse human health and ecologic effects. The actual cleanup of a site is the culmination of the Superfund remediation process. Remedial actions at a particular site may take years, depending on such factors as size of the site and the specific cleanup methods chosen. After remediation, NPL sites are removed from the NPL. According to the EPA, cleanup construction has been completed at 504 NPL sites and another 473 sites are in construction, as of January 1998. Moreover, the EPA states that more than 89% of nonfederal NPL sites are either undergoing cleanup construction (remedial or removal), or have been completed (Fields, 1998).

One criticism of the CERCLA program is the allegedly slow pace of NPL site remediation. The U.S. Department of Justice (DOJ, 1996) asserts that the rate of site remediation has quickened because of reforms in the EPA's administration of CERCLA. However, the GAO (1997) found that the time to complete remediation at NPL sites has increased over time (Figure 1.3). From 1986 to 1989, cleanup projects were completed, on average, in 3.9 years after sites had been placed on the NPL. By 1996, the time for

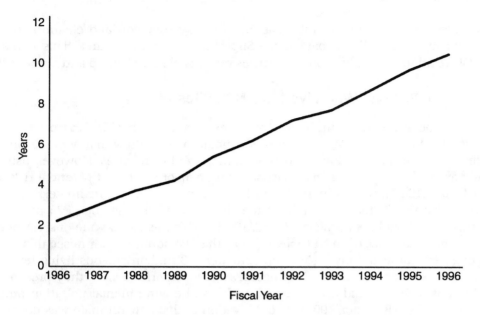

Figure 1.3. Average length of time to complete cleanup of NPL sites (GAO, 1997).

cleanup completions had increased to an average of 10.6 years. According to the GAO, the EPA attributed the increase to the growing complexity of sites, efforts to reach settlements with parties potentially responsible for the costs of site remediation, and resource constraints.

The data in Figure 1.3, which imply that the time to complete cleanups of NPL sites may continue to increase, have been challenged by the EPA (Fields, 1998). The EPA contends that the GAO analysis failed to acknowledge a backlog of hundreds of sites already in existence when CERCLA was enacted, and relied on analysis of operable units, not site completions. There is merit to the EPA's complaint about the GAO analysis; administrative reforms in how the EPA administers its CERCLA programs should decrease the time-to-cleanup for NPL sites.

States and territories have remediated non-NPL sites more quickly than the EPA has remediated CERCLA sites. Of 6,052 such remediated sites, 4,261 were completed in less than one year (ASTSWMO, 1994). This duration is less than for federal CERCLA sites, but the comparison is complicated. CERCLA defines duration of response actions as the time from the discovery of a site through the remedial action, whereas state/territories define it as the time from the start of the remedial investigation through the remedial action (ibid.). Moreover, it can be argued that the federal CERCLA program, which includes federal facilities, remediates sites that are more contaminated and complicated than are the state/territorial sites.

Given the high cost of remediating hazardous waste sites and the nature of assigning costs to PRPs, it is not surprising to find that remedy selection by the EPA and individual companies (where voluntary cleanup is undertaken) is a source of conflict. A GAO study found several alleged problems with the completeness and consistency of cleanup decisions at sites managed by both the EPA and PRPs. Problems noted were inadequate cleanup goals and insufficient justification for selecting a particular remedial action (GAO, 1992). The EPA's administrative reforms in implementing CERCLA programs have made improvements in several areas, but remedy selection and cleanup goals will remain subjects of active debate as the Superfund program matures. This stems from CERCLA's financial implications for parties responsible for cleanup and other costs.

Number of Persons Who Live Near NPL Sites

The number of persons in the United States at potential health risk from releases of hazardous substances from waste sites is difficult to estimate with any precision, given uncertainties in the frequency and geographic extent of releases. However, using currently available data, two estimates exist of numbers of persons at potential risk. Using data from 1,134 NPL sites in 1991, the EPA estimated that approximately 41 million persons live within four-mile radii of the sites, and an average of 3,325 persons live within one-mile radii of such sites (NRC, 1991b). This means that one of six Americans lives within four miles of an NPL site. In 1995, the EPA administrator raised this estimate in congressional testimony: "With approximately 73 million persons living fewer than four miles from one or more of the nation's active Superfund sites, they present some of the most complex and diverse of all health and environmental pollution problems facing us today" (Browner, 1995). The derivation of this new estimate was not stated.

A demographic[6] analysis of persons living near 1,200 NPL sites led ATSDR to estimate that approximately 11 million persons live within one-mile borders of those sites (Heitgerd et al., 1995). A separate analysis by ATSDR of 972 NPL sites found 949,000 children six years or younger lived within one mile of sites' borders, which represented 11% of the population. This is an average of 980 children six years or younger per NPL site. Given a current total of about 1,300 NPL sites, one can calculate that approximately 1.3 million young children live within one mile borders of NPL sites. This is a large number of young children at potential health risk. Actual health risk must be determined by such factors as patterns of release of substances from sites, degree of exposure of persons in contact with released substances, toxicity of the substances, and individual susceptibility to a substance's toxicity.

CHARACTERISTICS OF UNCONTROLLED WASTE SITES

As the number of NPL sites and the cost to remediate them have increased, interest has developed in characterizing the sites on the basis of demographics and socioeconomic features, influenced in part by environmental equity concerns, that is, concern that hazardous waste sites may be distributed inequitably in minority and low-income communities (Chapter 9). Statistical characterization of NPL sites is therefore important for both social and economic reasons.

Zimmerman examined the socioeconomic characteristics of 1,090 NPL sites, which was approximately the number of NPL sites at the time of his study (1993). He examined approximately 800 in depth. Several of his findings are useful for placing NPL sites in perspective:

- Population. The populations living near NPL sites varied widely. The mean population in 1990 was about 88,000, and the standard deviation was about 278,000.

---

[6] Demography focuses on population trends such as growth; for example, increases in percentage of the elderly in the U.S. population, and migration patterns (Detels and Breslow, 1991).

- Age. An average of 13% of the populations in communities with NPL sites in 1990 were elderly, comparable with the 12.4% of elderly in the nation.
- Population growth. NPL sites in 1990 were mostly located in communities that had modest growth rates of about 5% between 1986 and 1990. The national growth rate for the same period was 3.8%.
- Population density. NPL sites in 1990 were located in rather densely populated areas. The 1990 mean density of 2,658 persons per square mile is several times greater than the 1990 national average of 70.3 persons per square mile. Zimmerman notes that this large population density is an artifact of the NPL designation process (i.e., EPA's Hazard Ranking System), which uses population density as a criterion for listing a site on the NPL.
- House value. The median house value was $75,000 in communities with NPL sites in 1990, with a high degree of variability. This was about equal to the national median house value of $79,100 in 1990.
- Home ownership. Owner occupancy in communities with NPL sites in 1990 averaged about 60%, compared with the national average of 64 percent.

Zimmerman's data suggest that some key socioeconomic indicators for communities near NPL sites in 1990; that is, population density, house values, and home ownership, did not differ from corresponding national data. A key difference, however, was the finding that NPL sites were located in densely populated areas, which are given priority by the EPA's Hazard Ranking System.

Municipal Solid Waste Landfills

Because municipal solid waste landfills are known to receive some hazardous materials from households and industry, the GAO sought survey responses from 301 metropolitan and 322 nonmetropolitan landfills concerning various solid waste issues (GAO, 1995c). Their survey, like the Zimmerman examination of socioeconomic characteristics of populations living near uncontrolled hazardous waste sites, found several interesting characteristics of solid waste landfills of relevance to human health issues.

- Counties and municipal governments owned 69% of metropolitan landfills and 79% of nonmetropolitan landfills in operation in 1992.
- The average metropolitan landfill was about 191 acres in size, with a range of 1–2,000 acres. The average nonmetropolitan landfill was about 98 acres, with a range of 1–1,200 acres.
- The average metropolitan landfill received about 50% of its waste from the community where it was located and 36% from the remainder of the county. Comparable percentages for nonmetropolitan landfills were 61% and 34%, respectively.
- Typically, the waste sent to landfills was household waste, industrial nonhazardous waste, and construction-related debris.
- GAO estimated 7% of the metropolitan landfills had received hazardous waste from sources that generate small quantities of waste and less than 3% had received hazardous waste from sources that generate large quantities of waste. Among nonmetropolitan landfills, 8% had received hazardous waste from small-quantity generators, and 1% received from large-quantity generators.
- No protective liners were beneath any of the waste units for 47% of metropolitan landfills; 55% did not have leachate collection systems in place at any of their waste units. More than 90% reported that they had groundwater monitoring. About 16% of metropolitan landfills had caused groundwater contamination at some time.

- No protective liners were beneath any of the waste units for 66% of nonmetropolitan landfills; 80% did not have leachate collection systems in place; 67% reported they had groundwater monitoring in place. About 7% of nonmetropolitan landfills were reported as having caused groundwater contamination at some time.

The GAO survey (1995c) and the study of Wolf et al. (1997) indicate that small amounts of hazardous materials do reach metropolitan and nonmetropolitan landfills. The volume of these materials is not well characterized, but their presence in landfills justifies a human health concern.

SYNTHESIS

Public opinion polls indicate that the American public continues to place a high priority on the need to reduce hazardous waste and to repair existing environmental damage. An industry coalition's survey found hazardous waste cleanup and safe disposal of hazardous waste were top priorities for 27% of respondents, a figure that exceeded priorities ascribed by the survey's respondents for air and water pollution. Survey respondents also indicated that the CERCLA program had been unsuccessful and needs fundamental overhaul.

In the 1970s, discoveries of hazardous substances in communities became a concern to environmentalists, legislators, and public health officials alike. The legislative outcome was the Comprehensive Environmental Response, Compensation, and Liability Act of 1980 (CERCLA), which governs actions to remediate uncontrolled hazardous waste sites and responds to human health impacts.

In distinction to CERCLA, the Resource Conservation and Recovery Act (RCRA) covers the permitted, controlled management of hazardous and solid wastes. RCRA is the federal statute that regulates waste generators, waste transporters, and waste management facilities. Sites covered by RCRA include landfills, waste piles, surface impoundments, land treatment units, tanks, containment areas, and satellite accumulation areas (Ruttenberg et al., 1996).

The number of hazardous waste sites is large. At the end of December 1996, the EPA's inventory listed 41,266 uncontrolled hazardous waste sites (EPA, 1996a). However, the EPA has inactivated approximately 30,000 sites because they posed little or no threat to health or the environment (Browner, 1997). As of December 31, 1996, 1,296 sites were listed on or proposed for the EPA's National Priorities List (NPL), which are the sites posing the greatest threat to the public's health and the environment (EPA, 1996b,c).

Of particular import are uncontrolled waste sites that are the responsibility of the federal government. As of April 1995, federal agencies had placed 2,070 facilities on the federal facility docket, the EPA's listing of the facilities awaiting evaluation for possible cleanup (GAO, 1996). The EPA has placed 154 federal facilities on the NPL and, as of February 1996, had proposed another five facilities for NPL listing (GAO, 1996). Of the 154 federal facilities on the NPL, the largest number are DOD facilities (127), followed by the Departments of Energy (20), Interior (2), Agriculture (2), Transportation (1), and other departments (2).

In addition to the federal Superfund program, states and U.S. territories have major programs under way for removal and remedial actions at non-NPL sites that parallel those of the federal CERCLA. The GAO estimates 130,000 to 425,000 state sites might need to be evaluated for possible remediation. As of December 31, 1992, according to data from 39 states and one territory, 21,905 hazardous waste sites had been identified. Of these, 3,527 sites had undergone at least one completed state removal, 4,834 sites were in a state's version of the remedial investigation/feasibility study phase, 2,689 sites had completed construction through state remedial processes, and 11,000 sites were described as still active in some part of their state remedial process (ASTSWMO, 1994). The costs associated with 3,395 sites amounted to approximately $1.2 billion.

The cost of remediating hazardous waste sites and the impact on human health and natural resources is huge. Russell et al. (1991) projected that cumulative cleanup costs for all sites from the year 1990 through 2020, using current remediation practices, will be approximately $750 billion, with plausible lower bound at something less than $500 billion and upper bound at approximately $1 trillion (Russell et al., 1991). Their analysis included both federal and nonfederal sites and covered CERCLA NPL sites, state and private sector waste remediation programs, underground storage tanks, and RCRA sites requiring corrective action. Specific to CERCLA NPL sites, Russell et al. estimate cumulative costs through year 2020 to range from $106 billion (assuming 2,100 nonfederal sites) to $302 billion (assuming 6,000 nonfederal sites). Less stringent cleanup would lower these projected costs; greater stringency would increase them. Their "best guess" of cumulative remediation costs for federal facilities is $240 billion for DOE sites and $30 billion for DOD sites.

The number of persons at potential risk of exposure to substances released from hazardous waste sites is imprecisely known. Using data from 1,134 NPL sites in 1991, the EPA estimated that approximately 41 million persons live within four-mile radii of 1,134 NPL sites, an estimate that was later revised upward to 73 million persons. A demographic analysis of persons living near 1,200 NPL sites led the ATSDR to estimate that approximately 11 million persons live within one-mile borders of these sites. A separate analysis by the ATSDR of 972 NPL sites found 949,000 children six years or younger live within one mile, which represented 11% of the population. This extrapolates to approximately 1.3 million young children living near the current 1,300 NPL sites. This is a large number of young children at potential health risk. Actual health risk must be determined by such factors as patterns of release of substances from sites, degree of exposure of persons in contact with released substances, toxicity of the substances, and individual susceptibility to a substance's toxicity.

Because substances released from hazardous waste sites can harm human health and the environment, prevention should be the operative word, with action in any program intended to reduce the toll of hazards on human health and ecologic systems. The public health literature well documents the fact that preventing a disease is usually less costly than treating it. This approach applies equally well to environmental hazards.

Preventing the production of hazardous waste through industrial recycling or preventing contact with any generated hazardous waste will obviously lessen any adverse effects on human health or the ecosystem. Furthermore, strategies to prevent contact can be implemented even after hazardous waste has been produced and released into the environment. For example, prompt interdiction of contact with the released substances can help minimize or prevent subsequent human health effects, and surveillance of exposed populations can be an effective early indicator of any adverse health effects.

Chapter Two

Public Health Assessment

Are all hazardous waste sites equally threatening to the public's health? How does one differentiate between sites on the basis of human health hazard? This chapter describes the public health assessment of uncontrolled hazardous waste sites. It details the steps in a public health assessment and presents salient findings from assessments of CERCLA sites. The 30 hazardous substances found most often in completed exposure pathways are identified, and the national level of public health hazard of NPL sites is described.

Throughout life we are all concerned about our health and well-being. We know that disease can be debilitating and costly, reducing our quality of life. We recognize the importance of knowing about the health of our community, because epidemics of disease can spread to individuals and their families. This is certainly true of communicable diseases such as tuberculosis. Patterns of illness that a community associates with an environmental hazard can, like communicable diseases, be of great concern to individuals in the community. When a person or a community becomes concerned about health, some kind of health assessment is called for.

Assessing one's personal health is pursued through contact with medical practitioners. For many individuals this is handled by periodic medical checkups. The physician usually performs various diagnostic examinations that include laboratory tests of body fluids and exercises professional judgment when comparing the patient with what are known to be normal values for a healthy person. Medical science is melded with the physician's experience and judgment. This is one form of health assessment, personal assessment because it pertains to an individual.

Conducting an assessment of a group's or community's health status is a tradition of public health. This can take the form of administering medical diagnostic tests to examine the health status of each member of the group or of a statistically representative sample of a larger group. Another kind of health assessment involves examining environmental indicators that are known to be associated with adverse health effects. It is therefore important to distinguish between a personal health assessment and a group or community health assessment.

The ATSDR determined through discussions with the public that the term "health assessment" does not always accurately describe ATSDR activities when assessing the effects of substances released from a hazardous waste site. Often "health assessment" was assumed by the public to mean assessing personal health, rather than assessing the level of hazard to a population's health. For that reason, ATSDR changed to another name, public health assessments, to better distinguish its efforts to evaluate the health status of a population from the efforts of a health care provider to evaluate the health status of an individual (ATSDR, 1992a).

The practice of public health has been defined as the art and science of protecting and improving community health through preventive medicine, health promotion, disease control, and the application of social and sanitary sciences. A community may be defined as a group or class having common interests. The term "public health assessment" encompasses the art and science of dealing with the human health issues related to hazardous substance releases into the environment, and ensures community involvement in the process. These concepts of public health and community are consistent with the intent and use of what is called a "health assessment" in CERCLA and RCRA.

Persons who live near hazardous waste sites are often anxious about the sites' potential effects on their health and their children's health. As noted in Chapter 1, American public opinion polls report this anxiety. The U.S. Congress recognized the health implications of uncontrolled hazardous waste sites when the Comprehensive Environmental Response, Compensation, and Liability Act of 1980 (CERCLA, or Superfund) was enacted. The statute created the Agency for Toxic Substances and Disease Registry (ATSDR) and mandated that the agency conduct various activities that Congress perceived would address health issues of what became known as Superfund sites. A primary responsibility of the ATSDR is to conduct public health assessments of communities near individual uncontrolled hazardous waste sites, as described in a subsequent section of this chapter.

CERCLA AND RCRA MANDATES

The CERCLA statute of 1980 authorized the ATSDR to conduct public health assessments of sites, but operationally this was achieved only when the EPA referred the sites to the ATSDR. Moreover, the ATSDR was not structured as an agency by its parent department, the Department of Health and Human Services, until litigation was brought jointly in 1983 by the Chemical Manufacturers Association, the American Petroleum Institute, and the Environmental Defense Fund. The lawsuit was settled in 1984, and the ATSDR was established as a component of the Department.

The Department's reluctance to support the ATSDR's development and the EPA's uncertainty as to the value of ATSDR's public health assessments resulted in public health assessments being conducted for only a small number of CERCLA sites through 1984. This changed when CERCLA was reauthorized by Congress in 1986.

CERCLA requires congressional reauthorization every five years. The ATSDR's responsibilities were substantially increased under the amendments to CERCLA that constituted reauthorization of the statute in 1986 (Johnson, 1990). The most significantly increased mandate was the conduct by the ATSDR of [public] health assessments. Three very challenging new requirements were added. First, the ATSDR was required to conduct a public health assessment of each site within one year of its being placed, or proposed for placement on the NPL by the EPA. Second, the ATSDR was required by December 10, 1988, to conduct public health assessments of all NPL sites that existed on that date. Thus, 951 NPL sites required public health assessment. Third, an individual or licensed physician could petition ATSDR to conduct a public health assessment of a site of concern to the petitioner.

Furthermore, section 3019(b) of the Resource Conservation and Recovery Act of 1984 (RCRA), as amended, provides that when a landfill or surface impoundment poses a substantial potential risk to human health in the judgment of the EPA, or a state with an authorized RCRA program, or an individual, the EPA or the state may request the ATSDR to perform a public health assessment. However, this authority is tempered by this condition: "If funds are provided in connection with such request the Administrator (ATSDR) shall conduct the health assessment." This requirement has led to few RCRA facilities being submitted to the ATSDR for public health assessment.

Both CERCLA, as amended, and RCRA specify congressional intent as to what should constitute a public health assessment. Specifically, section 104(i)(6)(F) of CERCLA and section 3019(f) of RCRA state that, "...the term 'health assessments' shall include preliminary assessments of the potential risk to human health posed by individual sites and facilities subject to this section, based on such factors as the nature and extent of contamination, the existence of potential for pathways of human exposure (including ground or surface water contamination, air emissions, and food chain contamination), the size and potential susceptibility of the community with the likely pathways of exposure, the comparison of expected human exposure levels to the short-term and long-term health effects associated with identified contaminants and any available recommended exposure or tolerance limits for such contaminants, and the comparison of existing morbidity and mortality data on diseases that may be associated with the observed levels of exposure." RCRA 3019(f) adds, "The assessment shall include an evaluation of the risks to the potentially affected population from all sources of such contaminants, including known point or nonpoint sources other than the site or facility in question. A purpose of such preliminary assessment shall be to help determine whether full-scale health or epidemiological studies and medical evaluation of exposed populations shall be undertaken."

In response to these CERCLA and RCRA mandates, the ATSDR developed a framework and guidance for conducting health assessments (ATSDR, 1992b). The agency also has worked to fund state health departments in their conduct of health assessments of communities impacted by hazardous waste sites.

Although the ATSDR met the CERCLA mandate to conduct health assessments of 951 NPL sites by December 10, 1988, the agency was later reviewed and criticized by the General Accounting Office (GAO, 1991). The criticism was based largely on the ATSDR's reliance on pre-existing health studies or assessments conducted by other federal agencies as a surrogate for a site's health assessment. In particular, one set of health assessments for 165 sites was believed to need updating. The GAO also recommended that in the future a sample of health assessments be peer reviewed periodically. Furthermore, the GAO recommended closer coordination between the ATSDR and the EPA in developing and using health assessments. All three recommendations were accepted and implemented by the ATSDR.

OPERATIONAL DEFINITION OF PUBLIC HEALTH ASSESSMENT

A public health assessment is a form of hazard evaluation. It is conducted using weight-of-evidence and professional judgment. Public health assessments are the ATSDR's and many state health departments' primary diagnostic instrument to determine the level of hazard posed by individual waste sites. They are used to identify (1) CERCLA sites where environmental officials should interdict communities' current exposure to hazardous substances and (2) sites that need health investigations and other public health interventions. The following operational definition applies (ATSDR, 1992b):

> A [public] health assessment is the evaluation of data and information on the release of hazardous substances into the environment in order to assess any current or future impact on public health, develop health advisories or other recommendations, and identify studies or actions needed to evaluate and mitigate or prevent human health effects.

A public health assessment of an uncontrolled hazardous waste site or similar source of uncontrolled release of substances differs from a site-specific baseline risk assessment that the EPA conducts under its CERCLA authorities. As defined by the National Research Council (NRC), "...risk assessment means the characterization of the potential adverse health effects of human exposures to environmental hazards" (NRC, 1983). The Department of Health and Human Services (DHHS) adopted a slightly different definition, which is preferred for this book because exposure is given greater emphasis in the DHHS definition (DHHS, 1985). The DHHS defined **risk assessment** as "...the use of available information to evaluate and estimate exposure to a substance(s) and its consequent adverse health effects." According to the NRC, risk assessment consists of four elements:

- Hazard identification—The qualitative evaluation of the adverse health effects of a substance(s) in animals or in humans.
- Dose-response assessment—The process of estimating the relation between the dose (i.e., the amount of the substance received by the target organism) of a substance(s) and the incidence of an adverse health effect.
- Exposure assessment—The evaluation of the types (routes and media), magnitudes, time, and duration of actual or anticipated exposures and of doses, when known; and, when appropriate, the number of persons who are likely to be exposed.
- Risk characterization—The process of estimating the probable incidence of an adverse health effect to humans under various conditions of exposure, including a description of the uncertainties involved.

The principal differences between the ATSDR's public health assessment and the EPA's baseline risk assessment of a CERCLA site are contrasted in Table 2.1. First, public health assessments are quantitatively derived; risk assessments rely on statistical and biologic models to quantitatively calculate risk. Health assessments drive public health interventions (e.g., epidemiologic investigations); risk assessments are used for risk management actions (e.g., cleanup levels of contaminants in waste sites). Public health assessments focus primarily on the health consequences of current and past releases of hazardous substances; risk assessments focus on current and future releases of substances from sites. Public health assessments include human health outcome data where available (e.g., cancer data in state health departments) and community health concerns; risk assessments emphasize environmental contamination data and extrapolate these data in terms of import on sensitive populations.

Each of these two kinds of assessment serves the needs of their performing organizations. The Office of Technology Assessment stressed the importance of conducting site-specific public health assessments for determining public health actions in communities impacted by hazardous waste sites (OTA, 1995).

Table 2.1. Contrasts Between Public Health Assessments and Site-Specific Baseline Risk Assessments (adapted from Johnson, 1991; Harris, 1994).

Public Health Assessment (PHAs)	Baseline Risk Assessments (BRAs)
PHAs use quantitative estimates to provide qualitative judgments about 1) the relative hazard posed by contaminants at or near a site, 2) the need for follow-up health actions (Figure 2.4).	A BRA yields quantitative estimates of health risks. It offers a relative perspective of which contaminants, exposure routes, and media pose the most significant risks to each of the exposed groups selected for evaluation. Provides qualitative conclusions and quantitative estimates of human health risks and ecologic risk with supporting discussions on cleanup.
PHAs generally use the ATSDR's environmental media evaluation guides (EMEGs), which are derived from the ATSDR's Minimal Risk Levels (MRLs) (see Chapter 4) to assist in identifying hazardous substances to be further investigated at a site. Medical and environmental public health perspectives are weighed to assess human health hazards.	Statistical and/or biologic models are used to calculate the EPA-derived health criteria including the EPA's cancer slope factors, reference dose values, and reference concentrations. The EPA also examines whether Applicable, Relevant, Appropriate Requirements (ARARs) have been exceeded.
PHAs contain the Public Health Action Plan, which may include follow-up health investigations, recommendations for additional monitoring, or health education (Figure 2.4).	The EPA conducts site-specific risk assessments and determines cleanup levels and actions consonant with the level of assessed risk.
Public health assessments focus primarily on the health consequences of current and past releases of hazardous substances. In the absence of site-specific information, standard exposure values are used, such as those in the EPA's *Exposure Factors Handbook*.	The EPA examines current and future risks. Standard values derived from the EPA's *Exposure Factors Handbook* are used to calculate risk levels.
The ATSDR may collect exposure information (e.g., by analyzing biologic tissues and fluids) and health outcome data for communities near sites and may survey residents' health concerns.	The EPA collects and interprets environmental monitoring information not previously available for a site. It also collects information on populations exposed and sensitive environments and evaluates the potential for ecologic impacts.
PHAs are advisory.	BRAs bear regulatory weight of authority.

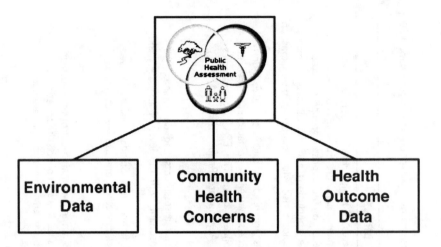

Figure 2.1. Foundations of public health assessments (ATSDR, 1992b).

Core Information and Data

Public health assessments are typically conducted by multidisciplinary groups of toxicologists, environmental engineers, epidemiologists, and other specialists. Their work involves analysis of three core sets of information: environmental data, community health concerns, and health outcome data that form the foundation of a public health assessment (Figure 2.1) (ATSDR, 1992b). A brief elaboration of each set follows.

Environmental data for a hazardous waste site or similar source of hazardous substances includes information on environmental contamination and environmental pathways. For example, environmental contamination data might include concentrations of benzene in groundwater samples taken on the site, and environmental pathways might include data on levels of contaminants in private wells that furnish water for residents' use. Contamination and pathway information is provided in site-specific remedial investigation reports, preliminary assessments, geologic surveys, and site inspection reports obtained from the EPA, other federal departments, and pertinent state and local environmental departments. Also, site visits by health assessors are important sources of environmental data because local residents may know of hazardous waste disposal previously unreported to environmental agencies.

Community health concerns associated with a waste site constitute the second key component of public health assessments. The community associated with a hazardous waste site includes the population living near the site, local health officials, other local officials, and the local news media. To obtain information on a community's health concerns, a health assessor must become an investigator through interviews in the community and other efforts to involve the community in conduct of the public health assessment. Local residents often know how hazardous substances were generated at a former industrial facility or how waste was managed at the facility. Getting the community's involvement is a first step in the risk communication process (Chapter 8). Woe will be the health assessor's lot who conducts a public health assessment without first hearing the community's health concerns.

Health outcome data are the third key source of data for conducting public health assessments. Health outcome data are community-specific and can include databases at the local, state, and national level, as well as data from private health care organizations and professional institutions and associations. Databases to be considered include medical records, morbidity and mortality data, tumor and disease registries, birth statistics, and surveillance[1] data. In some communities, biologic specimens are obtained from residents for evaluating exposure to hazardous substances or signs of adverse health conditions.

Relevant health outcome data can play an important role in assessing the public's health associated with a particular hazardous waste site and for determining health activities following the public health assessment. Of the three core sets of information necessary for conducting public health assessments, health outcome data are usually the least available to the health assessor and, when available, are the most difficult to relate to the site being assessed. This is because health outcome data are usually collected for other than environmental health use. For example, tumor registries are maintained for health care providers to use in assessing cancer prevalence and the cost and effectiveness of cancer prevention and treatment programs. (Environmental health is used here to denote the area of public health that is concerned with effects on human health of socioeconomic conditions and physical factors in the environment.)

The conduct of a public health assessment for a particular waste site requires the assessor to implement a structured analysis of the aforementioned three core sets of information. The process is described in a guidance manual the ATSDR developed in conjunction with state health departments (ATSDR, 1992b). Figure 2.2 shows a general schematic of the public health assessment process that is illustrated later in this chapter.

Findings of a public health assessment are contained in a report that has a structured format. A draft report is first released to relevant federal and state agencies to catch any errors of fact, analysis, or interpretation and then to the community and public at large. A review and comment period for this draft public health assessment report allows the public to use factual data to advise the health assessor of any errors or misinterpretations. The format of a public health assessment report is illustrated later in this chapter.

Uses of Public Health Assessments

The ATSDR is the federal agency primarily responsible for the public health activities of CERCLA. The ATSDR's initial activity at a hazardous waste site is usually a public health assessment, which is required for every site the EPA proposes or places on the NPL. Public health assessments can also be conducted in response to petitions from the public. By law, the ATSDR must respond to each petition and explain to the petitioner in writing the basis for not accepting a specific petition. Through September 1997, the ATSDR had received 378 petitions to conduct public health assessments. The agency accepted 220; another 90 petitions are awaiting decision. Thus, the ATSDR accepted 76% of 288 petitions for which decisions have been made, the remaining 24% were either denied and returned to the petitioner or converted by the agency into health consultations. A **health consultation** is ATSDR's response to a specific question or request for information pertaining to a hazardous substance or facility (which includes waste sites). It is a more limited response than is an assessment.

The sources of petitions are shown in Figure 2.3. More than half of all petitions come from community groups or individuals; 17% come from nonelected government offi-

---

[1] See glossary for definition.

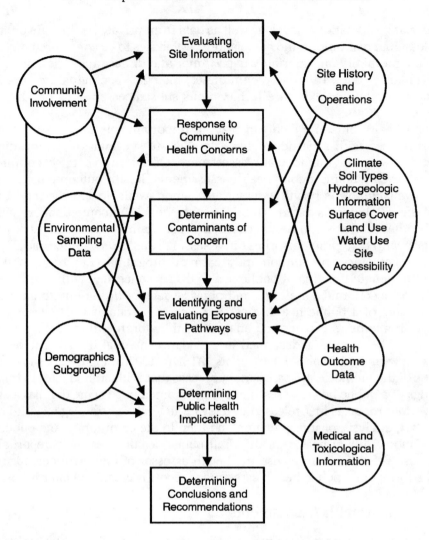

Figure 2.2. Factors in the health assessment process (ATSDR, 1992b).

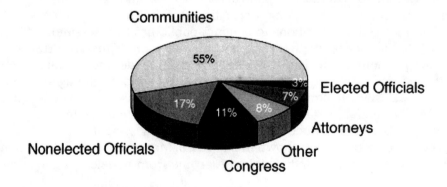

Figure 2.3. Sources of petitions for public health assessments.

cials, and 11% emanate from members of Congress. If the intent of Congress was to provide individuals with a mechanism in CERCLA by which they could have their health concerns addressed, the number and sources of petitions indicate that congressional intent is being realized.

In addition to public health assessments and health consultations, the ATSDR's initial response at sites can also include public health advisories (Glossary). While developing any of these investigations, the ATSDR engages community participation through a variety of avenues such as public meetings, public availability sessions, and Community Assistance Panels (CAPs), which are community-based panels formed to advise the ATSDR on issues of concern. The documents the ATSDR produces during the process are placed in a public repository, where the public is encouraged to review and comment on them. The public health assessments, health consultations, and public health advisories then undergo the ATSDR review to determine whether follow-up, health-related activities are needed for populations at risk in the affected communities.

When the outcome of a public health assessment indicates a need, the ATSDR and state health departments can initiate a variety of actions, including: pilot health-effects studies (usually in the form of symptom-prevalence studies, disease-cluster investigations, exposure studies), various kinds of epidemiologic studies, health surveillance systems, and exposure registries. Examples and the details of these kinds of health actions are described in Chapter 5. The principal health actions resulting from public health assessments are illustrated in Figure 2.4.

As commented previously, the public health assessment serves a diagnostic function. Each assessment results in recommendations and a plan of follow-up health actions. Public health assessments determine much of what the ATSDR and state and local health agencies will do in communities that have hazardous waste sites. Moreover, public health assessments are designed to identify: (1) knowledge gaps concerning the toxicity of substances identified at the facility or site being assessed, (2) communities near sites, facilities, or other sources of hazardous substances releases where biologic measurement of human exposure or medical investigations are needed, and (3) the need for additional health and environmental information (e.g., additional off-site environmental monitoring).

Case Example: Escambia Wood–Pensacola

The Florida Department of Health and Rehabilitative Services conducted a public health assessment (PHA) of the Escambia Wood-Pensacola site in Pensacola, Florida, in conjunction with the ATSDR. Some health risk communication issues related to this site are described in Chapter 8. The following summary is taken verbatim from the PHA report.

Report Summary

The Escambia Wood–Pensacola hazardous waste site (EWP), also known as the Escambia Treating Company, is in Pensacola, Escambia County, Florida. It is on Palafox Highway about one mile north of Fairfield Drive. The area around the site includes homes, light industries, and businesses. From 1942 to 1982, EWP treated wood poles and timbers. They used two chemicals to treat the wood: creosote and pentachlorophenol. The soil and groundwater on and around the site are contaminated. In 1991, EWP went out of business and abandoned

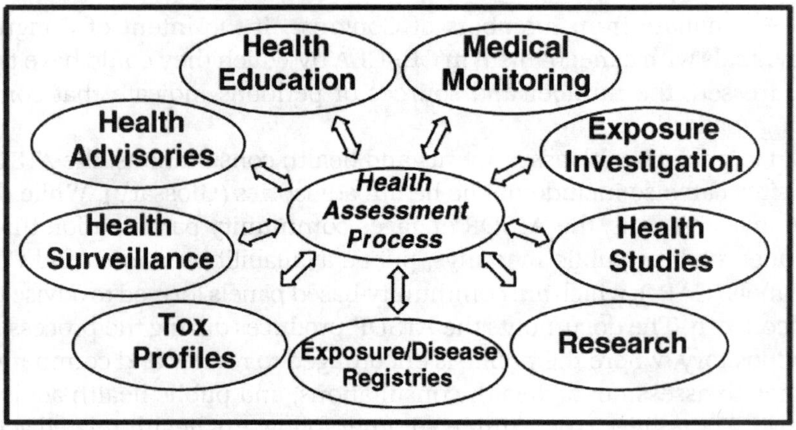

Figure 2.4. Uses of public health assessments (ATSDR, 1992b).

the site. From 1991 to 1993, the U.S. Environmental Protection Agency (EPA) dug up contami-nated soil and stored it under a secure high density polyethylene cover. During this time, nearby residents complained of odors coming from the site that irritated their eyes and skin and sometimes made it difficult for them to breathe.

Nearby residents believe that contamination released into the air during the excavation work has worsened their health problems. They are worried that breathing these contami-nants may cause cancer or make them sick in the future. Residents are also worried that future work at the site will expose them to more hazardous chemicals and cause more health problems.

We focused our public health assessment on the following chemicals: arsenic, benzene, dioxins/furans, pentachlorophenol (PCP), and polycyclic aromatic hydrocarbons (PAHs). Former workers at the plant and trespassers on the EWP site may have accidentally eaten contaminated soil or breathed contaminated dust in the air. Arsenic and PCP in this soil or dust may have caused skin irritation. Arsenic and PCP may have also caused liver, kidney, and nervous system damage; and blood-forming and immune system problems. Arsenic and benzo(a)pyrene (a PAH) in the soil or dust may have increased the risk of skin, lung, and blood-forming system (leukemia) cancers. Former workers at the EWP site and site trespass-ers may be at an increased risk of liver, spleen or adrenal cancer from PCP.

Former workers at the plant, trespassers on the site, and nearby residents may have acci-dentally eaten soil contaminated by dioxins/furans. They may have also breathed dioxins/furans in contaminated air. The U.S. Public Health Service and EPA are currently reviewing studies on the health effects of these chemicals to estimate their toxicity.

Groundwater under and to the southeast of this site is contaminated and is moving toward the east-southeast. This groundwater has already mixed with contamination from the nearby Agrico Chemical Company hazardous waste site. Contaminated groundwater, however, is unlikely to affect people since there are no public or private drinking water wells in the area.

Based on the information we have, this site is a public health hazard. We recommend EPA maintain site security and put up more hazardous waste warning signs. We also recommend EPA take more surface soil samples in the neighborhood north of the site. This is necessary to find out how much soil contamination exists and how far it extends. We recommend EPA make sure cleanup companies protect their workers from hazardous chemicals. Finally, we *recommend that the Agency for Toxic Substances and Disease Registry (ATSDR) conduct a comprehensive community health evaluation of residents near the site. This is necessary to find any illnesses that may be connected to hazardous chemicals from this site.*

Several public health features are evident from this summary of the public health assessment report for the Escambia Wood–Pensacola site and the digest of the report. First, a clear statement is made to the public that the site was found to be a hazard to public health. Second, the assessment examined both on-site and off-site contamination data and determined that substances of public health concern were present, including arsenic, benzene, dioxins/furans, pentachlorophenol, and PAHs. Third, a usually major exposure pathway, groundwater, was found not to be a current route of human exposure to hazardous substances here because potable municipal water was being used by local residents. Fourth, the public health assessment for this site provides a clear, unambiguous agenda for specific public health interventions, including the recommendation that the ATSDR conduct a health evaluation of local residents. Fifth, the public's comments submitted to the ATSDR during the public comment period were considered and responses were placed in the public health assessment report.

HAZDAT DATABASE

During the conduct of public health assessments, the ATSDR acquires site-specific environmental contamination data collected by the EPA, DOD, DOE, other federal agencies, and state agencies. The ATSDR compiles these data, along with toxicity data for individual hazardous substances and human exposure data, into a national database called HazDat. Many of this book's statistical data pertaining to NPL and other waste sites were derived from the HazDat database. HazDat is a comprehensive, on-line database for hazardous waste sites available to the public through computer information networks.[2]

As will be discussed in Chapter 4, the data aggregated in HazDat have enabled ATSDR analysts to make observations about (1) the sources of uncontrolled hazardous waste sites and chemical classes, (2) the priority hazardous substances associated with hazardous waste sites, (3) the environmental media that are most often contaminated and the contaminants found in completed environmental pathways, and (4) primary mixtures of contaminants.

HazDat contains environmental contamination and other data on more than 3,000 uncontrolled hazardous waste sites for which the ATSDR has conducted public health assessments, prepared health consultations, or provided responses to emergencies involving releases of toxic substances into community environments. It contains toxicity information abstracted from the ATSDR's Toxicological Profiles for more than 200 substances most frequently encountered at NPL sites.

As previously noted, the amount of hazardous waste left in the environment is enormous. To remediate the environment requires knowledge about what substances are found in waste sites. Data evaluated in HazDat from all public health assessments that ATSDR conducted through December 1996 for 1,391 uncontrolled hazardous waste sites (including 1,155 NPL sites), indicate that about 39% are waste storage/treatment facilities or landfills; 32% are abandoned industrial facilities, 8% are waste recycling facilities,

---

[2] HazDat is available through the Internet World Wide Web (http://atsdr1.atsdr.cdc.gov:8080).

5% are mining sites, and 5% are government properties (Susten, 1997). Figure 2.5 depicts the major categories of hazardous waste sites resulting from this analysis.

Noteworthy from an examination of Figure 2.5 is that more than two-thirds (71%) of uncontrolled hazardous waste sites are (1) storage/treatment facilities or landfills and (2) abandoned industrial facilities. This implies that waste management facilities used in the past were inadequate to contain the hazardous waste stored or treated in them. Furthermore, industrial facilities did not adequately manage on-site hazardous waste, leading to contaminated properties that in turn were abandoned.

PRIORITY SUBSTANCES IN COMPLETED
EXPOSURE PATHWAYS

One key step in preparing a public health assessment is to identify all completed exposure pathways at a site. A completed exposure pathway consists of the following five elements: a source of contamination, an environmental medium, a point of exposure, route(s) of exposure, and a receptor population (ATSDR, 1992b). Receptor populations include community residents and any relevant worker populations. All five elements must be present for a pathway to be considered completed (Table 2.2).

To illustrate a completed exposure pathway, consider an uncontrolled waste site known to be a **source** that is releasing contaminants. Environmental sampling indicates an **environmental medium**, groundwater, has been contaminated with trichloroethylene (TCE). Furthermore, the contaminated groundwater has caused a **point** of exposure, private wells, to be contaminated. A survey reveals that the private wells are used to supply drinking water, a **means** for exposure. Persons who use the well water constitute a **receptor** population. If any of these five events is missing, the waste site would not be considered to have a completed exposure pathway.

The ATSDR's evaluation of data in HazDat for the years 1993 through 1996 showed 168 sites to have completed exposure pathways. At 91% of these sites, the exposure occurred through multiple contaminated media. Where exposure occurred from only one contaminated environmental medium, the following media were contaminated at the indicated percentages of sites: water (18%), soil (5%), air (4%), and biota (3%). However, these contamination data need to be understood in the context of how the EPA collected them. When hazardous waste sites are evaluated, soil and groundwater are almost always sampled; however, for cost-containment reasons, air monitoring and sampling of biota are usually conducted only if evidence exists that contaminants in these media are likely to be of health import.

Knowledge of the hazardous substances found in completed exposure pathways is of great importance because persons are potentially exposed to these substances. An analysis of NPL sites with completed exposure pathways reveals 30 contaminants that are found in completed exposed pathways of at least 6% of these sites. The 30 substances will be called Completed Exposure Pathways Priority Substances (CEPPS) throughout this book. The choice of 6% is entirely arbitrary. These substances are listed in Table 2.3, which was derived from the HazDat database and represents data from 1,450 NPL sites addressed in public health assessments, advisories, or consultations. Of this number, 530 NPL sites have one or more completed exposure pathways. To illustrate how Table 2.3 was constructed, TCE was found at 213, or 40.2%, of the 530 NPL sites with completed exposure pathways.

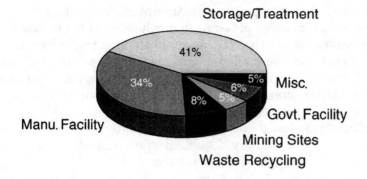

Figure 2.5. Categories of NPL sites.

Table 2.2. Elements of a Completed Exposure Pathway.

- A **source** of contamination
- An environmental **medium**
- A **point** of exposure
- **Mean**(s) to exposure
- A **receptor** population

Table 2.3. Hazardous Substances (n=30) Found in Completed Exposure Pathways at Indicated Percentages of 530 NPL Sites with Such Pathways (n=530).

Substance	%	Substance	%
Trichloroethylene	40	1,2-Dichloroethane	12
Lead	34	Methylene chloride	12
Tetrachloroethylene	30	Manganese	11
Arsenic	23	Toluene	10
Benzene	21	Copper	10
Cadmium	17	Nickel	9
Chromium	17	Carbon tetrachloride	8
1,1,1-Trichloroethane	16	Barium	8
Polychlorinated biphenyls (PCBs)	15	Polycyclic aromatic hydrocarbons (PAHs)	8
1,1-Dichloroethene	14	trans-1,2-Dichloroethene	8
Chloroform	14	DEHP [di(2-ethylhexyl)phthalate]	8
1,1-Dichloroethane	13	Antimony	7
Vinyl chloride	13	Benzo(a)pyrene[a]	7
Zinc	12	Beryllium	6
Mercury	12	Naphthalene	6

[a] Found in ATSDR's ToxFAQ for polycyclic aromatic hydrocarbons (PAHs).

Thus, about 36% of all sites assessed by the ATSDR have one or more completed exposure pathways, although the percentage varies from year to year. For instance, in fiscal years 1993–1994, 60% of sites had one or more completed exposure pathways (Johnson, 1995), a figure that grew to 80% for sites with public health assessments conducted from 1993 through 1996. The ATSDR attributes the increase in percentage of sites with completed exposure pathways to improvements in how the EPA ranks NPL sites and to the ATSDR's earlier presence in conducting its health assessments, which leads to the assessment of sites before exposure pathways might have been interdicted.

Identifying the substances found in completed exposure pathways is one way to set priorities among CERCLA hazardous substances. Therefore the 30 substances listed in Table 2.3 are a singularly important set of hazardous substances for CERCLA purposes, including human health purposes. The exposure assessment of these 30 substances is addressed in Chapter 3, and their toxicologic properties are described in Chapter 4. ToxFAQs, which are brief summary documents, are available to the public for the 30 substances in Table 2.3 through ATSDR's website.

The ATSDR and state health departments use data on the substances found in completed exposure pathways for several purposes. These include prioritizing health investigations, developing toxicologic databases, focusing toxicologic research, and refining the agency's list of priority hazardous substances. Knowing the priority substances gives focus. Toxicology databases can be developed and kept current, and risk assessments can draw on a coherent body of scientific information.

HEALTH HAZARD CATEGORIES

In its public health assessments, the ATSDR classifies the health hazard posed by individual uncontrolled hazardous waste sites according to the six categories shown in Table 2.4. (ATSDR, 1992b). A site is assigned one of these categories on the basis of professional judgment, using weight-of-evidence criteria; the assignments are not risk-based derivations. The ATSDR's site categories differ from the EPA's Hazard Ranking System for the same sites. For example, the EPA's ranking scheme takes into account ecologic effects and environmental hazard; the ATSDR's scheme focuses solely on human health impact. By classifying sites according to human health hazard, the ATSDR is able to direct program resources and effort to those sites believed to present the greatest hazard to the public's health. Some EPA regions use the ATSDR's hazard categories to refine their priorities for site remediation.

Some key distinctions that determine a site's hazard category are highlighted in Table 2.4. For example, the primary distinction between Urgent Public Health Hazard and Public Health Hazard is the duration of exposure to hazardous substances. In the former, short-term exposures are of concern; long-term exposures shape the latter hazard category. Implicit in this distinction is the potential level of exposure; short-term exposures are presumed of higher level than are the long-term exposures that characterize the Public Health Hazard category. Also noted in Table 2.4 is the category Indeterminate Health Hazard, which is the outcome when the health assessor has insufficient information to make a hazard determination. This often results from inadequate off-site environmental contamination data or inadequate toxicologic information.

Table 2.5 shows the percentage of sites placed in each hazard category by the ATSDR's public health assessments completed through December 1996 and in individual fiscal years (1992–1996). The number of public health assessments exceeds the number of waste sites assessed because assessments of some sites were updated when new environmental or health data became available.

The percentages of uncontrolled hazardous waste sites that the ATSDR has classified as Urgent Public Health Hazards or Public Health Hazards has varied, generally increasing over time. An increasing percentage of sites was categorized as health hazards after

Table 2.4. ATSDR Public Health Hazard Categories (ATSDR, 1992b).

Hazard Category	Description/Criteria
Urgent Public Health Hazard	This category is used for sites that pose urgent public health hazards as a result of **short-term exposures** to hazardous substances.
Public Health Hazard	This category is used for sites that pose public health hazards as a result of **long-term exposures** to hazardous substances.
Indeterminate Health Hazard	This category is used for sites with **incomplete information**.
No Apparent Public Health Hazard	This category is used for sites where human **exposure** to contaminated media is occurring or has occurred in the past but **at levels below those of health concern**.
No Public Health Hazard	This category is used for sites that **do not pose** public health **hazards**.
Unclassified	This category is no longer used, but was used circa 1985 for sites with very few data.

Table 2.5. Percentage of Sites in Each ATSDR Hazard Category According to Public Health Assessments (PHAs)[a] (Johnson, 1995; ATSDR, 1996a).

Category	All PHAs	FY 1992	FY 1993	FY 1994	FY 1995	FY 1996
	(n=1,826)	(n=233)	(n=66)	(n=70)	(n=37)	(n=75)
Urgent Hazard	2%	2%	5%	6%	5%	1%
Health Hazard	21%	35%	47%	50%	46%	33%
Indeterminate Hazard	64%	41%	35%	33%	24%	20%
No Apparent Hazard	9%	20%	11%	8%	22%	39%
No Hazard	4%	2%	2%	3%	3%	7%
Unclassified	<1%	0%	0%	0%	0%	0%

[a] Data reflect the status of the PHA documents through February 1997.

1990, when more extensive EPA environmental contamination data became available to the ATSDR, and when human health data from state health departments became more available to the ATSDR. In addition, the EPA made improvements in its Hazard Ranking System, its method for ranking hazardous waste sites in terms of their overall hazard to human health, ecologic impact, and environmental degradation. Furthermore, the ATSDR developed the capacity to investigate sites sooner after their proposed placement on the NPL.

In the years 1992 through 1996, the sites that the ATSDR classified as Urgent or Public Health Hazard together averaged 46% (Table 2.5). It is important to note that the actual hazard to health must be established by site-specific epidemiologic or other investigation. Nonetheless, about half the CERCLA sites the ATSDR assessed from 1992 onward pose threats to the health of persons living near the sites at the time ATSDR conducted the assessment.

It is important to stress that the ATSDR's public health assessments focus primarily on current or past releases of hazardous substances, therefore sites that currently represent lesser levels of health hazard should be understood as being possible hazards in the future should releases occur.

The category Indeterminate Hazard indicates sites for which information available to the ATSDR was not sufficient to permit a hazard classification. For the years 1992 through 1996, an average of 31% of the sites were placed in the Indeterminate Hazard category. However, the percentage of such sites steadily decreased after 1992, reaching 20% of sites in 1996. This decrease is due to a closer working relationship early in the site-inspection phase of the CERCLA process between the ATSDR and the EPA, resulting in improved collection of site contamination data.

The combined percentages for the categories No Apparent Hazard and No Hazard, averaged for the years 1992 through 1996, is 23%. Since 1995, the trend has been for an increased percentage of sites placed in both these categories. Here, too, the increase is largely a consequence of better ATSDR and EPA methods for identifying sites for which, in this case, the hazard to human health lies with the potential of future releases of substances from the sites.

The foregoing statistical data, drawn from the ATSDR's HazDat database for sites with completed health assessments, led to the following statement to Congress in 1995 (Johnson, 1995):

Of the 136 sites for which public health assessments were conducted and advisories were issued in fiscal years 1993 and 1994, ATSDR classified 54 percent as health hazards. Historically, 23 percent of 1,719 public health assessments for more than 1,300 Superfund sites represented a health hazard, according to ATSDR's public health assessment criteria.

The data in Table 2.5 update this statement to Congress.

PUBLIC HEALTH ADVISORIES

Health departments and agencies use public health advisories to warn the public of imminent, serious threats to their health. A public health advisory may be issued, for instance, when a local water supply is found to be contaminated by biologic or chemical agents under conditions that make the water unsafe to drink. Public health advisories are intended to effect personal and institutional actions. In the contaminated drinking water example, persons would be advised to boil the water (microbiologic contamination) or forego the water (chemical contamination). Institutions like local or state environmental groups would be expected to identify the source of contamination and eliminate it.

Acting under its CERCLA and RCRA authorities to conduct public health assessments and provide consultations, and in its capacity as a federal public health agency, the ATSDR issues a Public Health Advisory when an imminent, serious threat to the public's health is detected. An ATSDR Public Health Advisory is sent directly to the EPA Administrator to alert the EPA to a public health threat. Other government agencies—such as state and local health and environmental agencies—are also notified. The ATSDR's Public Health Advisories take into account the following:

- the levels (or concentrations) of hazardous substances in the environment,
- whether persons might be exposed and by what pathways of exposure,

- what harm the substances might cause to a person's health,
- whether working or living at or near the contaminated site puts persons at risk,
- whether other hazards; for example, physical hazards like unsafe buildings, are present.

Because a Public Health Advisory is meant to effect quick and serious actions by health and environmental authorities, several significant outcomes can result from the issuance of an advisory:

- temporarily or permanently relocating persons at health risk,
- EPA's placing sites on the NPL,
- removing contaminants from the site,
- providing residents with safe drinking water,
- educating local physicians and community residents.

Since 1987, the ATSDR has issued 24 Public Health Advisories. Each has achieved the intended purpose—to alert the public and government agencies of an imminent, serious risk to the public's health. A case study follows of a Public Health Advisory conducted under the ATSDR's CERCLA authorities.

Case Study: Hoboken, New Jersey, Mercury Contamination

In 1996, the ATSDR issued a public health advisory about a building in Hoboken, New Jersey. The advisory notified the EPA, the State of New Jersey, the Town of Hoboken, and the building's residents of an ongoing, imminent health hazard due to mercury contamination within the building. This case study illustrates the importance of conducting a thorough environmental assessment whenever older, industrial property is being renovated for residential use.

Background—In the mid-1990s an unused industrial facility was renovated by real estate developers. It was converted into a five-story structure containing 17 condominium units, an attached townhouse, and artists' studios (ATSDR, 1996b; Sasso et al., 1996). The industrial complex had been used by several companies from 1920 through 1993 as a site to manufacture such items as vapor lamps, electrical equipment, and tools and dyes. The building is of brick construction with wooden floors and solid wood floor supports. Renovation included the installation of a new elevator, and installation of new plumbing and electrical conduits throughout all floors. Residential occupancy began in mid- to late-1994.

Environmental Data—In early-1995, during renovation of one of the condominium units, pools of mercury were observed in the subflooring (ATSDR, 1996b). This triggered action by the building's tenants association. Consultants were retained later in 1995 to measure mercury levels in the building and to perform remediation. More than 200 gallons of debris contaminated with mercury were removed from the unoccupied fifth-floor unit, but mercury was still present in the building. One consultant measured mercury vapor levels ranging from 4 to 21 $\mu g/m^3$ in breathing-zone air. In some locations in the building, mercury vapor air levels ranged from 24 to 77 $\mu g/m^3$ at wall and floor openings (ATSDR, 1996b) and up to 888 $\mu g/m^3$ in areas where liquid mercury was visible (Sasso et al., 1996).

In October 1995, drops of elemental mercury were observed in fourth-floor condominium units, including on stove and countertop surfaces. Mercury vapor levels on the fourth floor, measured by a private consultant, were 7 µg/m³ to 26 µg/m³. In late November, urine mercury levels for five residents of the two fourth-floor units ranged from 11 µg/L to 65 µg/L of urine (normal range: 0–20 µg/L). In mid-December local health officials asked the New Jersey Department of Health (NJDOH) and the EPA for assistance. Maximum air mercury levels detected by the NJDOH ranged from 10 µg/m³ to 50 µg/m³ (Sasso et al., 1996). At the time the public health advisory was issued, the building had 32 occupants; six were children ages nine months to eight years (Sasso et al., 1996).

With assistance from the ATSDR, the Hoboken Board of Health and the Hudson Regional Health Commission, the NJDOH analyzed urine specimens from 29 of the building's 32 residents. Results indicated concentrations of mercury in urine ranging from 5.7 µg/L to 102 µg/L. Of the 29 persons, 20—including five of the six children—had urine mercury levels equal to or greater than 20 µg/L; eight of the residents had urine mercury concentrations equal to or exceeding 56 µg/L (Sasso et al., 1996). Thus, measurements of mercury vapor in air and total mercury in urine evidenced residents' exposure to mercury at levels of public health concern. These data led the ATSDR to issue a health advisory on January 22, 1996.

Health Issues—Mercury is a toxic substance at sufficiently high levels of exposure. The ATSDR's Minimal Risk Level[3] (Chapter 4) for chronic duration inhalation of mercury vapor is 0.014 micrograms of mercury per cubic meter of air. Total mercury levels in urine are typically 4–5 µg/L in the American population, ranging up to 20 µg/L (ATSDR, 1996b). The nervous system is very sensitive to all forms of mercury, although some forms are more harmful than others because of the body's reaction to them. Breathing metallic mercury vapors, or breathing or ingesting methyl mercury causes the most harm because more mercury in these forms reaches the brain. Exposure to high levels of metallic, inorganic, or organic mercury can permanently damage the brain, kidneys, and developing fetus. Effects on brain function can result in irritability, shyness, tremors, changes in vision or hearing, and memory problems. Short-term exposure to high levels of metallic mercury vapors may cause other effects including lung damage, nausea, vomiting, diarrhea, increases in blood pressure or heart rate, skin rashes, and eye irritation.

Communication Issues—Because the residents of the building themselves recognized the hazard of mercury contamination, many of the health risk communications focused on interpreting the environmental and biologic data. Local health officials met often with residents to explain their interpretation of the data. Federal and state officials coordinated an education effort on mercury toxicity for physicians in the Hoboken area. Newsmedia gave some attention to the mercury contamination problem, but this was not sustained coverage, probably because the mercury contamination was not widespread in the community.

Outcome—On January 4, the city of Hoboken condemned the building. Residents evacuated it on January 12, 1996. Through arrangements coordinated by federal, state, and local health officials, residents were referred to a clinic in New Jersey that specializes in environmental and occupational health problems (Fiedler et al., 1996). Twice the clinic measured, five to eleven weeks apart, urine mercury levels in adults and six children and administered a battery of neuropsychologic tests. A urine mercury level

---

[3] See Glossary.

of 50 µg/L is considered elevated. For adults, the mean urine mercury concentration, corrected for creatinine, was 24.8 µg/L for the first sampling period (conducted before evacuation of the building) and 31.9 for the second sample. For children, the corresponding mean values were 52.5 and 98.6 µg/L. (The increase in urine mercury concentrations is thought to be the result of laboratory error.) Although no significant neurobehavioral deficits were observed in the group of exposed subjects, urine and hair mercury levels were correlated with measures of impaired fine motor coordination. Because stress levels were elevated in the study group, special crisis counseling sessions were held to help residents deal with the consequences of evacuation.

In October 1997, the EPA announced that the condominium building would be demolished under the agency's CERCLA authorities. The cost of demolition and relocation of residents was $14 million, which does not include the fair market value of the residents' property (Smothers, 1997).

Lessons Learned—The public health advisory was an important statement that codified the concern of federal, state, and local health officials. The advisory was useful in mobilizing risk management actions, such as residents' evacuation of the mercury-contaminated building, a program of medical evaluations, and crisis counseling for persons who had to evacuate their residences.

PUBLIC HEALTH INTERVENTIONS

A unique and particularly significant part of any public health assessment report is the Public Health Action section. An ATSDR multidisciplinary team of physicians, toxicologists, engineers, health educators, and epidemiologists reviews each public health assessment. Public health recommendations are developed for each health assessment, and, when appropriate, are made part of a site's final public health assessment report. The Public Health Action section contains these recommendations, which fall into three categories:

- recommendations to cease/reduce exposure to hazardous substances,
- recommendations for follow-up health investigations,
- recommendations for follow-up educational actions.

Recommendations to cease or reduce exposure to hazardous substances are risk management actions and, under CERCLA, are the responsibility of the EPA. Such actions can include temporary relocation of persons, provision of an alternative supply of water, and restricting access to a site.

Follow-up health investigations can be of two different types: (1) studies using biologic markers of exposure; for example, studies of blood lead levels in children living near NPL sites; and (2) studies of biomedical effects; for example, investigations of patterns of birth defects in children whose mothers lived near a particular hazardous waste site.

Some grassroots environmentalists have stated their opposition to "more studies" of persons exposed to hazardous substances released from waste sites or similar sources of environmental pollution (Dunham, 1995). It is their contention that a person's proximity to a hazardous waste site is presumptive proof that harm to health has occurred. Therefore in their view, "health studies" are not needed. Rather, in their view, government and private industry (i.e., the potentially responsible parties) should be providing health and medical services to the site's residents. This view runs counter to traditional medical and public health practice. Health studies are crucial for determining the nature and extent of morbidity in a community or target population. Physicians use knowledge of morbidity to provide medical care for their patients. Public health practitioners use findings from health studies to prevent further health problems from hazardous substances released from a particular waste site.

Follow-up educational actions recommended in a public health assessment can take the form of educating both health professionals in a community and the population potentially impacted by the site. Professionals' health education would target the needs of local physicians and other health care providers. This could include updates provided by a state's health department, the ATSDR, or environmental medicine specialists on the toxicity of lead or the health hazard of dioxins. Community health education is intended to provide health information to a community or group that is at potential health risk because of hazardous substances released from the assessed site. Community health education includes basic information, developed for a lay audience, on principles of toxicology, epidemiology, and environmental health interventions; for example, methods to reduce one's personal exposure to hazardous substances. Community health education is addressed through public meetings and smaller individual sessions that involve personal interaction between health professionals and the public.

The ATSDR reviewed recommendations in 647 public health assessments it had provided to the EPA and states and found that about 60% of the assessments included recommendations to interdict or reduce current, ongoing exposure pathways (Lichtveld, 1997). These recommendations can include relocation of community residents (rare), provision of alternative drinking water, issuance of fish-consumption advisories, posting of warning notices, or restriction of access to the site. Second, at 61% of sites, follow-up public health actions were indicated. Of these actions, 55% involved the need to implement health education for communities and local health care providers; epidemiologic health investigations accounted for 24% of the recommendations.

Case studies in Chapter 8 describe professionals' and community health education activities in Stratford, Connecticut; Chattanooga, Tennessee; and Pensacola, Florida.

SYNTHESIS

Federal and state health agencies pursue the assessment of hazards to human health from uncontrolled hazardous waste sites and similar forms of substance releases through hazard evaluations called public health assessments. A public health assessment is a multidisciplinary staff's structured evaluation of environmental contamination data, community health concerns, and health outcome information. A public health assessment differs from a site-specific baseline risk assessment conducted by environmental agencies. A public health assessment is a form of hazard evaluation and does not quantify the level of risk. Health officials use it to determine what health actions to take for a given site after the health assessment. Environmental agencies use site-specific baseline risk assessment to help determine site remediation actions.

Public health assessments of uncontrolled hazardous waste sites have shown that all sites do not present the same level of hazard to the public's health. Because the mix of hazardous substances differs from site to site and different exposure pathways can lead to very different levels of human exposure to substances, it is not surprising that individual hazardous waste sites represent different levels of health hazard.

According to the ATSDR's public health assessments, from 1992 through 1996, about 46% of CERCLA sites represented public health hazards under the criteria that define Urgent or Public Health Hazard *at the time the sites were assessed.* Over the same years, another 31% were classified as Indeterminate Hazard, because environmental contamination data were inadequate to make more definitive classifications. This means that about 23% of CERCLA sites, when assessed by the ATSDR or state health departments, did not represent *current* hazards to the public's health. This figure should not be interpreted to mean that the sites are unimportant or that the EPA should not have placed them on the NPL. Indeed, sites are placed on the NPL for reasons that include not only human health risk but ecologic risk and environmental contamination potential as well.

The primary benefit of the public health assessment is to determine what public health interventions should be undertaken. Without a structured analysis like the public health assessment, each site might be viewed as needing health studies, or surveillance, or physician and community health education, or to be added to the ATSDR exposure registry system, or some other kind of intervention. This would be a waste of money and other resources, because the sites differ enough to warrant public health intervention plans that are specific to each site.

As a dividend from conducting public health assessments of all NPL sites, the ATSDR has used the accumulated data to construct HazDat, the only national database on NPL and similar uncontrolled waste sites. The HazDat database permits looking across NPL sites to conduct analyses of patterns of substances most often released. From 1993 through 1996, 80% of sites assessed by the ATSDR had one or more completed exposure pathways. The substances associated with sites that have completed exposure pathways are especially important, for those substances, by definition, come into contact with human populations living near such sites. Using information from HazDat, 30 Completed Exposure Pathways Priority Substances are found at 6% or more of these sites. Subsequent chapters address the exposure and toxicologic implications of these 30 substances.

Chapter Three

Exposure Assessment

What is meant by exposure assessment? How is exposure assessment conducted, and, in particular, what is a biologic marker? How does exposure differ from dose? In answering these questions, this chapter describes the public health importance of assessing the extent and nature of human exposure to hazardous substances released from uncontrolled hazardous waste sites. This chapter is organized around seven kinds of exposure assessment data and the value of each. Emphasis is given to the exposure properties of the 30 substances found most often in NPL sites that have completed exposure pathways. Reference data for these priority substances and key information gaps in knowledge are described.

Determining the effects on human health of hazardous substances released from waste sites requires a knowledge of the conditions of exposure; the toxicity of the substances, singly and in combination; and the susceptibility of persons exposed (e.g., children). This chapter discusses the assessment of exposure to hazardous substances. In general, environmental contaminant levels in communities near waste sites are often available, but individuals' exposure levels are almost always lacking.

Two key concepts in toxicology, **exposure** and **dose**, must be clearly distinguished. The National Research Council (NRC) defines **exposure** as: **"...an event that occurs when there is contact at a boundary between a human and the environment at a specific contaminant concentration for a specified period of time; the units to express exposure are concentration multiplied by time"** (NRC, 1991c). Thus, exposure means contact with a given concentration of an agent for a given time; that is, the amount of a substance available for a specified time. For example, a young child playing for a given time on soil contaminated with lead (e.g., 500 micrograms lead per gram of soil, often expressed as 500 parts per million) could potentially be exposed to lead through ingestion. Similarly, a person who lacks respiratory or skin protection and stands for a specified period near the exhaust pipe of an automobile with its engine running can be exposed to chemical substances for that period.

Specific contaminant concentrations and periods of time will vary for different hazardous substances and different environmental media, and exposure can vary greatly within geographic areas, thus diluting any health findings. Confounders, too, are often unknown or difficult to ascertain (Griffith et al., 1989), and methodologic difficulties exist for assessing a population's exposure to hazardous substances. For individuals, exposure is seldom known, thereby limiting any linkages between exposure and health outcomes.

For these reasons, exposure is often estimated from questionnaire data, employment records, and environmental contamination data for the area in which a study population lives. A resulting problem can be exposure misclassification, or placing persons in exposure categories without confirming their actual exposure. Such misclassifications can weaken any association between exposure and health outcomes.

Dose is different from exposure, although the two concepts are interrelated. Dose is a concept from pharmacology and medicine that measures the body's uptake of a given substance or the actual amount of the substance that enters a person's body (such as lead by all applicable routes of exposure). Physicians prescribe doses of drugs or other therapeutic agents. Dose is generally expressed as units of a substance per unit of body weight or body surface area. In the examples given above for exposure, if the child ingests lead-contaminated soil, this would result in a measurable dose of lead in body tissues (e.g., blood, bone). For the person standing near the auto exhaust, the volume of exhaust gases inhaled or absorbed through the skin during the exposure period would constitute the dose. If the person had worn respiratory and skin protection, the resultant dose would be zero.

As noted in Chapter 2, risk assessment has emerged as a key instrument to determine which environmental hazards, including uncontrolled hazardous waste sites, to manage and according to what priority. Risk Assessment is defined by the Department of Health and Human Services (DHHS) as "...the use of available information to **evaluate and estimate exposure** (emphasis added) to a substance(s) and its consequent adverse health effects" (DHHS, 1985). According to a paradigm the National Research Council described in 1983, risk assessment consists of four elements (NRC, 1983). Recalling that risk is defined as the probability of an adverse health effect resulting from exposure to a hazardous substance(s), it is worth repeating from Chapter 2 the four elements of risk assessment, but with emphasis on the role of exposure assessment (DHHS, 1985):

- Hazard identification—The qualitative evaluation of the adverse health effects of a substance(s) in animals or in humans.
- Dose-response assessment—The process of estimating the relation between **the dose (i.e., the amount of the substance received by the target organism)** of a substance(s) and the incidence of an adverse health effect.
- **Exposure assessment**—The evaluation of the magnitudes, time, and duration, and types (defined by routes and media) of actual or anticipated exposures and of doses, when known; and, when appropriate, the number of persons who are likely to be exposed.
- Risk characterization—The process of estimating the probable incidence of an adverse health effect in humans **under various conditions of exposure**, including a description of the uncertainties involved.

The use of risk assessment by governmental agencies and private sector organizations to characterize, estimate, and manage risk has grown in favor and therefore use. Although risk assessment, like a tulip bulb in early spring, holds the promise of something grand, it requires, in fact, further nurturing and cultivation. Because of data limitations, exposure assessment is an area of particular difficulty in conducting risk assessments for uncontrolled hazardous waste sites and other environmental hazards. As will be discussed in this chapter, research on exposure assessment is occurring and represents a high priority for improving risk assessment and public health assessment.

PURPOSES OF EXPOSURE ASSESSMENT

Materials known to be hazardous to human health have been found in ambient air, groundwater, surface water, soil, and agricultural produce. Although this sounds dire, and sometimes is, environmental contamination may not necessarily lead to adverse effects on human health (although other consequential effects may occur, for example, damage to ecologic systems).

The toxicity of a substance is first and foremost a matter of dose; that is, how much of the substance gets into the body and reaches vulnerable tissues (discussed in Chapter 4). How much exposure occurred and how much of a dose a person received is the relevant question and the essence of exposure assessment. It is often the critical element of risk assessment because too often the lack of relevant human exposure data constitutes the weakest or even the missing link in risk assessment.

Exposure assessment proceeds along the line of questions implied in its definition. What are the contaminants of concern? What environmental media are contaminated? Does any particular pathway of contamination predominate? What are the spatial and temporal characteristics of the exposure? Who and how many persons are exposed? Are persons of particular sensitivity (e.g., children) exposed to contaminants of concern? Answers to these and related questions constitute exposure assessment.

The main factors that must be considered in exposure assessment are summarized in Table 3.1. Programs of exposure assessment, as well as individual investigations of human exposure, must take these factors into consideration.

Briefly, any long-term program of exposure assessment must start with a knowledge of the personnel and monetary resources that will be required to ensure the quality of the program. Developing laboratories that can conduct exposure assessments of environmental media and analyze biologic samples from human populations is a sizeable commitment. National programs that monitor exposure of human populations at risk for adverse health effects that result from exposure to environmental contaminants are vital to protect public health. Considerable cost is associated with such programs.

APPROACHES TO EXPOSURE ASSESSMENT

The philosophy on the conduct of exposure assessment has changed with time. Although historically much attention has been placed on measuring or estimating exposure to a single or principal contaminant, especially for occupational exposures to workplace pollutants, this approach is changing. An important NRC study on exposure assessment is *Human Exposure Assessment for Airborne Pollutants* (NRC, 1991c). The report presents a framework and methods for assessing and analyzing the totality of an individual's exposure to air contaminants in the course of all activities over specified increments of time.

The NRC clearly articulates the necessity for **total exposure assessment**. Total human exposure accounts for all exposure a person has to a specific toxicant, regardless of environmental medium (air, water, soil, food) or route of entry into the body (inhalation, ingestion, dermal absorption) (Lioy, 1990). Although the NRC report focuses on air pollutants, the principles of exposure measurement and characterization described in the report are applicable to other contaminated environmental media. For example, a study of a community population exposed to hazardous substances from a waste site should also assess total exposure to any important source(s) of hazardous substances other than

Table 3.1. Principal Factors in Exposure Assessment (adapted from NRC, 1991c).

- Contaminant and potential biologic response [What is the contaminate(s) of concern and what is known about the toxicity and human health consequences of contact with the contaminant(s)?]
- Specification and selection of the target population [What is the population at risk from exposure to the contaminant? Does the population of concern contain persons with special sensitivity to the toxicants of concern?]
- Available technology for collecting and analyzing personal environmental and biologic marker samples [What equipment, analytical methods, laboratory support, and trained personnel are available to acquire samples from individuals within the population of interest?]
- Spatial and temporal variability of concentration distribution patterns [Is there a particular geographic area of concern? Does the environmental contamination change over time in ways that should be measured to adequately characterize human exposure to contaminants of concern?]
- Selection of the sampling period in appropriate relationship to the time scale of biologic effect [Is the biologic effect dependent on time-related factors of exposure, and if so how should the exposure assessment be conducted to circumscribe the biologic effect?]
- Frequency and intensity of exposure [How often does the exposure occur in the population of interest and how intense is that exposure?]
- Precision and accuracy requirements [How much precision and accuracy is required to have scientifically adequate data on the exposure of human populations, bearing in mind that increased precision and accuracy are accompanied by increased costs in terms of resources?]
- Costs and available resources [What are the financial and personnel resources available to conduct an exposure assessment of a population of concern? If adequate resources are not available, are alternative methods available to obtain the exposure information?]

those released from the waste site being assessed. For a public health assessment, the health assessor must consider the toxicants in air, water, soil, dust, and food and relate these various routes of exposure to the body burden (dose) of the hazardous substances. Also, from the perspective of public health practice, these exposures must be considered in the broader context of nutritional status, tobacco and alcohol use, medications, and other host factors. This has been referred to as **Integrated Exposure Assessment**.

In the lexicon of the EPA this calls for the use of Total Exposure and Assessment Methodology (see, e.g., Lioy et al., 1991). Total exposure assessment should be considered the ultimate and ideal approach to determining human exposure. But practically speaking, total exposure assessment will probably remain as a goal because of the high cost of conducting total exposure assessments and the lack of available analytical methods (Johnson, 1992).

NRC divided the approaches to conducting exposure assessments into two broad classes, direct methods and indirect methods (NRC, 1991c). Each class is further divided into currently available methods, as demonstrated in Figure 3.1.[1] To illustrate these various methods, consider as a hypothetical example an exposure assessment of a community living near a hazardous waste site. The site is part of an operating chemical production plant. Assume that the local health department requested an exposure assessment by federal health officials. Following discussions with community represen-

---

[1] Reprinted with permission from *Human Exposure Assessment for Airborne Pollutants: Advances and Opportunities.* Copyright 1991 by the National Academy of Sciences. Courtesy of the National Academy Press, Washington, DC.

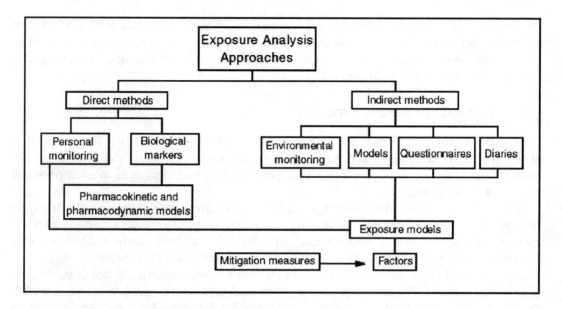

Figure 3.1. Possible approaches for analysis of contaminant exposure (NRC, 1991c).

tatives and local and state health department officials, what exposure assessment methods might the investigators use?

Indirect Methods to Assess Exposure

Environmental monitoring—Following a visit to the community putatively impacted by a hazardous waste site, a review is conducted of records kept by the local environmental waste management authority. The investigators suspect that arsenic and trichloroethylene (TCE) are the primary environmental contaminants at the hazardous waste site. Soil samples collected by the EPA are analyzed for arsenic content and other metals of concern, using a standard analytical method that it developed and quality controlled. It is unclear whether TCE should be sampled in groundwater.

Models—The state environmental agency uses an environmental fate and migration computer model to determine whether TCE has migrated into groundwater and in what direction. Using the model's predicted migration of substances in groundwater, federal and state authorities decide to drill a limited number of sampling wells in a southerly direction from the waste site. An EPA standard method for TCE analysis indicates no migration of TCE into groundwater.

Questionnaires—The area of the community that might be at greatest threat of exposure to arsenic needs to be circumscribed. The local health department designs a questionnaire, drawing on state health department resources, and administers it to area residents to determine whether children play on or near the waste site. Questionnaire responses indicate that a large number of children live in the area and some do play on the waste site. Young children become the investigators' principal group of health concern.

Diaries—After the questionnaire is used to identify persons of presumed greatest risk of exposure, the investigators ask selected families to keep daily records of their children's play and other activities related to contact with the waste site. Information from the diaries will be used to determine spatial (i.e., geographic) and temporal (i.e., time) patterns of children's activities that could help interpret biologic measurements of exposure and possible estimation of dose.

Direct Methods to Assess Exposure

Personal Monitoring—The potential problem of residential exposure to arsenic as a result of parents' workplace exposure is of concern. Investigators (who have gained the cooperation of the parents' employers) ask potentially exposed parents to wear personal monitors at the workplace.

Biologic Markers—Because investigators believe children are the group most likely to be exposed to arsenic, the local health department arranges to collect urine samples for measuring arsenic levels. State laboratories perform the analyses, using analytical methods developed by the federal Centers for Disease Control and Prevention.

Parents' workplace monitoring data reveal that arsenic exposure is not occurring in the chemical plant. The investigators find elevated urinary arsenic levels in some children, compared with health guidelines developed by the World Health Organization. (If reference values do not exist for some contaminants, exposure measurements could be taken from a population comparable to the community under study.) Parents of the affected children are notified by local health officials. A comprehensive plan is implemented to restrict the children's further exposure to arsenic-contaminated soil. Local health authorities agree to monitor the children's urinary arsenic levels to confirm decreased contact with the waste site.

To complete the hypothetical example, local health department officers meet with the parents of children who had elevated arsenic levels. They are told about the arsenic exposure and advised to restrict their children's contact with the waste site. The site is fenced, and local agencies urge parents to monitor the area.

This "hypothetical" example is, in fact, not fiction. It represents the general approach that the ATSDR, state, and local agencies took when investigating the Crystal Chemical CERCLA site in Houston, Texas. The exposure assessment identified several children whose urinary arsenic levels were at the upper extreme of the distribution of urinary arsenic values. After federal, state, and local authorities implemented recommended risk management actions, the ATSDR and local health officials re-evaluated the children who had had elevated urinary arsenic levels. All levels had decreased to acceptable values (ATSDR, 1989a).

The basic approach for determining exposure to substances released from waste sites begins with a knowledge of the nature and extent of the waste in a given site. This is typically learned during a site investigation that the EPA or a state environmental agency conducts. Next, the nature and level of contaminants in environmental pathways is determined. The environmental contamination assessment will depend on several factors, including the structure and integrity of the waste site, the kind and quantity of substances found, the method of containment of these substances, the years of disposal and the amount of waste, the biologic persistence of the site's substances, and the geologic and meteorologic characteristics of the site and its surrounding area (Buffler et al., 1985).

The foundation for assessing the nature and extent of human exposure is knowledge of which environmental pathways are contaminated with hazardous substances. The extent of human exposure depends on the population size, proximity of communities to the waste site, and degree of persons' contact with the site and the site's released substances.

The importance of human exposure measurement or estimation in conducting site baseline risk assessments and public health assessments was recognized early after passage of the CERCLA legislation in 1980. For example, Corn and Breysse (1983) argued that estimating the health risk of residents in communities adjacent to waste sites necessitates estimates of exposure and dose. They note that each route of entry of toxicants into the human body must be considered if the total dose to a person is to be measured. Knowing the concentrations of various hazardous substances in various environmental media facilitates a determination of total body dose. The end result of exposure to a hazardous substance can be a disease, but several steps lie in a continuum of events between exposure and disease, as shown in Figure 3.2.

This simplified scheme is applicable to environmental and other kinds of hazards to health. The first steps in the environmental contamination-to-disease continuum are described in this chapter. Biologic change and disease are covered in Chapters 4 and 5, respectively. Environmental contamination must be present at hazardous waste sites for a risk to one's health to exist. Moreover, hazardous substances must enter the body (i.e., biologic uptake) and come into contact with target organs for biologic change to occur. With some hazardous substances, metabolism occurs, and the metabolite makes contact with target organs.

Biologic change does not necessarily produce disease. For example, in acute exposure of healthy individuals to carbon monoxide (CO), CO molecules attach to hemoglobin in red blood cells, and the hemoglobin has less affinity for attaching to oxygen molecules. Although this biologic change means less oxygen is available to the cells of the body, no disease occurs unless exposure to CO is very great.

The dose at the target site of biologic effect is called the **biologically effective dose**. It is that fraction of the hazardous substance or its metabolite identified at the site of biologic action (e.g., liver) from which the ensuing health effect derives (e.g., liver tumor). Measurement of biologically effective doses (BEDs) is seldom possible because specific measurement methods are generally lacking (NRC, 1991b). In the absence of BEDs, researchers and health assessors will measure internal dose, such as the concentration of lead in bone tissue.

Measuring levels of contaminants in environmental media (e.g., groundwater) is an essential step in assessing human exposure. When biologic indices (e.g., levels of lead in blood) are available for individual hazardous substances, doses of toxicants can be measured. Biologic exposure indices offer several advantages over environmental samples: they give a measure of the internal dose rather than the exposed dose, they account for all routes of exposure, and they can, in some instances, provide an integrated measure of exposure over a period of time (Sim and McNeil, 1992). Biologic fluids or tissues that have been measured for persons' exposure to hazardous substances include blood, urine, breast milk, exhaled air, and adipose tissue (i.e., fat). Regarding Figure 3.2, health investigators look for specific and sensitive biologic indicators, be they hazardous chemical levels, metabolic products, diagnoses of disease states, or measurements of altered physiologic states that predispose or increase susceptibility to disease. If such biologic indicators can be identified and quantified, they become extremely valuable tools for assessing exposure to hazardous substances.

Corn and Breysse suggest that in the presence of dozens or even hundreds of hazardous substances released from hazardous waste sites, a surrogate group of substances

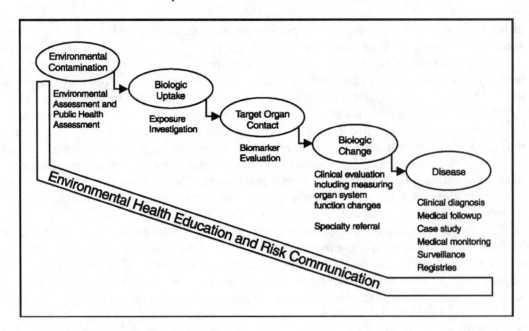

Figure 3.2. Relationship between exposure and disease (Lichtveld and Clinton, 1997, with permission of authors).

could be selected to indicate human exposure associated with the site. This suggestion has not been followed, because indicator substances have not been identified and because health officials are uncertain about which specific indicators would suffice to characterize a complex mixture of substances.

In the absence of exposure data or biologic indices of exposure, distance of residence from the site can be used as a crude surrogate of exposure, although as will soon be explained, this approach is not preferred.

HUMAN EXPOSURE ASSESSMENT

Several approaches are available for assessing human exposure to hazardous substances. These cover the full gamut of possible exposure scenarios, ranging from assessment of past exposures to predicting future exposures. Moreover, exposure assessment includes a very broad spectrum of methods, ranging from simulation models that can predict levels of contaminants in environmental media, to using incidence rates of human disease that estimate probable environmental exposure levels.

The NRC has provided valuable advice on the hierarchy of data quality for exposure assessment (NRC, 1991b). Because all data used for exposure assessment of populations around waste sites are not of equal worth, knowing their strengths and limitations is important, as shown in Figure 3.3. The different kinds of exposure data increase in value as one ascends the pyramid. Exposure data of greatest value are from quantified individual measurements; the least valuable are from use of residence or employment in a geographic area as surrogates for quantified measurements.

Direct measurement includes personal monitoring (i.e., measuring a person's exposure over a designated period of time) and the use of biologic markers. All other

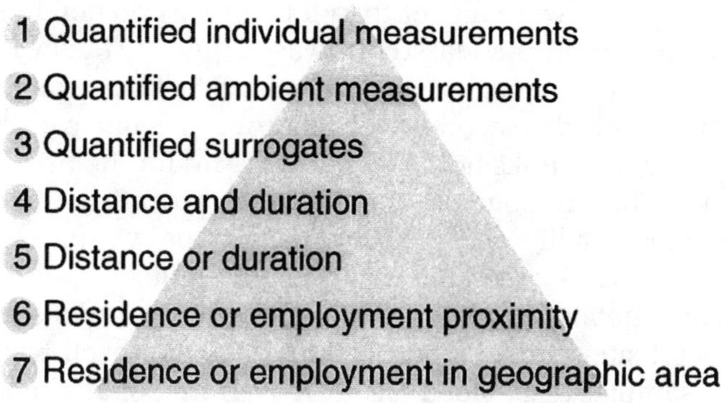

1 Quantified individual measurements

2 Quantified ambient measurements

3 Quantified surrogates

4 Distance and duration

5 Distance or duration

6 Residence or employment proximity

7 Residence or employment in geographic area

Figure 3.3. Hierarchy of data values for exposure assessment (NRC, 1991b).

types of exposure data shown in Figure 3.3 are surrogates because they are indirect measurements of exposure. The sections that follow will elaborate and give examples of the seven levels of the exposure data hierarchy shown in Figure 3.3.

Quantified Individual Measurements

Level 1 Value—Quantified measurements of substance levels in individuals are the most compelling exposure information. **Biologic markers** are the usual means for measuring levels. According to the National Research Council (NRC, 1991b), "...a biological marker is any cellular or molecular indicator of toxic exposure, adverse health effects, or susceptibility." Although the NRC divided biologic markers into three broad classes: (1) markers of exposure, (2) markers of effect, and (3) markers of susceptibility, none of these constitutes a distinct, discrete category, but each is part of a continuous spectrum of events in the human body (Figure 3.2).

For most communities around hazardous waste sites, few quantified data exist on individuals' exposure levels to substances released from the sites. Many reasons exist for this paucity of data. For some sites, releases have been interdicted and the ability to measure any past exposures has been lost because the substances (or their metabolites) do not remain in body tissues long enough to be measured. (Notable exceptions include lead in bone and dioxin in adipose tissue for which the toxicants remain in tissue for several years.) For some sites, exposure, as measured by environmental contamination, may be ongoing, but biologic measurement methods may not be available to assess actual human uptake.

A small body of literature contains quantified individual human exposure data for substances released from waste sites. These investigations (exposure assessment studies) have focused on measuring exposure to specific hazardous substances in human populations thought to be at elevated risk of exposure. Some examples follow.

Amsterdam, The Netherlands: In a study from The Netherlands, tetrachloroethene (tetrachloroethylene, perchloroethylene, or PCE) was measured in the exhaled air of persons working or living near pollution sources in Amsterdam (Monster and Smolders, 1984). Six children (four to five years old) and two teachers in a school located near a factory and 29 elderly persons living in an "old folks home" near a former waste dump were evaluated. The PCE mean concentrations (24 $\mu g/m^3$) in exhaled air of children living near the factory were higher than for comparison children (2.8 µg/m³). PCE ambient air concentration in the classroom was 13 µg/m³. Mean PCE concentrations in the exhaled air of elderly persons living on the first floor were higher (7.8 µg/m³) than for persons living on the second floor or above (1.8 µg/m³). PCE concentrations in the ambient air of the old folks home were 8.2 µg/m³ for the first floor and 1.6 µg/m³ for the second floor. The investigators concluded that breath analysis was an effective measurement method readily accepted by two groups, young children and elderly persons, for whom exposure assessment is often difficult.

Exposure to heavy metals found in and released from hazardous waste sites has been the subject of several site-specific investigations. The approach has typically involved collecting biologic samples (e.g., blood, urine, hair) from populations presumed to be at greatest likelihood of exposure and comparing the levels of metals with those in a reference group. Several examples of this approach follow.

Anaconda, Montana: In response to residents' health concerns, a survey was conducted of children living near a former copper smelter in Anaconda, Montana (Binder et al., 1987). The primary environmental concern was arsenic in soil and house dust. Levels of arsenic ranged up to 1,950 ppm in soil and 386 ppm in house dust in the Mill Creek area. Two other areas near the Anaconda site and a comparison area in Livingston, Montana, were included in the study. Urine samples were obtained from 232 children and measured for total arsenic concentration, adjusted for urinary creatinine concentration. The mean age of the study's participants was four years. Mill Creek children had mean urinary arsenic levels (approximately 55 µg/L) that were significantly different from the comparison children's mean level (approximately 15 µg/L). The investigators speculated that excess levels of urinary arsenic were occurring where soil arsenic levels exceeded 100 ppm.

Indiana and Massachusetts: Stehr-Green et al. (1988) conducted a multisite evaluation of human exposure to polychlorinated biphenyls (PCBs) found at hazardous waste sites. Levels of PCBs in blood serum were measured in persons living near 12 waste sites. Risk of exposure was based on residents' self-reported activity patterns and measured levels of PCBs in soil, groundwater, and surface water. The investigators considered a mean serum PCB value of 6.0 ppb as a population-based background level. At 10 sites, serum PCB levels were at or below the background level, even though PCB contamination levels were as high as 330,000 ppb in soil and 2.5 ppb in well water samples. At two sites (Monroe County, Indiana; Newport County, Massachusetts), serum PCB levels were elevated over the background level. For the Monroe County site, community residents (n = 51) had a geometric mean PCB serum level of 9.0 ppb; for Newport, Massachusetts, the geometric mean was 12.9 ppb (n = 42 persons). This study showed that elevated exposure to PCBs could occur from a waste site, but investigators could not differentiate on a site-specific basis the factors that contributed to uptake.

Leadville, Colorado: An exposure assessment study that looked comprehensively at sources of lead and the association with children's blood lead levels was conducted of residents in Leadville, Colorado, the site of an historic mining district (Cook et al., 1993). Smelting in the community ended in 1961, leaving high levels of soil contaminants. More than 60% of the surface soils had lead concentrations greater than 1,000 ppm. Several sources of lead were identified and measured: yard soil and indoor dust, lead in paint, and tap water. Blood lead levels were measured in 150 children ages 6 to 71 months, and behavioral characteristics were obtained through interviews with a parent or guardian. The arithmetic mean blood lead level was found to be 10.1 µg/dL (SD = 5.6 µg/dL).

Approximately 41% of the children had blood lead levels $\geq$ 10 µg/dL, the public health action level recommended by the federal Centers for Disease Control and Prevention (CDC, 1991). Results showed that three sources of exposure to lead were associated with children's elevated blood lead levels: lead in the backyard soil, lead brought home on miners' clothing, and lead from soldering in the home. Two pathways of exposure were associated with elevated blood lead levels: swallowing things other than food, and taking food or a bottle outside to play. This well-conducted study indicates that lead from past smelting operations had been absorbed by children's bodies, and soil contamination was an important source of exposure to lead.

Denver, Colorado: Investigators in Colorado measured blood lead levels in 443 children age six months to six years living in four lower socioeconomic neighborhoods of Denver (Gottlieb and Koehler, 1994). Two neighborhoods were adjacent to a cadmium refinery. Findings showed that 8% of the children had blood lead levels $\geq$ 10 µg/dL; 45% were below the detection limit of 4 µg/dL. Children living near the smelter had a slightly higher probability of having blood levels $\geq$ 5 µg/dL. This study illustrates one difficulty of relating site-specific waste site releases with body burden levels of toxicants. Removing lead from gasoline contributed significantly to lowering blood lead levels in young children, perhaps overriding any increase in uptake of lead attributable to releases from the smelter and soil contaminated by the smelter.

Silver Valley, Idaho: The Bunker Hill NPL site is located in the Silver Valley of Idaho. The site was mined for lead and other metals and had a smelter that operated for over 60 years until it closed in 1981. The smelter's operations led to massive releases of lead into the environment. Since 1974, lead exposure has been assessed among the children of the Silver Valley through blood lead screening. Mean blood lead levels in children decreased annually until 1992 when a slight increase occurred. Nogueras et al. (1995) conducted a case-control study of matched pairs of Silver Valley children to determine factors that might explain the increase in the mean blood level. The case participants were 69 children whose blood lead levels were $\geq$ 10 µg/dL, matched for sex and age with 69 children whose blood lead was < 10 µg/dL. Information on soil lead levels was provided by the Idaho Department of Health and Welfare. A questionnaire was administered in person to the parent or legal guardian of each study participant to obtain risk factor information, including demographic, mouthing, behavioral, recreation, and occupational information. Regression analysis showed yard soil remediation protected against elevated blood lead levels in children, but pets having free access to and from houses showed increased blood lead levels.

Galena, Kansas: Located in the southeast corner of Cherokee County, Kansas, the Galena NPL site is one of six former smelter subsites in the county. Mining and smelting operations left the area contaminated with lead and cadmium. On-site soil samples collected in 1985–1986 contained lead in maximum concentrations of 500 ppm in surface soil, and 3,800 ppm in surface mine waste and tailings. Maximum concentrations of cadmium were 12 ppm in surface soil and 60 ppm in surface mine wastes. Off-site sampling also revealed lead and cadmium contamination. In 1990 the Kansas Department of Health and Environment conducted a study to measure blood lead and urine cadmium levels in Galena community residents (Feese et al., 1996). Biologic and environmental data and questionnaire information were collected from 167 Galena residents and a comparison group of 283 residents from an area of Missouri. Findings showed young children (6–71 months) in the Galena group had a mean blood lead level (4.2 µg/dL) that was statistically significantly greater than that for the comparison group's young children (3.1 µg/dL). Up to 93% of participants in both study areas had blood lead levels less than 10 µg/dL. Statistically significant correlations between blood lead levels and environmental lead levels in soil (r = 0.49, p < 0.006), dust (r = 0.67, p < 0.0001), and paint (r = 0.29, p < 0.1) were found in the Galena group. Urine cadmium levels did not differ between the Galena group and the Missouri comparison group.

Southwestern Missouri: Heavy metals found in mining waste can pose a health hazard to residents living nearby. A study evaluated blood lead and urine cadmium levels in persons living near an abandoned lead mining, milling, and smelting area in southwestern Missouri, which had become the Jasper County NPL site (Murgueytio et al., 1996). Soil lead levels on the site ranged up to 7,300 ppm and cadmium levels up to 250 ppm. Access to the site was not restricted; some persons used it for recreation. Biologic samples were obtained for lead ascertainment from 391 study participants living near the NPL site and from 271 persons in a comparison community; urine cadmium levels from these two groups were measured in 356 exposed and 249 comparison persons. Children from ages 6–71 months formed about 60% of the exposed group and 50% of the comparison group. Results indicated that children living in the study area had greater exposure to lead but not to cadmium than did comparison children. A significantly higher prevalence of elevated (i.e., > 10 µg/dL) blood lead levels was found in the study group (14%) than in the comparison group (0%). Mean blood levels were 6.25 µg/dL in the study group and 3.59 µg/dL in the comparison group. All children with elevated blood lead levels lived in areas with soil lead levels more than 150 ppm, but lead in paint also may have contributed to the elevation.

Ottawa County, Oklahoma: The Tar Creek NPL site is located in the northeastern corner of Oklahoma in Ottawa County (Herrman et al., 1997). This site, which is part of a larger mining district, extends more than 40 square miles. The mining activities resulted in a large volume of tailings and other mining wastes. Households in the mining district had soil lead levels above 500 µg/kg. Primary contaminants in soil were lead, cadmium, and zinc. Exposure of children to lead was a major concern to state and county health officials. Health officials conducted blood lead screening of 230 children ages 6 to 72 months throughout the mining district. Findings showed that 28% of children had blood lead levels ≥ 10 µg/dL, 9% had ≥ 15 µg/dL, and 2.6% had ≥ 20 µg/dL. These figures indicate that a significant percentage of children were exposed to excessive amounts of lead in their environment. Investigators attribute the source of lead exposure to children's contact with soil and dust and lead-based paint in some households.

Recommended Biologic Markers

Although not described in detail here, the NRC conducted studies of biologic markers specific to five toxicity endpoints: (1) reproductive, (2) pulmonary, (3) nervous system, (4) immune function, and (5) urinary tract. These five areas were selected to correspond to the seven health conditions ATSDR had identified as priority health concerns at CERCLA sites (Table 5.9). The NRC's recommendations on biologic markers are found in five monographs: *Biologic Markers in Reproductive Toxicology* (NRC, 1989a), *Biologic Markers in Pulmonary Toxicology* (NRC, 1989b), *Environmental Neurotoxicology* (NRC, 1992a), *Biologic Markers in Immunotoxicology* (NRC, 1992b), and *Biologic Markers in Urinary Toxicology* (NRC, 1995). These five studies are an important, but underutilized, body of scientific work that merits greater incorporation into exposure assessment investigations.

Another NRC study resulted in the monograph entitled *Animals as Sentinels of Environmental Health Hazards* (NRC, 1991d). This study was designed to "...determine how animals could be used for ecologic and human health risk determinations as well as provide an early-warning system for risk assessment and management." This project bridges the gap between environmental media analysis and human analysis. The NRC concluded, "Domestic and wild animals can be used to identify and monitor a wide range of environmental hazards to human health and ecosystems." Government agencies have apparently not responded to instituting an animal sentinel surveillance system, probably because of the cost of establishing and maintaining this kind of surveillance system.

Quantified Ambient Measurements

Level 2 Value—Quantified ambient measurements of hazardous substances in environmental media are an important surrogate for quantified individual measurements (i.e., biologic markers). Measurement of contaminant levels in soil, air, water, and food can be used in exposure assessment investigations and epidemiologic studies to predict human populations at health risk. Where possible, such investigations should be complemented by measurement of individual exposures. Examples follow of studies in which quantified ambient measurements were essential for relating environmental exposure to biologic uptake of contaminants at hazardous waste sites.

Woburn, Massachusetts: Researchers from the Harvard School of Public Health investigated a cluster of childhood leukemia cases in Woburn, Massachusetts (Lagakos et al., 1986). They obtained information on 20 cases of childhood leukemia (children's ages 19 and younger) diagnosed in Woburn from 1964 through 1983. In 1964, water from two municipal wells contaminated with VOCs from an industrial waste site had been pumped into the municipal water system. All 20 cases were identified in tumor registries maintained by the state or hospital tumor registries. For exposure assessment, the investigators estimated the percentage of contaminated water delivered annually to each household in the sample of households surveyed, using water distribution records and other data. Two exposure metrics were used: (1) cumulative exposure, from birth, to contaminated water from the two contaminated wells and (2) a binary indicator of exposure (no exposure vs. some exposure). Using either exposure metric, the findings, as described in Chapter 5, showed a positive association between exposure to water from the contaminated wells and the incidence of childhood leukemia.

Illinois, Kansas, Missouri, Pennsylvania: Multisite studies are designed to evaluate total exposure patterns when the same hazardous substance is found at multiple sites. These studies afford the opportunity to evaluate composite health outcomes at several sites where persons are exposed to the same hazardous substance. Linking data from multiple sites increases the population size and as a result the statistical power, thus enhancing the ability to detect meaningful associations between disease and exposure to hazardous substances.

In 1991, the ATSDR initiated a multisite study of four NPL sites to evaluate possible adverse health effects from chronic, low-level exposure to lead and cadmium (Stallings and Jones, 1995). The NPL sites were smelting and mixed mining and smelting areas in Illinois, Kansas, Missouri, and Pennsylvania where on-site soil lead levels ranged from 1,500 to 48,000 ppm at the Illinois secondary smelter site, while off-site samples ranged from 27 to 9,493 ppm. Biomedical tests were used to detect any subtle pathophysiologic changes in the renal, immune, and hepatobiliary organ systems. No comparison group was included in the study; the investigators relied on multivariate analysis of the four exposed groups.

Combining populations from all four sites yielded 2,208 study participants: 1,701 constituted the exposed group, 507 were in the comparison group. Of this number 1,465 had household environmental samples taken to measure concentrations of lead in yard soil, house dust, interior paint, and drinking water. From households where environmental samples were collected, 552 (38% of households) had yard soil lead concentrations greater than 500 mg/kg (500 ppm), and 544 (43%) households had house dust lead concentrations greater than 500 mg/kg.

Multivariate analysis was used to evaluate whether area of residence, behavior, socioeconomic factors, and levels of lead and cadmium in biologic samples are associated with concentrations of lead and cadmium in environmental media. Findings showed 93% of participants had blood lead levels < 10 µg/dL. The geometric mean blood lead level in the target population was 4.3 µg/dL (n = 1645 persons), which was significantly higher than in the comparison group (3.4 µg/dL, n = 493). The geometric mean blood lead level among participants from smelting areas (4.5 µg/dL) was slightly higher than for persons from mining areas (3.9 µg/dL). Of the environmental factors evaluated, only yard soil lead concentrations > 500 mg/kg were associated with a mean blood lead level ≥10 µg/dL in children 6 through 71 months of age. These children were more than twice as likely to have a blood lead level ≥10 µg/dL as were children not exposed to mean soil lead concentrations ≥500 mg/kg. Regarding cadmium, 95% of all study participants had urine cadmium excretion levels < 1 µg/g creatinine. The mean urine cadmium excretion was higher in target areas than in comparison areas (0.18 vs. 0.11 µg/g creatinine).

This exposure assessment study is remarkable in terms of the large number of subjects, elaborate environmental characterization, and multivariate analysis. Although the blood lead levels were generally below the health level of concern, 10 µg/dL, the study makes a valuable contribution toward a better understanding of the relationship between soil lead levels and children's blood lead levels.

Quantified Surrogates

Level 3 Value—In addition to quantified ambient measurements of hazardous substances in environmental media, other surrogates of exposure can be developed and quantified. Some examples of quantified surrogates follow.

Mellary, Belgium: An interesting example of a quantified surrogate is the use of cytogenetic measurements as a surrogate for measuring individuals' exposure to releases from hazardous waste sites. Use of cytogenetic biomarkers to assess exposure has been rare in studies of environmental toxicants. Belgium investigators used such markers to examine the extent of exposure of persons living near chemical waste sites (Vleminckx et al., 1997). Ambient air concentrations of several hazardous substances, benzene, toluene, ethylbenzene, xylenes, TCE, and tetrachloroethylene, were documented and thought to be associated with the chemical dumps. Vleminckx et al. used sister-chromatid exchange (SCE), refined by high frequency cell (HFC) analysis, as a sensitive bioassay to detect potential exposure to low levels of genotoxic chemicals. An HFC displays a number of SCEs higher than the 95th percentile of the SCE-distribution in the reference group.

Using cytogenetic biomarkers, the investigators conducted a cross-sectional study of children living near chemical waste dumps in Mellary and Hensies, Belgium, and compared them with children living in Mellery, but distant from the dump, and Wavre, a semirural town. Not only were the SCE mean frequencies higher in the exposed group in 1990, but all individuals in the group exceeded the acceptable number of HFCs. A follow-up in 1992 showed the same outcome. However, in 1993 mean SCE frequencies and HFC incidence in populations of "exposed" children did not differ from those of the comparison groups. The investigators attribute the decreased SCE and HFC rates to closure of the waste site. Unfortunately, the investigators did not have measured environmental levels of contaminants to correlate with the quantified surrogate (SCE) used in their investigation.

Karnes County, Texas: A cross-sectional study (Chapter 5) was conducted of persons living in Karnes County, Texas, near uranium mining activities and attendant waste to ascertain effects on chromosomes (Au et al., 1995). Although data are not cited, the investigators assert that uranium mill tailings in the area of study contained most of the radionuclide species produced in the decay of uranium. The investigators conducted household surveys of radon gas levels. The study population consisted of 24 target (i.e., living within one mile and downwind of the uranium facility) and 24 reference subjects. The reference subjects also lived in Karnes County, but 10 miles from the uranium site.

Blood samples were collected for chromosome assay. Findings showed that individuals who lived near the uranium site had a significantly higher frequency of cells with chromosome aberrations and deletion frequency compared with the reference group. The latter findings are indicative of abnormal DNA repair response in the target group.

Fullerton Hills, California: A different kind of quantified surrogate was part of research to fulfill a graduate school degree. A study of persons living near the McColl NPL site in California assessed the presence of illness symptoms and odor complaints among current residents (Lipscomb, 1989). Results from a study conducted seven years earlier by the California Department of Health Services were used to select three zones in the neighborhood on the basis of odor complaints: high rate of odor complaints (57 persons), lower odor complaint zone (66 persons), and a comparison zone with few odor complaints (70 persons). The questionnaire survey provided information on environmental worry and perception of odors for a one-year symptom period. Prevalence rates were calculated by exposure area.

Findings showed that elevated symptom reporting was strongly associated with several measures of environmental worry. When the association between symptom reports and "exposure" was stratified by level of worry, the association was strongest and remained significant only in the high-worry stratum. The investigator concluded, "...these findings suggest that the concern and worry associated with living in a community with a waste disposal site, rather than the exposure to chemicals from the site, is responsible for the excess symptoms reporting observed in this study." As with the Belgian study, no environmental contamination data were available to correlate with the odor-perception threshold data.

Environmental Contamination Models—Some health assessors have used mathematical and physical models to estimate the concentrations of hazardous substances in environmental media. Because groundwater contamination is frequently the source of exposure for communities near waste sites, the application of groundwater migration models is relevant to assessing the health of populations potentially exposed to hazardous substances. The following example illustrates the use of a model to estimate the distribution of contaminated municipal water to households.

Southington, Connecticut: Aral et al. (1996) modeled a municipal water system to estimate the distribution of hazardous substances that had reached the water system from an NPL site. Assessing the distribution of contaminants in a municipal water system is a complex problem. Aral et al. note that a typical analysis may involve these components: (1) analysis of pumping stress placed on groundwater resources by extraction wells, (2) fate and transport of contaminants in the subsurface and their movement, (3) hydraulic characteristics of the water distribution system, (4) distribution of contaminants within a water-supply network, and (5) spatial analysis of census subdivisions as they relate to contaminant distribution in the waste-supply network.

Aral et al. used hydraulic and chemical-fate modeling to estimate the distribution of contaminants released from the Solvents Recovery Services of New England NPL site in Southington, Connecticut. The water distribution system consisted of 1,700 pipeline segments, 186 miles of pipes, nine groundwater extraction wells, and three municipal reservoirs. Because the water distribution system was so large, Geographic Information Systems (GIS) methodology was used to integrate water system contaminant estimates and census data. GIS refers to computer hardware and software that marries geographic information (i.e., maps) with environmental or other data (e.g., census).

In 1976–1977 groundwater samples from three wells that supplied part of the city's municipal water supply were found to be contaminated with VOCs and possibly heavy metals. By overlaying the computational model and GIS with census data, the exposure to VOCs showed significant spatial variation from one census block[2] to another. For the NPL site under evaluation, the computational model showed that census blocks served by contaminated water were generally east of the contaminated wells, and that areas west of the wells were relatively free of contaminated water. Thus, modeling permitted health assessors to more precisely locate the population of the city at greatest health risk.

Distance and/or Duration

Levels 4 & 5 Value—As Chapter 5 will show, epidemiologists have often used residential distance and/or duration as surrogates for quantified measurement of individuals' exposure to substances released from hazardous waste sites. This is because individual exposure data were lacking or could not be reasonably acquired during the course of investigation. Duration of residence could be ascertained through questionnaires administered to the population being studied, and distance from the waste site could be obtained from maps or similar means. Examples of this approach follow.

New York State: Chapter 5 shows a study conducted in collaboration with the New York State Department of Health that examined the association between congenital malformations in children and pregnant women's proximity to hazardous waste sites in the state (Geschwind et al., 1992). An "exposure risk index" was created for each respondent, incorporating the distance from and the hazard ranking score for each site

---

[2] Census blocks are small areas bounded on all sides by visible features such as streets, roads, streams, and railroad tracks, and by invisible boundaries such as city, town, township, and county limits, property lines, and short, imaginary extensions of streets and roads (BOC, 1992).

within a one-mile radius of the birth residence. Findings indicated that maternal proximity to uncontrolled hazardous waste sites was associated with an increased risk of approximately 12% of bearing children with any kind of congenital malformation (i.e., birth defect).

United Kingdom: Knox and Gilman (1997) conducted a required-reading study that uses distance from environmental hazards as a surrogate for measured levels of environmental contaminants. Although not expressly concerned with hazardous waste sites, this study is important because it advances the examination of disease clusters in community populations. In Chapter 5, the study of disease clusters is presented as a challenge to epidemiologists because small numbers of cases of disease often constitute a specific cluster.

Knox and Gilman examined relationships between birth and death addresses of 22,458 children who died from leukemia and cancer in the United Kingdom between 1953 and 1980. The addresses were studied in relation to proximity to areas of possible environmental hazards. Their study measured relative case densities of leukemia and cancer close to, and at increasing distances from, different kinds of environmental hazards. The sources of environmental hazards included oil refineries and allied industries, motor car factories, major users of petroleum products, users of kilns and furnaces; and airfields, railways, motorways, and harbors.

Home address postal codes (PCs) and their map coordinates were identified at birth and at death for children who died from cancer. Industrial and other sources presumed to have released hazardous substances into the environment, identified from business and other directories, were located on map coordinates and linked to PCs. Railway lines and motorways were located with use of digitized geographic survey maps. Numbers of births and deaths at successive radial distances from each environmental hazard were counted and compared with expected values. The expected values were based on counts in all PCs at similar distances. Relative case density ratios at successive distances from the hazards were calculated from observed and expected numbers, aggregated over similar sites (e.g., oil refineries). The inclusion of case density is a key element of this investigation; the procedure advances how disease clusters are evaluated in epidemiologic studies.

Findings showed childhood cancers were geographically associated with two groups of industrial atmospheric effluents: (1) petroleum-derived volatile compounds and (2) kiln and furnace smoke and gases, and effluents from internal combustion engines.

Residence or Employment Proximity

Level 6 Value—Residence in an area thought to be contaminated by substances released from a hazardous waste site, but with no measure of distance from the site or duration of possible exposure, is another surrogate for exposure. This approach is exemplified by collecting health data from persons living in areas of known contamination. These kinds of data are useful for developing hypotheses about exposure and health outcome associations.

Montreal, Québec, Canada: The risk of giving birth to low-birth-weight infants was examined among persons living near the Miron Quarry municipal solid waste landfill in Montreal, Québec (Goldberg et al., 1995a). The landfill is the third largest in North America and is located in a densely populated metropolitan area. Potential exposure to ambient air pollutants from the landfill was defined by residential proximity to the site. From 126,655 births available for analysis, case-control analyses (Chapter 5) were conducted to evaluate the risk of low birth weight (less than 2500 grams), very low birth weight (less than 1500 grams), preterm birth (fewer than 37 weeks), and small-for-gestational age (less than third percentile). As described in Chapter 5, findings indicated low birth weight was significantly elevated in the exposure zone proximal to the landfill.

National Exposure Registry—CERCLA directs the ATSDR not only to perform health assessments and health studies at CERCLA sites but also to maintain a national registry of persons exposed to toxic substances (Johnson, 1990). This is a mandate that is too sweeping to implement. The ATSDR's National Exposure Registry (Chapter 5) is closely linked to both its public health assessment program and the agency's epidemiologic studies program. It comprises substance-specific subregistries. The ATSDR defines an exposure subregistry as a list of persons and their health-related characteristics who have had documented exposure to a specific hazardous substance of health concern. Documented exposure means that the ATSDR can vouch for the presence of a contaminant being delivered to a household. The level of individuals' exposure, if any, within the household is unknown for purposes of the exposure subregistry. Each subregistry contains information about the registrants' self-reported health status and other vital information.

The ATSDR has developed several exposure subregistries for persons exposed to various substances. In 1996, the TCE subregistry listed 4,883 persons from 13 CERCLA sites in Michigan, Illinois, and Indiana; the dioxin subregistry listed 250 persons from four different sites—all in the Times Beach, Missouri area; and the benzene subregistry listed 1,127 persons from sites in Texas. A subregistry is planned for persons exposed to radioactive iodine (^{131}I) released from the Hanford facility of the Department of Energy.

Subregistries are valuable resources for advancing scientific understanding of associations between disease and exposure to specific hazardous substances. This is because large numbers of persons constitute each subregistry and this permits epidemiologic studies that are not restricted by small numbers of participants. The subregistries provide many of the same advantages that multisite analysis brings to health studies (Chapter 5).

Residence or Employment in Geographic Area

Level 7 Value—Residence in an area putatively contaminated by releases from hazardous waste sites, but without necessarily having environmental contamination data, is another surrogate for quantified exposure data. In the context of Figure 3.3, place of residence represents the least valuable surrogate of exposure assessment data. Examples follow.

Clinton County, Pennsylvania: The Drake Chemical NPL site in Lock Haven, Pennsylvania, was contaminated with beta-naphthylamine, a known human bladder carcinogen. Other contaminants included benzidine and benzene. To evaluate the prevalence of bladder cancer, investigators obtained county-wide age-adjusted, sex-, race-, and site-specific cancer mortality rates for the years 1950 through 1979 (Budnick et al., 1984). No environmental or personal exposure data were available to the investigators, nor was there information on bladder cancer risk indicators (e.g., occupational history) for persons who died of bladder cancer. During the 1970s, bladder cancer deaths were significantly increased among white males in Clinton County, and a significantly greater number of other cancer deaths occurring in the general population of Clinton and three adjacent counties would be expected. The trend of increased bladder cancer mortality among Clinton County white males was not evident in the county's white female population, reducing the likelihood that a general environmental exposure to carcinogens had occurred in the county. The investigators concluded that county-wide cancer mortality rates were useful as a screening method for more in-depth investigation of specific hazardous waste sites. No association between the Drake Chemical site and county-wide cancer mortality rates was possible with this study.

Tucson, Arizona: In another site-specific study, an investigation of births in Tucson, Arizona, was conducted to evaluate the association between congenital cardiac malformations and parental contact with drinking water contaminated for several years, primarily with TCE released into groundwater from an NPL site (Goldberg et al., 1990). Using the investigators' own case registry, 707 children with congenital heart disease born between 1969 and 1987 were identified. These children were conceived in the Tucson Valley, and their parents had spent both the month before the first trimester of pregnancy and the first trimester of the case pregnancy in the Tucson Valley. Of the 707 children, 246 lived in the area with contaminated water; the remaining 461 served as a case-comparison group. According to the investigators, "The odds ratio for congenital heart disease for children of parents with contaminated water area contact during the period of active contamination was three times that for those without contact ($p < 0.005$) and decreased to near unity for new arrivals in the contaminated water area after well closure."

CHILDREN'S EXPOSURE PATHWAYS

Children who live near hazardous waste sites are at potential risk of exposure to substances released from these sites. As noted in Chapter 1, approximately 1.3 million young children live within one mile of the borders of the current total of about 1,300 NPL sites. Health and risk assessors must include children as an at-risk population in their analyses. Knowledge of children's pathways of exposure is particularly important for these assessments. Two such pathways are particularly important for assessing children's health risk: breast milk from nursing mothers and play activities.

Human Breast Milk

In 1988, approximately 54% of mothers nursed their babies at the time of discharge from the birth site (DHHS, 1991). Because breast milk is the nutritionally optimum food for babies and also bestows passive immunity to nursing infants, government agencies recommend that mothers nurse infants. Although this section will highlight the potential problem of toxicants that are sometimes passed to nursing infants in breast milk, the many benefits of nursing must always be considered in relation to any health risk conveyed by these toxicants. It is prudent public health policy to include nursing infants in any site-specific public health assessment and baseline risk assessment. Assessments must also include the many positive benefits of nursing of infants.

Golding (1997) reported that breast milk can contain various unnatural constituents derived from the mother, including medications, addictive drugs (e.g., nicotine, caffeine, alcohol, heroin, cocaine), viruses, and toxicants. Breast milk may become a major route of excretion for persistent lipophilic chemicals that generally diffuse through cell membranes and concentrate in tissues with high fat content (e.g., adipose tissue). Sim and McNeil (1992) note that "Lipophilic compounds without a halogen group are usually rapidly metabolized and eliminated from the body and do not accumulate in fat." Because such compounds, as well as inorganic compounds (e.g., metals), do not accumulate in fat tissues, breast milk concentrations generally reflect recent exposure to the substances. The presence of lead in breast milk is an exception. This can occur when lead stored in bone is mobilized along with calcium during pregnancy and lactation.

By contrast, when substances that have high lipid solubility and are slowly metabolized appear in breast milk, they may constitute a measure of past exposure to toxicants that include polyhalogenated biphenyls (e.g., PCBs), polyhalogenated dibenzodioxins (i.e., dioxins), and organochlorine insecticides (Rogan and Gladen, 1983; Sim and McNeil, 1992).

Several investigators have measured levels of hazardous substances in human breast milk (e.g., Schecter et al., 1996; Koopman-Esseboom et al., 1996; Schlaud et al., 1995). ATSDR investigators reviewed the scientific literature on breast-feeding exposure of infants to cadmium, lead, mercury, chlorinated dibenzodioxins (CDDs), and chlorinated dibenzofurans (CDFs). Abadin et al. (1997) concluded that under normal conditions breast milk contains concentration ranges of < 1 µg/L cadmium, 2–5 µg/L lead, and 1.4–1.7 µg/L mercury. They proposed that breast milk concentrations of 5 µg/L cadmium, 20 µg/L lead, and 3.5 µg/L mercury are adequate health screening levels. To determine these screening levels, they used the Minimal Risk Levels (MRLs) for cadmium and mercury and a slope value to estimate the contribution of lead intake to blood lead concentration.

Pohl and Hibbs (1996) reviewed the literature on CDDs and CDFs and used risk assessment to estimate MRLs for these substances. Intermediate-duration (15 days–12 months) oral MRLs were derived from information in the ATSDR Toxicological Profiles. When converted to dioxin equivalents, or TEQs, the MRL for CDFs is 15 pg/kg/day; the corresponding MRL for CDDs is 7 pg/kg/day. Health assessors and health care providers can use these screening levels for metals, and MRLs for CDDs and CDFs can be used when advising lactating mothers on any hazards to their nursing infants from these substances.

Play Activities

Young children's play activities can increase their contact with hazardous substances. Children place materials and objects in their mouths that can contain hazardous substances, which are then ingested, thereby contributing to the elevation of contaminants in body tissues. Health and risk assessors must include these routes of exposure in their assessments.

When young children play outdoors, ingestion of soil that contains contaminants is a real possibility. The EPA assumes that most children ingest relatively small amounts of soil, with the upper 95th percentile of children ingesting an average of 200 mg/day (Calabrese et al., 1997); however, the degree to which children's soil ingestion is incorporated into site-specific, baseline risk assessments is unclear, which emphasizes the importance of studying children's soil ingestion rates and other routes of exposure to hazardous substances.

Stanek et al. (1998) investigated the prevalence of mouthing and ingesting outside soil among children aged one to six years. Through face-to-face interviews with 533 parents or guardians of children who attended three clinics in western Massachusetts, 28 behaviors were investigated in healthy young children. Approximately 32% of children were one year, and about 15% each were ages two, three, and four years. No significant differences in the distribution of children's age by race, age by sex, or race by sex were found. Findings showed that 38% of children put soil in their mouths at least monthly; 24%, at least weekly; and 11%, daily. Children one year old showed the greatest rates of mouthing objects outdoors; high-risk behavior decreased quickly for children aged two years or more. These data help health assessors more accurately characterize age groups of children who are at increased risk of exposure to soil contaminants.

Children's soil ingestion data that have acute toxicity implications are especially important to health assessors. Calabrese et al. (1997) assessed the risks posed by soil contaminants at currently accepted guidance concentrations when soil pica (i.e., profound soil ingestion by children) is assumed to reach 5 g soil/day. For 13 substances, doses reported to produce acute toxicity were compared with doses that would occur if a small child ingested 5, 25, or 50 g soil containing each substance at the EPA screening concentration. Substances included in the analysis were: antimony, arsenic, barium, cadmium, copper, cyanide, fluoride, lead, naphthalene, nickel, pentachlorophenol, phenol, and vanadium. For four of these substances (cyanide, fluoride, phenol, vanadium), soil pica episodes were found to result in contaminant doses approximating or exceeding acute human lethal doses; for five of the remaining substances (barium, cadmium, copper, lead, nickel), doses were within reported dose ranges for nonlethal toxicity in humans. These data suggest that acute soil pica episodes may be greater hazards than previously thought and that health and risk assessors should adopt more protective exposure and dose assumptions for children who have pica behavior.

Children's play with toys can also result in exposure to hazardous substances if the toys bear contaminants. Gurunathan et al. (1998) evaluated nondietary ingestion and dermal contact with room surfaces and children's toys after two apartments were treated with Dursban, which contained approximately 42% chlorpyrifos. Measurements included chlorpyrifos concentrations in air, on toys, and in dust on smooth surfaces. Air concentrations of chlorpyrifos were negligible, but total chlorpyrifos dose from children's playing with toys contributed to 39% from dermal contact and 61% from nondietary oral doses. The investigators concluded, "Chlorpyrifos deposited on and in toys and other absorbent surfaces following a typical broadcast application and reentry period were found at levels that could yield substantial doses to a child playing in the treated residence."

REFERENCE DATA

Reference data are at the heart of any exposure assessment of populations of health concern. Assuming a health investigator has collected a particular set of exposure data, what do the data mean in terms of human health? Do the population's data indicate excessive exposure to a contaminant(s) known to cause, for example, adverse reproductive effects? Is there evidence of any trend in the data over time, such as increased groundwater contaminant levels? These questions illustrate the need to compare exposure data with reference data. That, in turn, raises the questions, "What kinds of reference data are advisable, and what sources of such data are available?"

In conducting a health study, the frequent practice is to collect or estimate a study group's exposure to the hazardous substance of concern. When a comparison group is included, that group's exposure data can serve as a reference. When no exposure data from a comparison group are available, the most readily available reference data are regulatory or advisory guidelines provided by national or international health and environmental agencies. Such data consist of published levels of environmental and biologic limits that indicate how much exposure would produce adverse health or environmental effects. For example, the World Health Organization provides exposure guidelines for some hazardous substances as part of its environmental criteria documents series. Federal agencies like the EPA also issue regulatory statements about limits of contamination for specific contaminants in various environmental media.

On a larger scale, the establishment and maintenance of exposure-monitoring systems is quite important for public health purposes. Regular monitoring of environmental contamination levels and of human biologic indicators of exposure provides essential data for risk assessment and risk management. In the United States, the EPA conducts a program of national air monitoring and maintains a database on contaminants found in biologic samples provided to the EPA. Another valuable database was developed by the National Center for Health Statistics in conducting its National Health and Nutrition Examination Surveys, which periodically conduct U.S. population-based statistical sampling. Among the samples collected are biologic specimens that were evaluated for exposure to lead, cadmium, and 36 volatile organic compounds (e.g., Ashley et al., 1994).

Similarly, as described in Chapter 2, the ATSDR maintains a national database called HazDat that contains all environmental contamination data that the EPA collected for sites on its National Priorities List. The ATSDR uses its HazDat system to determine in which communities health studies, exposure registries, or other health services should be performed or established. For example, the ATSDR used the HazDat system to retrospectively identify hazardous waste sites for which additional health actions should be initiated, after recent knowledge that lower than previously supposed levels of exposure to lead can cause consequential adverse effects on children's intelligence and development (ATSDR, 1993b).

The importance of population-based exposure data is best illustrated by the effect that reduced levels of lead in gasoline had on blood lead levels in the American population. Lead in gasoline that was emitted in vehicle exhaust was recognized in the United States and elsewhere as a significant source of environmental contamination. This was then acknowledged as a major public health problem because of lead's toxicity to young children, and lead was gradually phased out as an additive to gasoline. As the quantity of lead added to gasoline diminished, so did the average blood lead levels, as verified through analysis of blood lead data acquired from a representative sample of the American population (EPA, 1986). This finding showed the remarkable effectiveness of a specific environmental risk management action (elimination of lead in gasoline) in making a positive contribution to improving the public's health (as evidenced by less human uptake of lead).

The importance of population-based exposure data relevant to specific, known environmental risk factors is evident in the federal government's national health-promotion and disease-prevention objectives (DHHS, 1991). The 16 national objectives in environmental health include reducing asthma morbidity; health risks from contaminated drinking water; and exposure to lead, criteria air pollutants, radon, solid waste, hazardous waste, infectious agents, and toxic agents. Other objectives address environmental health services and data collection. In particular, national objectives give high priority to improved health surveillance systems (Chapter 5), which could be linked to national environmental contamination databases like the Toxics Release Inventory maintained by the EPA of substances released by industrial sources.

Analytical Methods

Analytical methods must be at the core of any exposure assessment. Methods must obviously be available if measurement of exposure is to occur. Moreover, these analytical methods must be reliable and specific to the assessment of interest. The substances found in sites with completed exposure pathways (Table 2.3) are particularly important for public health purposes, because these substances are released from sites that have the greatest potential for human exposure. A review of the ATSDR Toxicological Profile for each CEPPS was conducted to determine how many analytical methods are available to measure biologic and environmental exposure. All CEPPS were found to have analytical methods to measure the substances' concentrations in biologic and environmental media, although most methods require the availability of sophisticated laboratory instrumentation such as gas chromatographs and mass spectrometers.

A health investigator must do two things to determine whether an individual's exposure data are of health concern. First, the investigator should obtain a biologic sample of a person's exposure dose, preferably with coincident measurement of ambient environmental contaminant levels. This requires informed consent from the person before collecting the biologic sample. Informed consent means telling the individual the purpose and procedures for collecting data and obtaining written permission to proceed. Samples can be collected from urine, blood, hair, and body tissues, and laboratory protocols should be followed in collecting all samples. For example, measuring lead concentration in blood is a complex laboratory procedure that can easily be compromised if good sample collection and laboratory procedures are not carefully followed. Second, the investigator must compare or reference the person's biologic level of a hazardous substance or related biomarker to normative data. Therefore, an essential component of any exposure assessment is developing laboratory reference data for individual toxicants.

The National Center for Environmental Health (NCEH) of the Centers for Disease Control and Prevention (CDC) has developed **reference range** concentrations of several hazardous substances found in nonoccupationally exposed populations. By their definition, "Reference range concentrations are those concentrations of a toxicant, its residue, or its metabolite that are found in the general population. We assume that the general population is not occupationally exposed to the parent compound and is not otherwise overtly exposed" (Hill et al., 1995). Reference range data provide vital information on "background" levels of toxicants in the general population.

The NCEH's reference range data cover only volatile organic compounds (VOCs) and pesticides. Reference range data are available for 31 VOCs and 12 pesticides. The substances were selected from the ATSDR's CERCLA priority list of hazardous substances (Table 4.2). VOCs and pesticides or their analytes were measured in whole blood from 1,000 individuals using laboratory methods described elsewhere (Needham et al., 1995). Blood samples were obtained as part of the Third National Health and Nutrition Examination Survey (NHANES), which is a national, population-based health survey conducted periodically by the National Center for Health Statistics, CDC. The NCEH randomly selected 1,000 individuals from the NHANES survey to include in a reference range study. The reference range population sampled represents a broad spectrum of the American population in terms of geography, race, gender, and urban/rural communities.

The NCEH's reference range data for 31 VOCs measured in blood are listed in Table 3.2. The substances shown in bold type are CEPPS; 13 VOCs are among the 30 CEPPS. Having reference range data for 12 of the 13 VOCs is a significant exposure data re-

Table 3.2. Blood Concentrations (in ppb) of Selected Volatile Organic Compounds in a Group of Nonoccupationally Exposed, U.S. Adults (Ashley et al., 1994; Sampson, 1996).

Analyte (Substances in bold are CEPPS)	Detection Limit	Number	Minimum	5th %tile	Median	Mean	95th %tile	Maximum
Acetone	200	1013	340	650	1800	3100	>6000[a]	>6000[a]
Benzene	0.030	873	ND	ND	0.061	0.13	0.48	1.9
Bromodichloromethane	0.009	1049	ND	ND	ND	ND	0.020	0.28
Bromoform	0.027	603	ND	ND	ND	ND	0.034	0.41
2-Butanone	0.50	1078	0.54	1.9	5.4	7.0	17	>24[a]
Carbon Tetrachloride	0.019	1036	ND	ND	ND	ND	ND	0.056
Chlorobenzene	0.007	1002	ND	ND	ND	ND	0.013	0.092
Chloroform	0.021	966	ND	ND	0.203	0.043	0.12	>1.8[a]
Dibromochloromethane	0.013	1013	ND	ND	ND	ND	0.023	0.088
Dibromomethane	0.044	639	ND	ND	ND	ND	ND	0.053
1,2-Dichlorobenzene	0.044	808	ND	ND	ND	ND	ND	ND
1,3-Dichlorobenzene	0.019	1019	ND	ND	ND	ND	ND	0.041
1,4-Dichlorobenzene	0.073	1014	ND	0.077	0.33	1.9	>8.0[a]	>8.0[a]
1,1-Dichloroethane	0.009	1060	ND	ND	ND	ND	ND	0.034
1,2-Dichloroethane	0.012	1062	ND	ND	ND	ND	ND	0.026
1,1-Dichloroethene	0.018	1056	ND	ND	ND	ND	ND	0.056

cis-1,2-Dichloroethene	0.013	ND	ND	ND	ND	ND	0.023
trans-1,2-Dichloroethene	0.014	ND	ND	ND	ND	ND	0.038
1,2-Dichloropropane	0.008	ND	ND	ND	ND	ND	0.026
Ethylbenzene	0.020	ND	ND	0.060	0.11	0.25	>2.5[a]
Hexachloroethane	0.079	ND	ND	ND	ND	ND	0.12
Methylene Chloride	0.089	ND	ND	ND	ND	ND	1.6
Styrene	0.019	ND	ND	0.041	0.074	0.18	>1.7[a]
1,1,2,2-Tetrachloroethane	0.008	ND	ND	ND	ND	ND	0.016
Tetrachloroethylene	0.030	ND	ND	0.063	0.19	0.62	>4.4[a]
Toluene	0.092	ND	0.11	0.28	0.52	1.5	>4.0[a]
1,1,1-Trichloroethane	0.086	ND[b]	ND	0.13	0.34	0.80	>5.0[a]
1,2-Trichloroethane	0.016	ND	ND	ND	ND	ND	0.027
Trichloroethylene	0.010	ND	ND	ND	0.017	0.021	>0.85
m-/p-Xylene	0.033	0.034	0.074	0.19	0.37	0.78	>8.0[a]
o-Xylene	0.040	ND	0.046	0.10	0.14	0.28	3.5

ND=not detected; [a]Results are greater than highest linear standard; [b]Results are below detection limit.

source. The case study that follows illustrates how reference data were used to help develop a public health response to an episode of illegal pesticide application.

Case Study: Methyl Parathion Contamination

The use of reference range exposure data is illustrated in a study of widespread, illegal contamination of homes by methyl parathion, a pesticide approved only for use outdoors. Because methyl parathion was inappropriately applied and left as a contaminant in houses, it became a hazardous waste issue with attendant ramifications for public health authorities. Reference range data for methyl parathion developed by the NCEH were instrumental in identifying the severity of methyl parathion exposure and the need for attendant actions by health and environmental officials.

Severe human exposure to methyl parathion can lead to convulsions, unconsciousness, cardiac arrest, and death. At lower exposure levels, over time, persons may have a persistent lack of appetite, muscle weakness, and malaise. Persons can be exposed by swallowing, inhaling, and having skin contact with methyl parathion. Children, the elderly, and persons with pre-existing health conditions are particularly at risk of methyl parathion toxicity.

In 1994, state officials in Ohio discovered that methyl parathion had been used in inappropriate ways for pest control in houses in Lorain County (Hill et al., 1995). Methyl parathion is a highly toxic pesticide used principally, under controlled conditions of application, to spray cotton. The EPA has established re-entry standards for persons who work in fields sprayed with methyl parathion. In the Ohio episode, an unlicensed pesticide applicator had illegally used methyl parathion for pest control in several hundred houses. As a result, several hundred persons living in the methyl-parathion "treated" homes were exposed to the pesticide (Hill et al., 1995).

Local, state, and federal health and environmental authorities joined in a collaborative effort to identify the geographic area of methyl parathion contamination (Kavanaugh, 1996). Efforts were included in these actions to identify individuals at risk of exposure to methyl parathion at levels of health concern. As Hill et al. described, exposure was determined by measuring methyl parathion in air and dust samples in the houses treated by the exterminator. Biologic exposure was determined by measuring p-nitrophenol (PNP), a metabolite of methyl parathion, in urine.

Samples from 131 residents whose urine was measured for PNP concentration showed levels in excess of the reference range for nonoccupationally exposed persons (Hill et al., 1995). The mean PNP concentration in the methyl parathion group was 240 µg/L, compared with 1.6 µg/L in the reference range sample. The median PNP value was 28 µg/L, compared with < 1.0 µg/L in the reference range. In Lorain County, 78% of the methyl parathion-exposed residents had PNP concentrations greater than or equal to the 95th percentile concentration of the reference range.

Other episodes of methyl parathion contamination occurred in 1996 in Mississippi, Louisiana, Tennessee, and Illinois. They parallel what happened in Ohio (ATSDR, 1996c). In the largest of these contaminations, more than 1,100 homes were contaminated over a two-year period in Jackson County, Mississippi. The total number of homes in Mississippi alone will exceed 2,000. The EPA estimates the incident in Mississippi will cost more than $50 million in remediation and allied costs.

These unfortunate episodes of illegal use of methyl parathion for household insect control have led to changes in the pesticide's control and administration. The Danish manufacturer of methyl parathion has agreed to redesign the substance's containers to warn against improper use and to add an offensive odor to the pesticide to discourage its use in homes. From a public health perspective, these changes are desirable if they help stop the illegal application of methyl parathion.

The biologic exposure data, together with the household air and dust samples, became critical information for subsequent environmental and public health interventions. The widespread geographic contamination opened the potential for the EPA to temporarily relocate large numbers of persons from their homes and places of work. In response, the ATSDR and the EPA jointly developed criteria for relocation based on both PNP levels in urine and in-home environmental levels of methyl parathion. Using both kinds of exposure databases, the difficult decision about which families to relocate became more firmly based in science. For instance, relocation is not necessary for pregnant women or infants younger than 12 months when the PNP concentration in urine is less than 25 ppb *and* methyl parathion in in-home wipe samples is less than 50 $\mu g/100$ cm^2. Relocation is, however, indicated for these at-risk groups when PNP levels exceed 50 ppb.[3] Moreover, if adjusted PNP levels are $\geq 1,000$ ppb for adults or ≥ 500 ppb for children, the individuals should be seen immediately by physicians.

The EPA declared the geographic areas of methyl parathion contamination in Ohio and Mississippi for cleanup under its CERCLA authorities. Criminal charges have been

---

Case Example: Lorain County, Ohio

In November 1994, a medical epidemiologist at the Centers for Disease Control and Prevention in Atlanta was contacted about infants in Cleveland, Ohio, suffering from a rare pulmonary disease. Reports were also received of pesticides having been illegally sprayed in homes in Lorain County, Ohio. Although the infants' disease proved unrelated to the pesticides, what ensued was one of the largest responses ever by government agencies to a toxics problem in private residences.

Methyl parathion, which is restricted to outdoor use, had been sprayed in homes by an unlicensed pest exterminator. The situation eventually involved 20 government agencies working in cooperation under CERCLA's emergency removal authorities. Over a 16-month period, 869 persons were temporarily relocated while 233 houses were cleaned of the pesticide. A sample of persons whose houses had been contaminated by methyl parathion showed elevated urine levels of the pesticide. The long-term cost of remediating the houses is estimated to be in excess of $21 million (Kavanaugh, 1996).

---

[3] The complete relocation decision logic is available from ATSDR's Division of Health Education and Promotion.

brought against the pesticide applicators who violated federal pesticide laws by spraying homes and other structures with methyl parathion (BNA, 1996). The unlicensed Ohio pesticide applicator was found in violation of the Federal Insecticide, Fungicide, and Rodenticide Act. He was sentenced to serve 37 months in prison and two years of supervised release. In Mississippi, two unlicensed exterminators were each sentenced in July 1997 to prison for more than five years. Health authorities have conducted extensive community and physician education and are evaluating the possibility of implementing a long-term health surveillance program for the most severely exposed persons.

Without the availability of reference range data for methyl parathion, the extent of pesticide misapplication would have been much less certain.

EXPOSURE ASSESSMENT RESEARCH

The setting of an agenda for research on environmental hazards has often lacked a sense of priority, including research priorities for acquiring exposure and toxicity data relevant to human health. CERCLA prescribes an applied research program to fill key data gaps for priority substances. First, the ATSDR and the EPA must jointly rank hazardous substances released from CERCLA sites. A list of 275 hazardous substances has resulted from the ranking process (Chapter 4). Second, using the list of priority substances, the ATSDR is mandated to prepare, distribute to the public, and periodically update a Toxicological Profile for each substance. (Each profile contains information on how to assess exposure to the substance; for example, analytical methods to measure the substance in biologic media.) Third, using the profiles, the ATSDR is required to initiate a program of research to fill key gaps in knowledge about the substances' effects on human health (Johnson, 1990). This applied research program fills key gaps in both exposure and toxicity databases. The toxicity data gaps are described in Chapter 4.

Substances found most often in hazardous waste sites that have completed exposure pathways are an especially important subset of the 275 priority CERCLA substances (Table 2.3). The Completed Exposure Pathways Priority Substances (CEPPS), by definition, are substances posing the greatest risk of exposure for humans. Table 3.3 lists in alphabetic order the 30 CEPPS and their key exposure data needs. Research to improve exposure measurement is needed for most ranked substances. A review of Table 3.3 shows that 73 gaps in exposure data have been identified for the 30 CEPPS; 37 of these (51%) are currently being filled through applied research or other means.

Data gaps are filled through a combination of ATSDR-funded studies, private industry voluntarism, and the EPA's authorities in the Toxic Substances Control Act and the Federal Insecticide, Fungicide, and Rodenticide Act that require chemical producers to conduct research specified by the EPA (Auer, 1994). The data gaps in Table 3.3 highlighted in bold are being filled through one or more of these mechanisms.

Some general comments about exposure assessment research are also in order. Findings and data from exposure assessment research are important for improving and advancing understanding of the effects that hazardous substances have on human health and for improving risk assessments of hazardous waste sites. Several specific areas of research are important. First, continued development of biological markers is needed. The artificial division between markers of exposure and markers of disease should be

Table 3.3. Key Exposure Assessment Data Needs for the 30 CEPPS (ATSDR, 1996d; ATSDR, 1997a).

Substance	Exposure Data Gaps
Antimony	To be determined
Arsenic	(1) Comparative toxicokinetic studies, (2) Half-lives in surface water, groundwater, (3) Bioavailability from soil, (4) Exposure levels in humans[a]
Barium	To be determined
Benzene	**(1) Dose-response data in animals for acute- and intermediate-duration oral exposure, (2)** Exposure levels in humans[a]
Benzo(a)pyrene	To be determined
Beryllium	**(1) Dose-response data in humans for acute-and intermediate-duration inhalation exposures. The subchronic study should include extended reproductive organ histopathology, (2) Environmental fate in air; factors affecting bioavailability in air,** (3) Analytical methods to determine environmental speciation, (4) Exposure levels in humans
Cadmium	(1) Dose-response data in animals for chronic oral exposures, (2) Half-life in soil, (3) Exposure levels in humans,[a] **(4) Potential candidate for subregistry of exposed persons[b]**
Carbon tetrachloride	(1) Dose-response data in animals for intermediate-duration oral exposure, (2) Exposure levels in humans,[a] **(3) Potential candidate for subregistry of exposed persons[b]**
Chloroform	(1) Dose-response data in animals for intermediate-duration oral exposure, (2) Exposure levels in humans,[a] **(3) Potential candidate for subregistry of exposed persons[b]**
Chromium	**(1) Dose-response data in animals for acute-duration oral exposure to chromium (VI) and (III) and for intermediate-duration oral exposure to chromium (VI),** (2) Exposure levels in humans[a]
Copper	To be determined
1,1-Dichloroethane	**(1) Dose-response data in animals for acute- and intermediate-duration oral exposures. The subchronic study should include an evaluation of immune and nervous system tissues, and extended reproductive organ histopathology, (2)** Dose-response data in animals for chronic inhalation exposures. The study should include an evaluation of nervous system tissues, **(3) Potential candidate for subregistry of exposed persons.**
1,2-Dichloroethane	**(1) Dose-response data in animals for acute- and intermediate-duration oral exposures. The subchronic study should include an evaluation of immune and nervous system tissues, and extended reproductive organ histopathology, (2)** Dose-response data in animals for chronic inhalation exposures. The study should include an evaluation of nervous system tissues, **(3) Potential candidate for subregistry of exposed persons.**
1,1-Dichloroethene	To be determined
trans-1,2-Dichloroethene	To be determined
Di-(2-ethylhexyl)phthalate	(1) Dose-response data in animals for acute- and intermediate-duration oral exposures, **(2) Comparative toxicokinetic studies,** (3) Exposure levels in humans,[a] **(4) Potential candidate for subregistry of exposed persons[b]**
Lead, inorganic	(1) Analytical methods for tissue levels, **(2) Exposure levels in humans[a]**
Manganese	(1) Evaluate existing data on concentrations of Mn in contaminated environmental media at hazardous waste sites, **(2) Exposure levels in humans,[a]** (3) Potential candidate for subregistry of exposed persons,[b] (4) Relative bioavailability of different Mn compounds and bioavailability of Mn from soil, (5) Dose-response data for acute- and intermediate-duration oral exposures, (6) Toxicokinetic studies in animals

Table 3.3. Key Exposure Assessment Data Needs for the 30 CEPPS (ATSDR, 1996d; ATSDR, 1997a) (continued).

Substance	Exposure Data Gaps
Mercury, metallic	(1) **Dose-response data in animals for chronic-duration oral exposure,** (2) **Exposure levels in humans,**[a] (3) **Potential candidate for subregistry of exposed persons**[b]
Methylene chloride	(1) **Dose-response data in animals for acute- and intermediate-duration oral exposure,** (2) Exposure levels in humans,[a] (3) **Potential candidate for subregistry of exposed persons**[b]
Naphthalene	To be determined
Nickel	(1) Dose-response data in animals for acute- and intermediate-duration oral exposures, (2) Bioavailability of nickel from soil, (3) Exposure levels in humans,[a] (4) **Potential candidate for subregistry of exposed persons**[b]
PAHs	(1) **Dose-response data in animals for intermediate duration oral exposures. The subchronic study should include extended reproductive organ histopathology and immunopathology,** (2) **Dose-response data in animals for acute- and intermediate-duration inhalation exposures. The subchronic study should include extended reproductive organ histopathology and immunopathology,** (3) Exposure levels in humans, (4) **Potential candidate for subregistry of exposed persons.**
PCBs	(1) **Dose-response data in animals for acute- and intermediate-duration oral exposures,** (2) Dose-response data in animals for acute- and intermediate-duration inhalation exposures, (3) Biodegradation of PCBs in water; bioavailability of PCBs in air, water, and soil, (4) **Exposure levels in humans,**[a] (5) **Potential candidate for subregistry of exposed persons,**[b] (6) **Aerobic PCB biodegradation in sediment**
Tetrachloroethylene	(1) **Dose-response data in animals for acute-duration oral exposure,** (2) Dose-response data in animals for chronic-duration oral exposure, (3) Exposure levels in humans,[a] (4) **Potential candidate for subregistry of exposed persons**[b]
Toluene	(1) **Dose-response data in animals for acute- and intermediate-duration oral exposures,** (2) **Comparative toxicokinetic studies,** (3) **Exposure levels in humans,**[a] (4) **Potential candidate for subregistry of exposed persons**[b]
1,1,1-Trichloroethane	To be determined
Trichloroethylene	(1) **Dose-response data in animals for acute-duration oral exposure,** (2) Exposure levels in humans[a]
Vinyl chloride	(1) **Dose-response data in animals for acute-duration inhalation exposure,** (2) Dose-response data in animals for chronic-duration inhalation exposure, (3) Exposure levels in humans,[a] (4) **Potential candidate for subregistry of exposed persons**[b]
Zinc	(1) **Dose-response data in animals for acute- and intermediate-duration oral exposure,** (2) Exposure levels in humans,[a] (3) **Potential candidate for subregistry of exposed persons**[b]

Notes: Data gaps shown in bold print are being filled through research (ATSDR, 1996d for details). Dose-response data needs are listed both in this table and in Table 4.6 because dose-response involves both exposure and toxicity.

[a] Exposure levels in humans living near hazardous waste sites and other populations, such as exposed workers.

[b] Refers to substance-specific subregistry in the ATSDR's National Exposure Registry program.

eliminated because biologic markers truly encompass a continuous spectrum of events. Specific chemical metabolites, particularly those with stable half-lives or those that lodge in or bind to tissues, must be identified and measured. Also, identifying subtle, chemically-induced tissue changes is important, particularly those shown to be pre-clinical indicators of disease. Moreover, establishing the relationship between these preclinical indicators and subsequent disease would greatly increase the ability to esti-mate the probability of future disease. This would provide an unparalleled opportunity to assess the effectiveness of public health intervention strategies.

Closely related is the need for more information on chemical-specific toxicokinetics. If chemical uptake, distribution, metabolism, and elimination cannot be accurately predicted, truly predicting subsequent health consequences with any degree of accu-racy is impossible.

Efforts in the study of Quantitative Structure Activity Relationships (QSAR) hold promise. QSAR evaluates the relationship between physical and/or chemical properties of substances and their ability to cause particular effects or enter into specific chemical reactions. The primary goal of QSAR studies is to develop procedures for predicting the toxicity of a compound from its chemical structure by analogy with other toxicants whose structure and toxic properties are known (Hodgson and Levi, 1987). Although QSAR is usually associated with mechanism-of-action toxicology studies, applying the method to exposure assessment is equally important.

Approximately 75,000 chemicals were produced in the United States in 1996, accord-ing to the EPA's inventory established under the Toxic Substances Control Act (EDF, 1997). In theory all these chemicals could be candidates for causing environmental exposure because many are in commerce. Many eventually enter the waste stream and can constitute hazardous waste. Evaluating the environmental and health effects of each chemical individually would obviously be impractical, but QSAR studies, such as those used effectively for years in pharmacology and medicinal chemistry, could help predict not only the pharmacokinetics but also the health outcomes for entire chemical classes. In addition, the proper application of structure-activity principles has been and should continue to be a valuable tool for environmental modeling.

Another important need is for accurate, normal, baseline data, such as those given in Table 3.2. This need includes background levels in environmental media, normal tissue levels, and normal physiologic values, whether they be chemical concentrations, ingestion amounts, breathing rates, or enzyme levels. The range and distribution of these normal background values should be accurately determined and population-based; that is, determined from data on a statistically valid sample of a population.

Closely tied to this need for accurate assessment of background data is the need for probabilistic approaches for assessing exposure (NRC, 1991b). Today, exposure is some-times still estimated on the basis of single-scenario-point estimates. For example, expo-sure for an average 70 kg human is calculated assuming residence at a site for 70 years and daily ingestion of 2 liters of water contaminated at a specific chemical concentra-tion. However, each estimate—body weight, age, years residence near the site, water consumption, and even chemical concentration in the water—has a range and distri-bution. Appropriate use of these various distributions through probabilistic models would provide much more accurate and meaningful population-based estimates of minimum, maximum, and average exposures. These scenarios should, to the extent feasible, be tailored on a site- or context-specific basis.

SYNTHESIS

Exposure assessment is a critical element of public health and risk assessments. Unfortunately, lack of actual exposure data can constitute the weakest link in these assessments. To obtain exposure data for a human population of concern requires careful planning and rigorous execution.

Current trends in assessing exposure of human populations to hazardous substances indicate: (1) the need for total exposure assessment (i.e., dose determination) where possible, (2) the importance of evaluating all exposure data for evidence of excess exposure in persons most at risk (e.g., children, the elderly), (3) the need for greater cooperation among federal, state, and local government agencies as exposure assessment become more complex, and (4) the availability of too few population-based exposure databases.

This chapter describes a framework and methods for assessing and analyzing the total exposures of an individual to hazardous substances in all activities over specified increments of time. The NRC clearly articulates the need for **total exposure assessment**. Total human exposure includes all exposure a person has to a specific toxicant, regardless of environmental medium (air, water, soil, food) or route of entry into the body (inhalation, ingestion, dermal absorption) (Lioy, 1990). Public health assessments (Chapter 2) consider all relevant pathways of exposure to hazardous substances that might impact a community. Total exposure assessment is conducted in this context. However, to ascertain the body burden, or dose, of a substance in terms of a total exposure assessment requires biological measurements. These are seldom performed for communities near hazardous waste sites because of cost and other considerations.

The concern that hazardous waste sites and similar uncontrolled releases of hazardous substances can cause adverse health effects is in reality a concern that exposure to toxicants is occurring or has occurred in the past. In a public health context, the absence of exposure to a hazardous substance translates into the absence of possible harm to one's health. No exposure—no health problem. Given this obvious principle, the health official must determine whether indeed any exposure of a population is occurring or has occurred. And, if exposure has occurred, in what amount, and who was exposed? These questions represent the essence of exposure assessment.

Seven kinds of exposure assessment data are described in this chapter, although great differences exist in their value. Quantified individual measurements of exposure, using biologic markers, are the most useful and powerful kind of exposure data. Less valuable, but nonetheless useful, are surrogates for measuring individuals' exposure. These include quantified ambient measurements of contaminants in environmental media, distance from a site and/or duration of residence, and residence in an area of known or suspected contamination. Biologic markers can indicate a person's internal dose of a hazardous substance or related metabolite. Greater use of biologic markers is advocated in conducting epidemiologic studies and exposure assessments of populations at risk of adverse health effects.

Exposure assessments must be rooted in good laboratory science. Analytical methods exist for the 30 substances (CEPPS) found most often in NPL sites that have completed exposure pathways, although sophisticated laboratory instrumentation is usually required. The application of analytical methods has resulted in reference range data for 10 of the 13 volatile organic compounds on the list of 30 CEPPS. This set of data is essential for comparing a person's or a group's exposure with a particular hazardous substance to what is found in the general population.

Health and risk assessors must include children in their assessments in order to reduce the health risk from exposure to hazardous substances. Children's play activities that involve contact with contaminated soil and toys present the possibility of exposure to hazardous substances. Similarly, substances in human breast milk consumed by infants can present a source of exposure to some hazardous substances. A thorough characterization of children's exposure pathways therefore becomes paramount for accurate and protective health assessments.

Research on methods to measure exposure to individual hazardous substances released from hazardous waste sites is under way. These data, when acquired, will strengthen public health assessments and site-specific risk assessments.

Three generalizations are important. First, each hazardous waste site and community study is different in ways that must be factored into the exposure assessment. For example, children may be the sole population of concern in certain communities because of exposure and behavioral factors. Second, total exposure assessment should be the goal of any investigation (Johnson, 1992). For instance, a study of children exposed to lead released from a CERCLA site also must consider other possible sources of lead exposure; for example, lead in paint in older houses. Third, contaminant levels in environmental media are usually inadequately correlated with contaminant (or metabolite) levels in biological samples.

Chapter Four

Toxicology of CERCLA Hazardous Substances

What substances at NPL sites are the most hazardous to human health? What key principles of toxicology are relevant to the properties of substances released from hazardous waste sites and during chemical spills? The toxicologic information in this chapter draws heavily on the list of 30 substances found most often in NPL sites that have completed exposure pathways. The toxicology of mixtures of substances released from hazardous waste sites is also discussed.

Determining the health effects of substances released from uncontrolled hazardous waste sites requires knowing their toxicity. Similarly, site remediation cleanup levels must be established for contaminants using knowledge of their toxicologic properties. Using data in HazDat, more than 2,000 unique substances have been identified in the environmental media that the EPA sampled during site investigations. This large number reflects the magnitude and diversity of the American industrial sector. Each hazardous waste site is unique, but many share common properties; for example, many hazardous substances are found at multiple waste sites. Moreover, waste sites usually contain mixtures of substances, and this complicates public health assessment because scientific data on the toxicology of mixtures is sparse.

Substances released from hazardous waste sites are not equally hazardous to human health. For that reason, the ATSDR and the EPA have jointly prioritized hazardous substances released from NPL sites in terms of their threat to human health. As will be discussed later, CERCLA provides that placing a substance on the priority list of hazardous substances leads to the ATSDR's development of a Toxicological Profile for the substance. These documents contain summary information about the toxic and human health properties of substances profiled. Each profile must contain an analysis of gaps in scientific knowledge about a substance's toxic and physical properties, which in turn, results in a program of research mandated by CERCLA to fill key toxicologic and exposure data gaps.

This chapter identifies the priority substances released from uncontrolled hazardous waste sites, discusses the mixtures of substances most frequently released from sites, and describes salient toxicologic properties and data gaps for priority hazardous substances. A summary of the key principles of toxicology will put the content of this chapter into perspective. An in-depth description can be found elsewhere (e.g., Klaassen, 1996; Hodgson and Levi, 1987).

KEY PRINCIPLES OF TOXICOLOGY

Toxicology is the study of poisons. Toxicologists study substances for their toxic properties; that is, the capacity to poison an organism. Naturally occurring substances like animal or plant substances that have toxic properties are called toxins. Toxicants are technologic-derived substances with toxic properties. Snake venom is a toxin. Methylene chloride is a toxicant. These definitions are not universally followed; it is still common to hear mercury referred to as a "toxin" or a "toxic," but the better term is toxicant.

Pharmacology is the study of substances that promote an organism's health. Toxicology, which evolved as a specialty area within pharmacology because both sciences have common underpinning in chemistry, physics, and biology, focuses instead on assessing and understanding how substances can harm an organism's health. This section summarizes key principles of toxicology that are relevant to the health implications of uncontrolled releases of substances from hazardous waste sites and chemical spills. The clear distinction that was made in Chapter 3 between **exposure** and **dose**, two key concepts in toxicology, should be reviewed here before proceeding.

The dose makes the poison. This is the first and most fundamental principle of toxicology. For example, small amounts of zinc in the diet are essential, but large amounts of zinc are toxic. This fundamental principle is often misunderstood or misused in regard to hazardous waste issues. Toxicologists distinguish between applied dose and absorbed dose. **Applied dose** is the amount of a substance given to a human or test animal, especially through dermal contact. Applied dose is a measure of exposure because absorption is not taken into account. **Absorbed dose** is the amount of a substance that penetrates across the exchange boundaries of an organism through either physical or biologic processes after contact (exposure).

Because concentrations of hazardous substances found in environmental media are often "low," some scientists and special interest groups have averred that no health effects would occur. They assert that health effects occur only at higher levels of exposure. Their assertion is based in part on occupational health studies of workers who did not manifest adverse health effects at higher levels of exposure in the workplace. Other groups assert that these "low" levels *must* be associated with the health problems they have observed in neighbors or family members because they think the substances are toxic at any level of exposure. Where is the truth? An understanding of the difference between exposure and dose and a knowledge of basic elements of toxicology are fundamental to getting at the "truth."

Routes of Exposure—As discussed in Chapter 3, persons living near hazardous waste sites are potentially at risk of exposure to hazardous substances. The routes, or pathways, of exposure include air, surface water, soil, groundwater, and biota. The route of exposure must be viewed as the initial event in a multistage process, which was shown in Figure 3.2. In a toxicologic sense, these environmental pathways lead to the major routes by which hazardous substances enter the body. These routes are through the gastrointestinal tract (ingestion), lungs (inhalation), and skin (topical contact). Although placement of toxic substances into the blood stream, through intravenous injection, is not relevant to hazardous waste site releases, it generally elicits the quickest and greatest physiologic effect. Eaton and Klaassen (1996) give an approximate descending order of effectiveness for the other routes: inhalation, intraperitoneal, subcutaneous, intramuscular, intradermal, oral, and topical.

Duration of Exposure—A single high-dose exposure to a hazardous substance may result in toxic effects different from those following repeated low-dose exposures. Therefore, toxicologists must consider the duration, the intensity, and the frequency of exposure when assessing a person's exposure to a hazardous substance. When assessing duration of exposure, the ATSDR defines an acute exposure as lasting 14 or fewer days, intermediate exposure as lasting 15–364 days, and chronic exposure as persisting for 365 or more days. These definitions are specific to the ATSDR and are used in the ATSDR Toxicological Profiles. The ATSDR defines the intensity of exposure as dose rate or total dose, depending on the circumstances of exposure. The frequency of exposure is defined by the ATSDR as being either continuous or intermittent.

Susceptible Populations—Members of a population exposed to a given hazardous substance under the same nonlethal exposure conditions will not all express the same reaction. For several reasons, some persons within a population will be more susceptible to a substance's toxicity than will other persons. Susceptible groups include children, elderly persons, fetuses, persons with certain genetic disorders, and persons with infirmities.

Children's susceptibility to hazardous substances derives from their physiology and behavior. For instance, young children absorb more of a given amount of lead through their small intestine than do adults. Infants and young children's play activities and mouthing behaviors can increase their contact with hazardous substances. Fetuses are susceptible because many hazardous substances can pass from the mother into the developing fetus. Because fetal tissues and organs are changing rapidly, a toxic substance can have a major effect. Elderly persons and persons with infirmities are susceptible because they may not possess the strength and organ functioning needed to resist the impact of an exposure. For instance, a person with emphysema, a disease that reduces oxygen transport in the lungs, is more susceptible to carbon monoxide toxicity, which lessens the amount of oxygen carried by hemoglobin in the blood's red blood cells.

Persons with genetic disorders or genetic anomalies can also be susceptible to the effects of a specific toxic agent, depending on the kind of anomaly. For example, anemia can occur if the rate of red cell destruction in peripheral blood exceeds the normal rate of cell production in bone marrow. Some chemicals (e.g., naphthalene) have direct hemolytic effects (i.e., red-blood cell destruction) *in vivo*. Other chemicals, such as primaquine, only produce hemolysis in red blood cells that are congenitally deficient in glucose-6-phosphate dehydrogenase (Smith, 1996). Persons who have this genetic trait are therefore at increased health risk if exposed to certain hazardous substances.

Genetic testing includes genetic monitoring and genetic screening. In genetic monitoring, specific laboratory tests assess the risk associated with exposure to genotoxic substances. In genetic screening, tests for inherent genetic characteristics determine susceptibility for particular diseases. Both genetic monitoring and screening present ethical challenges (Van Damme et al., 1995). For example, screening tests used as a pretext to avoid risk management actions like eliminating workplace genotoxic substances would be unethical. Similarly, genetic test results used to discriminate in hiring practices raise serious ethical concerns.

HUMAN ORGAN SYSTEM TOXICITY

Effects of toxic substances on organ systems ultimately determine the nature and extent of the health impact (Figure 3.2). Different organs can be affected quite differently by the same toxic substance. Indeed, some substances have organ-specific toxicity, as for example the effect asbestos fibers have on the lung, causing asbestosis. A substance's damage to the tissues of an organ can alter the organ's function. For instance, substances that damage one or more liver tissues can change or halt the many physiologic functions the liver serves.

Knowledge of the effects that individual toxicants have on specific human organs has accrued over time, although significant gaps still exist for many substances. Much of this knowledge comes from laboratory-based programs that test the toxicity of individual substances on laboratory animals and then extrapolate the results to humans. In addition, investigations of workers' health have shed important light on the toxicity of some substances found in occupational environments. Furthermore, extreme exposure to some toxicants has led to clinical reports of organ toxicity.

The sections that follow highlight some toxic effects that hazardous substances have on human organ systems associated with the ATSDR's seven priority health conditions (Table 5.9). Human health effects of hazardous substances released from waste sites are discussed in Chapter 5.

Birth Defects and Reproductive Disorders (Teratogenicity, Developmental Toxicity)

The reproductive system is structurally and functionally designed to procreate a species. Males produce sperm to fertilize the eggs, or ova, that females produce. In humans, a fertilized egg, or embryo, grows into a fetus. The period before birth is called prenatal, that after birth is postnatal. Results from toxicology studies indicate that many substances can adversely affect the complex process of reproduction. One source notes that more than 900 agents are known to cause congenital anomalies in laboratory animal studies; 30 of these cause physical or structural malformations in humans (Hicks et al., 1993). Unfortunately, the list of environmental toxicants known to adversely affect male or female reproductive systems is limited. This area has high research priority, especially for substances found in hazardous waste sites that have completed exposure pathways.

The complex reproduction process can sometimes fail to produce offspring (e.g., miscarriage, sterility), or produces offspring that have congenital malformations (birth defects) or other abnormalities, or that develop abnormally. Because these three kinds of reproductive disorders have been associated with exposure to hazardous substances, toxicologists group the effects into three categories.

Reproductive system toxicity from exposure to certain substances can occur in both males and females. For example, the pesticide, 1,2-dibromo-3-chloropropane, causes the testes to atrophy and reduce their capacity to produce sperm, thereby lessening male fertility (Rodricks, 1992). Lead, chlordecone, methylmercury, and many chemotherapeutic drugs have been shown to be toxic to male and female reproductive systems.

Teratogens are substances that cause malformations and other effects in the fetus. The fertilized egg, produced by fertilization of an ovum by a single sperm, first enters the embryonic period, or 14 weeks in which different body organs are formed at different times and rates. Next comes the fetal period, which ends at birth. Many substances are know to produce **teratogenicity**. For example, the drug thalidomide, taken by pregnant women in the 1960s, caused devastating congenital malformations, including missing or malformed limbs in infants. Hazardous waste concerns are raised by epidemiologic studies in Chapter 5 that associate parental proximity to some kinds of CERCLA sites with particular birth defects and lower birth weight of infants.

Following birth, infants continue their growth and development toward becoming adults. Impaired or delayed development of infants and young children is the subject of **developmental toxicity**. For example, exposure of the fetus by maternal consumption of fish containing PCBs has been associated with decreased intelligence and delays in certain cognitive processes in infants (e.g., Jacobson and Jacobson, 1996). Similarly, fetal exposure *in utero* to lead from maternal lead exposure will cause effects on infants' development. The effects include reduced intelligence, delayed development of memory and other cognitive processes, and problems in social adjustment (e.g., Needleman, 1994; Needleman et al., 1996).

Physicians and psychologists have a variety of tests to assess reproductive system toxicity, teratogenicity, and development toxicity. Tests include measuring levels of male and female hormones required for normal reproductive-system functioning, sperm count and quality, and female ovulation. Genetic tests and imaging techniques (e.g., sonograms) can detect some teratogenic effects. Neonatal and infant development can be assessed by observational techniques and measured by various tests. To assess developmental effects, psychologists use batteries of tests that observe or measure cognitive, motor, and sensory functions. As one example, the ATSDR produced a battery of neurobehavioral tests to assess the effects on infants' development of parental exposure to hazardous substances (ATSDR, 1996e).

Cancer (Carcinogenicity)

The cells of the human body are genetically programmed to replace themselves. This is necessary because of the normal cycle of organ growth, tissue damage, cell death, and the need for increased numbers of certain cells to ward off infection or disease. Under normal conditions, cell division (mitosis) is orderly, in harmony with good health, and results in normal growth. A fertilized ovum grows into a fetus; at birth the fetus becomes an infant; infants grow into adults. This wondrous progression from one cell to a biologically complex individual occurs because individual cells divide and clone themselves. In the process, cells differentiate into different types based on selective activation/deactivation of different segments of the cells' genetic material that governs all cellular activities. Liver and blood cells, for example, replicate themselves under normal conditions, but cells of the musculature and nervous system do not.

Sometimes the orderly, genetically programmed process of cell division and replication goes awry. In one case, irretrievably damaged cells undergo programmed death (apoptosis) by a process involving the fragmentation of DNA. In another case, mutation can occur, causing cells to proliferate and lose their functional relationship to other cells. Tissue masses, called tumors, can form and occupy space in the body; for example, in the brain or lung, causing effects that include: compressing adjacent tissue, impairing an organ's normal function, and consuming an excess of energy that should have gone to normal cells.

Tumors are therefore masses of tissue consisting of abnormally replicating cells. Tumors can be benign or malignant. A benign tumor, although an abnormal mass of tissue, does not have the genetic property of uncontrolled growth that is characteristic of malignant tumors. **Cancer**, another term for **malignant tumor**, is also called a **malignant neoplasm**. Cancer cells can break away or be released from a malignant tumor and be transported in the blood stream or lymphatic system to other areas of the body. These migrating, malignant cells can form secondary tumors, or **metastases**, the effect of which is to spread the cancer into other parts of the body.

Malignant cell proliferation has many known causes, including viruses, radiation, chemicals, and other factors. Chemicals and physical agents that cause cancer are called **carcinogens**, and their capacity to cause cancer is referred to as **carcinogenicity**. Laboratory animal research and human epidemiologic studies have identified many individual chemicals and mixtures that can cause cancer.

Cancer researchers have found carcinogenicity to be a multistage process. At least three stages are commonly recognized: initiation, promotion, and progression. The first stage is **initiation**, which occurs when a carcinogen reaches the cell's nucleus and initiates a change in the cell's genetic material. The initiation stage is an alteration in the cell that "primes" it for subsequent carcinogenic development. The carcinogen can be a chemical to which a person was exposed or a metabolite of that chemical. A chemical reaction occurs between the carcinogen and DNA. As Rodricks (1992) stated well, "This reaction constitutes DNA damage, an unwelcome event because this magnificent molecule [DNA] controls the life of the cell and the integrity of its reproduction." Although cells possess the ability to repair DNA damage, damage that is severe enough is permanent.

The second stage of the carcinogenic process is **promotion**. "Promoting agents are chemicals that are not in themselves carcinogens, but which, when given after a low dose of an initiating agent, increase cancer incidence" (Hodgson and Levi, 1987). Promoters can either increase the number of tumors or decrease the latency period of tumor formation. As an example, tryptophan is a promoter for urinary bladder tumors in dogs treated with an initiating dose of 4-aminobiphenyl or 2-naphthylamine. The third stage, **progression**, is the development of full-fledged malignancies. Promotion and progression are the two stages involved in creating a neoplasm, a population of neoplastic cells, from a single neoplastic cell.

Physicians have available various methods and tests to diagnose and treat cancer. Pathologic evaluation of abnormal body tissues is the core of cancer diagnosis. Malignant tumors are differentiated from nonmalignant tumors on the basis of pathologic characteristics of cells in the tumors. Clinicians also use imaging methods; for example, CAT scans, to locate tumors and to assess their mass. Laboratory tests in addition to pathologic examinations of abnormal tissues can include hematologic and endocrine-function tests.

Immune Function Disorders (Immunotoxicity)

Humans are constantly assaulted by hosts of organisms, particles, and substances that are foreign to the body. Many of these alien materials; for example, viruses, have the potential to cause great harm to a person's health unless the body's defense mechanisms are mobilized. Although often overlooked, physical barriers of the body are the first line of defense against alien materials. These barriers include intact skin and secretions from the ears, nose, and eyes. When these barriers fail, the immune system becomes the body's foremost defense system. A compromised immune system can be fatal, as occurs with Acquired Immune Deficiency Syndrome (AIDS).

The immune system is a highly sophisticated recognition system in which the host is able to detect, and respond to, foreign materials, called **antigens** (Swanborg, 1984). As he stated concisely,[1] "When an antigen is introduced into the body, that antigen, as is true of most macromolecules entering the system, is taken up by macrophages. The macrophages, in turn, present the antigen to the lymphocytes, which are the cells involved in the immune system. The antigen and lymphocytes interact, and this interaction triggers the immune response."

The lymphatic system consists of vessels and organs specific to immune function. Lymph vessels are found throughout the body and circulate lymphatic fluid to lymph nodes, which can trap materials foreign to the body. Lymphatic system glands include the spleen, tonsils, thymus, and lymph nodes. Lymph glands produce one class of white blood cells called **lymphocytes**. These cells produce **antibodies**, which are proteins that are specific to neutralizing or destroying antigens. Lymphocytes are of two kinds: T lymphocytes, which are produced in the thymus gland, and B lymphocytes, which are of nonthymus origin. Both T and B lymphocytes have subsets of cell types that play different roles in an immune response.

Immunotoxicology is the study of adverse effects of substances on the components and functions of the immune system. As Rodricks (1992) comments, "The consequences of exposure to immunotoxic agents range from suppression of immunity, which can lead to reduced resistance to infection and certain diseases including cancer, to mild allergic responses." Immunotoxicants can affect the body's immune response in two ways: immunosuppression and immunopotentiation (Swanborg, 1984). Silica suppresses the immune response, presumably by damaging macrophages and impairing their ability to deliver the antigen (i.e., silica particles) to lymphocytes. Other immunosuppressive agents include benzene, PCBs, and some therapeutic drugs.

In immunopotentiation, the immune response is enhanced, often because a substance can combine with host proteins and alter them as they become antigenic. Swanborg describes the immunopotentiation effect of substances that cause contact hypersensitivity, a skin condition that results from T lymphocyte-mediated eczematous reactions to chemicals. Contact hypersensitivity occurs because a chemical, such as nickel, can bind to host proteins, thus becoming antigenic. Another immunopotentiating chemical, toluene diisocyanate, is known to cause respiratory allergic reactions.

Immunotoxicology is a relatively new specialty in the field of toxicology. Laboratory methods are emerging to screen chemicals for immunotoxicity. In the meantime, physicians use laboratory tests to measure the number, kind, and ratio of T lymphocytes and B lymphocytes and their subsets to diagnose the status of immune function. Tests, such as skin patch tests, can assess allergic responses to antigens.

Kidney Dysfunction (Renal Toxicity)

The kidneys are the body's primary means for removing wastes from blood. As part of the urinary system, they filter blood as it passes through and release the waste as urine. Urine then flows to the urinary bladder by way of two tubes called ureters, one

---

[1] Excerpted with permission from *Toxicology: Principles and Practice, Vol. 2*, Sperling, F., Ed. (1984), p. 109. Copyright 1984 John Wiley & Sons, Inc.

attached to each kidney. Urine stored in the bladder is released from the body through a tube called the urethra. Loss of kidney function is fatal unless kidney dialysis (artificial filtering of blood) or kidney transplant occurs.

Nephrons are the primary filtering units of the kidney. Each nephron consists of a renal corpuscle and a unit called a tubule. Blood containing metabolic wastes moves through a part of the renal corpuscle called the glomerulus, forming a filtrate. The kidney resorbs a portion of the minerals in the filtrate to maintain the body's mineral balance.

Many substances are known to be toxic to the kidneys, causing **nephrotoxicity**. Several metals, notably cadmium, chromium, lead and mercury, and chlorinated hydrocarbon compounds like carbon tetrachloride and chloroform are particularly nephrotoxic, causing severe damage to the tubules. The effect is to impair the removal of wastes from the body, an outcome that can be fatal.

Damage to the kidneys is assessed using clinical tests that measure elevated levels of proteins and glucose in urine and excess levels in blood of unexcreted waste products such as urea and creatine. Other clinical tests that measure kidney function include placing dye in the blood stream and using imaging techniques to observe distribution of the dye within the kidneys.

Liver Toxicity (Hepatotoxicity)

The human liver is a large, wonderfully complex organ whose normal functioning is essential for good health. The liver plays a key role in digestion (by secreting bile), regulating sugar levels in blood, storing iron and vitamins, and synthesizing proteins. The functional integrity of liver cells, or hepatocytes, is necessary for normal metabolism and the metabolism of foreign substances. The liver's metabolism of substances, such as ethanol ("alcohol"), detoxifies them by converting them into less toxic metabolites.

The liver has been the subject of extensive examination by toxicologists, both because of its vital functions and because of its role in detoxifying substances. Moreover, the liver is a prime target for toxicity, called **hepatotoxicity**, because all substances absorbed into the body through ingestion are routed through the liver after absorption in the digestive system.

Toxicologists classify substances toxic to the liver according to the type of injury produced (Rodricks, 1992). The health of the liver and its physiologic functions are crucial to the health of the individual. Sometimes substances found in the workplace and community environment can cause acute and chronic toxicity to the liver, depending on the dose delivered (Wilson and Straight, 1993). For example, carbon tetrachloride is a well known **hepatotoxicant**. Liver disorders can range in consequence from asymptomatic disorders in liver function to progressive chronic hepatitis, cirrhosis, liver failure, or fatal hepatic necrosis. Human health investigations and laboratory animal research have provided information about which substances cause liver toxicity and at what doses and the kind of damage caused

Several clinical tests are used to assess liver functioning. For example, certain liver enzymes found in blood following injury to liver tissue can be measured by laboratory tests.

Lung and Respiratory Diseases (Pulmonary Toxicity)

Without oxygen a person will die. This simple molecule (O_2) is vital material for energy processes that occur in the cells of the body. These same processes produce carbon dioxide (CO_2) as waste that must be removed from the body. The lung and nasal passages constitute the respiratory system, through which air enters the body by inhalation and waste products are released via exhaled air. Under ideal conditions, only oxygen and physiologically-inert gases (like nitrogen) would be inhaled, and CO_2 exhaled. In reality, the air breathed contains many kinds of chemicals, some of which have toxic properties; small particles, some of which can harm lung tissue; and microorganisms, some of which can cause disease.

The lungs can be simplistically visualized as a stalk of very small grapes, with air moving through the stalk (trachea) to branches (bronchi) to smaller branches (bronchioles) to the individual grapes (alveoli). The adult human lung contains about 500 million alveoli, with a total surface area of 500 ft^2 (Rodricks, 1992). In alveoli, oxygen enters the blood stream and CO_2 is removed as gas molecules diffuse between alveoli and capillary blood.

Toxicologists have divided damage to the lung caused by hazardous substances into four categories: irritation, fibrosis, cellular damage, and allergic reactions (Rodricks, 1992).

Many substances cause **irritation** of the respiratory system, typically resulting in constriction of the bronchi, or air passages leading into the lungs, and difficulty in breathing occurs. Formaldehyde, sulfur dioxide, and chlorine are examples of respiratory irritants. In **fibrosis**, the second category of pulmonary damage, some very small particles, dusts, and fibers inhaled over prolonged periods of time, typically years, cause changes in lung tissues that make the exchange of oxygen and CO_2 more difficult.

A third kind of pulmonary toxicity, **cellular damage**, occurs when cells along the respiratory tract are injured, causing the cells to release fluids. This results in an accumulation of fluid, or **edema**, in open spaces of the respiratory tract. The fourth category of pulmonary toxicity, **allergic reactions**, result from the body's response to many kinds of foreign agents, including protein molecules, chemical molecules, and microorganisms.

Physicians have several tests available to assess abnormal pulmonary function, including measuring how well the lungs exchange gases across the alveoli, the maximum volume of air that can be forcibly exhaled, enzyme activity of lung tissue, and allergic responses to inhaled agents.

Neurotoxic Disorders (Neurotoxicity)

Because the nervous system includes the brain, it is an incredibly complex system, superior in complexity to even the most sophisticated digital computers. Its functions are essential for the well-being of all living organisms above the level of one-cell microorganisms. The nervous system is classified according to both structure and function: the central nervous system (CNS) consists of the brain and spinal cord; the peripheral nervous system includes nerves that control the muscles, senses, and other control functions.

Nerves are bundles that include nerve fibers, connective tissue, and blood vessels. Each nerve fiber is part of a neuron, which is the individual unit of the nervous system. Neurons transmit electrical impulses along their length (axons), resulting in the release of chemicals, called neurotransmitters, into spaces, called synapses, that lie between neurons. Neurotransmitters released from neurons into synapses will cause electrical impulses to occur (or not occur) in adjacent neurons.

Many substances are toxic to the nervous system, causing **neurotoxicity**. Neurotoxicologists study toxic damage in the nervous system. Neurotoxicity takes several forms, depending on the mechanism of toxicity and location of damage or diminished function. Neurotoxicologists generally categorize toxic damage to the nervous system according to the primary site of anatomic or functional action.

Some **neurotoxicants** exert their effect on the brain. For example, carbon monoxide reduces the supply of oxygen to and in brain tissue, producing anoxia. When inhaled, carbon monoxide readily combines with hemoglobin in red blood cells and thereby reduces the capacity of cells to deliver oxygen to body tissues. Lethal levels of carbon monoxide cause pathologic damage to brain tissue.

Many hazardous substances are toxic to the central nervous system, including mercury, lead, manganese, gold, carbon disulfide, and a range of pesticides, among others. Neurotoxic effects of these and other substances range from mild to very severe, depending on the dose of the particular neurotoxicant, the anatomic location of any pathologic damage to the CNS, and susceptibility factors that include the age and health status of the exposed individual.

Three examples of CNS toxicity illustrate the range of adverse effects. First, severe, chronic exposure to manganese can cause effects that resemble Parkinson's disease. The pattern of chronic manganese toxicity includes impaired speech, incoordination, unsteady gait, speech difficulties, and lassitude. Second, high exposure to mercury can cause a psychologic condition called erethism (Rodricks, 1992). Signs of this toxicity include excessive timidity, self-consciousness, inability to concentrate, and loss of memory. Third, toxic exposures to carbon disulfide (CS_2) have occurred in the rayon and rubber industries where the solvent was used in production processes. Very high exposures produced severe psychoses, but at much lower levels, no significant changes were found in the psychologic function of rayon workers (Putz-Anderson et al., 1983).

Another kind of neurotoxicity occurs when substances like lead and hexachlorophene reduce the amount of myelin, the sheath of tissue that surrounds the axon and other parts of neurons. Loss or reduction of myelin changes the normal electrical properties of axons. Severe toxic damage to peripheral nerves from myelin loss or other changes can cause a condition called peripheral neuropathy, which manifests as impaired motor and sensory functioning. For instance, peripheral neuropathy in the leg can result in difficult walking and loss of sensation in that limb. Substances known to cause peripheral neuropathy include carbon disulfide, lead, and methyl n-butyl ketone. In a study of viscose rayon workers chronically exposed to 20 ppm or less of CS_2 during working hours for more than one year, nerve conduction velocities of motor and sensory nerves were reduced (Johnson et al., 1983).

There are many clinical tests available to physicians and psychologists to assess the effects of neurotoxicants on the human nervous system. Methods and tests from neurology include measuring the electrical activity of brain tissue, muscle activity, rate of conduction of electrical impulses by nerves, and computerized imaging techniques. Neuropsychologists use batteries of tests drawn from neurology and experimental psychology, including measuring a person's mood, sensory and motor functions, and cognitive processes such as memory.

PRIORITY SUBSTANCES

An examination of the HazDat database reveals more than 2,000 substances that have been identified in samples the EPA obtained in their remedial investigation of NPL sites. The primary chemical classes of contaminants most frequently reported in the

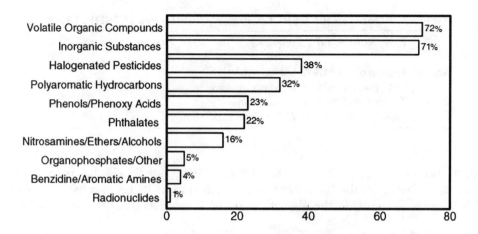

Figure 4.1. Classes of contaminants identified in public health assessments and advisories.

ATSDR's public health assessments and advisories (N = 1,719) conducted through 1996 are shown in Figure 4.1. These classes are: volatile organic compounds (VOCs) (72%), inorganic substances (71%), halogenated pesticides (38%), polyaromatic hydrocarbons (32%), phenols/phenoxy acids (23%), phthalates (22%), nitrosamines/ethers/alcohols (16%), and organophosphates (5%).

Within these chemical classes, substances differ widely in toxicologic and human exposure properties. These differences point to the need to prioritize individual hazardous substances. Information in the figure sheds light on what American society has discarded as chemical waste, and some waste likely reflects changes in regulations covering commercial uses of chemicals. For instance, when clean air regulations restricted the use of TCE in dry cleaning operations, many TCE stocks were apparently discarded as chemical waste.

The ATSDR has made observations about the contaminants released from NPL sites, the environmental media most affected by released contaminants, and the pathways by which persons are exposed to contaminants. Using HazDat, the ATSDR found that as many as a hundred or more different chemicals can be found in a single waste site (ATSDR, 1993a). These chemicals were found in widely varying combinations in water, soil, and air, and some combinations may be much more hazardous than any of the chemicals individually. As an example, the ATSDR found that more than 80% of the time, soil with copper, chromium, arsenic, cadmium or mercury at levels of health concern also contained lead in high levels. This kind of information is important for conducting site-specific, baseline risk assessments and public health assessments and for designing human epidemiologic investigations and toxicologic studies.

Priority Substances at NPL Sites

Of the more than 2,000 substances found in NPL sites, approximately 700 have been detected in contaminated environmental media at three or more NPL sites (an arbitrary number selected for comparison purposes). From this roster of 700 chemicals, the ATSDR and the EPA jointly developed a list of priority substances that pose the greatest

Table 4.1. Algorithm for CERCLA Priority Substances (ATSDR, 1996f).

Total Score = NPL Frequency + Toxicity + Human Exposure

where: Total score = 1800 points maximum
NPL frequency = 600 points maximum
Toxicity = 600 points maximum
Human exposure = 600 points maximum

hazards to human health (ATSDR, 1992c, 1994; ATSDR, 1996f). The list is mandated by CERCLA and is based on the three criteria prescribed in CERCLA and listed in Table 4.1 (ATSDR, 1996f). Elements of the algorithm are calculated as follows:

- *Frequency of Occurrence at NPL Sites*—ATSDR's HazDat database is the source for frequency of occurrence of substances at NPL sites or facilities. Presence of a substance at an NPL site constitutes one occurrence of release, whether the substance occurs in one or more medium.
- *Toxicity*—For a given NPL site, Reportable Quantities (RQs) are used when available, to represent the toxicity of hazardous substances. The EPA established the RQ numbers for regulatory purposes and based each on a substance's toxicity. If no RQ is available for a substance, the RQ methodology is applied to derive a toxicity/environmental score.
- *Potential for Human Exposure*—The exposure component of the ranking algorithm has two parts: concentration of the substance in environmental media and number of persons estimated to be at risk of exposure.

CERCLA directs the ATSDR and the EPA to jointly rank the hazard potential of the top 275 hazardous substances. To illustrate the list's content, the top 25 priority substances are listed in Table 4.2, along with the score for each substance, derived using the ranking algorithm. Arsenic is the hazardous substance ranked first, followed by lead, mercury, vinyl chloride, benzene, PCBs, cadmium, and so forth. Table 4.2 was derived using data from 1,450 NPL sites (ATSDR, 1997b).

Prioritizing hazardous substances is important because CERCLA requires the ATSDR to prepare a Toxicological Profile for each priority substance and to make these profiles available to the public. As of 1997, 224 Toxicological Profiles were available to the public, including those for each substance listed in Table 4.2. Toxicological Profiles have been excerpted to develop other documents, like the ToxFAQ sheets available through the ATSDR website. CERCLA requires that the ATSDR review each profile at least every 3 years and revise and reissue it as needed. The result is a credible, contemporary profile of each substance's toxicity available to both the public and specialists.

Completed Exposure Pathways Priority Substances (CEPPS)

Priority substances listed in Table 4.2 differ somewhat from those listed in Table 2.3, which is based only on frequency of appearance of substances at sites with completed exposure pathways. Although both lists are useful, this book emphasizes the 30 hazardous substances released most often from NPL sites that have known completed exposure pathways. In a public health context, these substances have the greatest potential for human contact by virtue of completed exposure pathway analysis.

Table 4.2. Top 25 Ranked CERCLA Hazardous Substances Found at NPL Sites (ATSDR, 1997b).

Rank	Substance	Score	Rank	Substance	Score
1	Arsenic	1627	14	Aroclor 1260	1183
2	Lead	1522	15	Trichloroethylene	1161
3	Mercury	1492	16	Chromium, hexavalent	1152
4	Vinyl chloride	1398	17	Dibenzo(a,h)anthracene	1142
5	Benzene	1373	18	Dieldrin	1130
6	PCBs	1340	19	Hexachlorobutadiene	1129
7	Cadmium	1315	20	Chlordane	1121
8	Benzo(a)pyrene	1286	21	Creosote	1121
9	Benzo(b)fluoranthene	1258	22	p,p'-DDE	1117
10	PAHs	1251	23	Benzidine	1116
11	Chloroform	1250	24	Cyanide	1113
12	Aroclor 1254	1187	25	Aldrin	1108
13	p,p'-DDT	1183			

Having identified the 30 CEPPS, the question arises about their toxicities. Although significant toxicology data gaps exist, a useful body of information is available on the carcinogenic (Table 4.3) and noncarcinogenic (Table 4.4) toxicities of the 30 CEPPS (Johnson and DeRosa, 1997). The toxicity data for each CEPPS are drawn from the ATSDR's series of Toxicological Profiles. More detailed information about each CEPPS is found in ToxFAQs, which include information on toxicity, exposure limits, and health effects. The following sections summarize the toxicologic properties of the 30 CEPPS.

Carcinogenicity

The carcinogenic potential of the 30 CEPPS is listed in Table 4.3. Cancer is a major health concern and a leading cause of death. The extent to which environmental factors cause cancer is a matter of debate, shaped largely by one's definition of "environmental factors." If personal risk factors, like tobacco smoking, are included as environmental factors, then the "environment" is a major contributor to cancer mortality rates because cigarette smoking is strongly associated with lung cancer mortality. If, however, "environmental factors" are more narrowly defined to include only toxicants released into the general environment (excluding the workplace environment), then some investigators have indicated that environmental factors account for only a small percentage of cancer mortality (Doll and Peto, 1981).

Regardless of the definition of "environmental factors," prudent public health policy advocates preventing exposure to carcinogenic substances, whether tobacco in the home, asbestos in the workplace, or vinyl chloride in groundwater. For this reason, several government agencies critique evidence for the carcinogenicity of substances and classify them according to stated criteria. The Department of Health and Human Services (DHHS), the EPA, and the World Health Organization's International Agency for Research on Cancer (IARC) classify carcinogens according to each organization's own criteria. Each organization reviews findings from laboratory animal studies, human epidemiologic evidence, and knowledge of basic carcinogenic mechanisms of toxicity to derive their categories.

Table 4.3. Percentage of 530 NPL Sites at Which 30 CEPPS Are Found, and Carcinogenicity Category of Each (ATSDR, 1997b; Johnson and DeRosa, 1997).

Substance[a]	%	Category	Substance[a]	%	Category
Trichloroethylene	40	2	**1,2-Dichloroethane**	12	2
Lead, inorganic	34	3	**Methylene chloride**	12	2
Tetrachloroethylene	30	2	Manganese	11	3
Arsenic	23	1	Toluene	10	3
Benzene	21	1	Copper	10	3
Cadmium	17	2	**Nickel**	9	2
Chromium	17	1	**Carbon tetrachloride**	8	2
1,1,1-Trichloroethane	16	3	Barium	8	3
PCBs	15	2	**PAHs**	8	2
1,1-Dichloroethene	14	2	trans-1,2-Dichloroethene	8	3
Chloroform	14	2	**DEHP [di-(2-ethylhexyl)-phthalate]**	8	2
1,1-Dichloroethane	13	3	Antimony	7	3
Vinyl chloride	13	1	**Benzo(a)pyrene**	7	2
Zinc	12	3	**Beryllium**	6	2
Mercury, metallic	12	3	Naphthalene	6	3

[a] Substances in bold print are known human carcinogens or reasonably anticipated to be human carcinogens. Carcinogenicity categories: 1–Designated a human carcinogen by DHHS, EPA, or IARC; 2–Designated as reasonably anticipated to be a human carcinogen by DHHS, EPA, or IARC; 3–Not classified by DHHS, EPA, or IARC.

Carcinogens released from hazardous waste sites have import for human health. The IARC, the DHHS, and the EPA classifications of the 30 CEPPS are listed in Table 4.3. Arsenic is shown in bold print to indicate it is a known or suspected human carcinogen; copper is not known to have carcinogenic properties and is therefore not shown in bold print. Of the 30 substances, four are known human carcinogens (arsenic, benzene, chromium, vinyl chloride), and 14 are reasonably anticipated to be carcinogens (benzo[a]pyrene, beryllium, cadmium, carbon tetrachloride, chloroform, 1,2-dichloroethane, 1,1-dichloroethene, di[2-ethylhexyl]phthalate, methylene chloride, nickel, PAHs, PCBs, tetrachloroethylene, TCE). Thus at least 18 of the 30 CEPPS represent carcinogenic hazards to community populations exposed to them.

As will be described in Chapter 5, the extent of carcinogenic hazard presented nationwide by uncontrolled hazardous waste sites is unknown. Few epidemiologic studies have had sufficient statistical power to adequately investigate cancer mortality in populations living near individual hazardous waste sites (Johnson, 1997). One study reported that U.S. counties with hazardous waste sites had higher mortalities from cancer of the bladder and gastrointestinal tract than did counties not having such sites (Griffith et al., 1989). A similar finding was reported for 20 of the 21 counties of New Jersey (Najem et al., 1983). Mortality from bladder and gastrointestinal cancers was statistically significantly elevated in comparison with national rates; the presence of hazardous waste sites was one of several statistically significant factors in multiple regression analysis of cancer rates. In one site-specific investigation, public access in Woburn, Massachusetts, to municipal well water contaminated with trichloroethylene and other volatile organic compounds was associated with an increased incidence of leukemia in children (Lagakas et al., 1986; MDPH, 1996).

The relatively sparse data on cancer rates in populations living near hazardous waste sites is somewhat surprising, given that analysis of sites with completed exposure pathways shows the presence of several known or suspected human carcinogens. Possibly the generally low exposures that characterize populations subject to completed exposure pathways would account for only small increases in cancer rates, which would be difficult to detect in cancer surveillance systems (Chapter 5). Given the 20 year latency typically associated with the onset of cancer, sufficient time may not have elapsed since releases of carcinogens into completed exposure pathways to cause elevations in cancer rates. Regardless, the presence of large numbers of carcinogens in completed exposure pathways is a concern for protecting the public's health. Further cancer studies are warranted for populations living near hazardous waste sites that have completed exposure pathways, and surveillance of these populations for any early signs of increased rates of cancer would seem purposeful.

The ATSDR's position on science and science policy issues pertaining to carcinogenicity of hazardous substances is given in the agency's Cancer Policy Framework (DeRosa et al., 1993). It outlines the ATSDR's practices and describes the roles that conclusions on carcinogens derived by other groups, professional judgment, and emerging scientific principles play. A central theme in the cancer policy framework is the use of risk analysis (CEQ, 1989) as the organizing construct on which to base biomedical and other scientific judgment when assessing the human health risk of carcinogens. Although risk analysis encompasses the traditional elements of risk assessment, pre-eminent emphasis is placed on biomedical and other scientific judgment, rather than on single numerical conclusions that often convey an artificial sense of precision, despite substantial uncertainties in such estimates.

Noncarcinogenic Toxicity

Besides considering the carcinogenic potential that some CEPPS possess, health officials also must be concerned about other toxicity endpoints. The term **noncarcinogenic toxicity** is used here to refer to toxic endpoints other than carcinogenicity. The noncarcinogenic toxicities of the 30 CEPPS are summarized in Table 4.4, using information from each substance's ATSDR Toxicological Profile (Johnson and DeRosa, 1997). Each CEPPS is described according to its toxic effects on 10 organ systems and functions: lethality, liver, kidney, lung, reproduction, nervous system, cardiovascular system and heart, immune system, skin, and constitutional complaints (e.g., nausea, diarrhea, gastrointestinal upset).

Table 4.4 includes exposure information because "the dose makes the poison." The exposure information is given in relative terms; for example, VH denotes Very High exposure according to route of exposure (i.e., oral, inhalation, dermal). To ascertain more precise reference levels for "Very High" exposure and the other designations, reference must be made to each substance's ATSDR Toxicological Profile. Animal toxicity findings are shown in bold type to distinguish them from toxicity data on human populations, either epidemiologic or clinical observations.

To illustrate noncarcinogenic toxicities, Table 4.4 indicates that tetrachloroethylene at very high levels of exposure by inhalation is lethal to humans, and at lower levels of inhalation is toxic to the nervous system and causes constitutional complaints. Moreover, tetrachloroethylene in laboratory animal studies is toxic to the liver and kidney by both oral and inhalation administration. Thus integrating exposure assessment across

Table 4.4. Noncarcinogenic Toxicities of the 30 Completed Exposure Pathways Priority Substances (Johnson and DeRosa, 1997).

Substance	Death	Liver	Kidney	Lung	Reproductive System	Nervous System	CV, Heart, Hematologic Effects	Immune System	Skin	Constitutional
Antimony	VH(i,o,d)	VH(o) H(i)	H(i)	H(i)	VH(i)	VH(o)	VH(o),L(i) H(i),L(i,o)		VH(d)	VH(o)
Arsenic	VH(o)			H(i),M(i) H(i)		H(o) L(o,i)	L(o,i)		H(o) L(i)	L(o,i) L(o,i)
Barium		H(o)	H(o)	H(o)		H(o)	H(o)			H(o)
Benzene	VH(i)				H(o)	H(o) H(o)	H(o) L(i)	L(i)		H(o)
Benzo(a)pyrene	H(o) H?(i)	M(o)	M(o)	H(i)	M(o)		M(o)	H(o,i)	M(o)	
Beryllium	VH(i,o)			H(i)			VH(i)	H(i),L(o) L(d)	H(d) L(d)	VH(i)
Cadmium	VH(i)		L(o,i)	H(i)						
Carbon tetrachloride	VH(o,i)	H(o,i)	H(o,i)	L(o,i)		H(o,i)				VH(o)
Chloroform	VH(i)	VH(o,i)	VH(o,i)	VH(i)	L(i)	VH(o,i) H(i)			H(d)	
Chromium	VH(o)	VH(o) H(o)	VH(o)	H(o) H(i)	H(o)	VH(o) H(o)	H(o)			VH(o,i)
Copper	VH(o)	H(i) H(o)	H(o)	H(i)		H(i)	M(i) H(o) VH(i)		H(i)	H(o) H(o)
1,1-Dichloro-ethane	VH(i) VH(o)					VH(i)				
1,2-Dichloro-ethane		H(o,i)	H(o,i)	H(o,i)		H(o,i) H(o,i)	H(o,i)	H(o,i)	H(d)	
1,1-Dichloro-ethene		H(i) H(i) H(i)	H(o,i)	H(i) H(o,i) H(i)	H(i)	VH(i) H(i) L(i)			H(d)	
trans-1,2-Dichloroethene	VH(o)					VH(o)	H(i) L(o)	M(i)		

Substance	1	2	3	4	5	6	7	8
DEHP	VH(o,i)	H(o)						
Lead, inorganic		H(o) L(o,i)	**H(o)** L(o,i) H(i) **H(it)** H(i)	H(i)	L(o,i) H(i)	L(o,i)	H(o,i)	H(d)
Manganese			H(i)	H(i)			H(i)	H(i)
Mercury, inorganic	**VH(i)**			H(i)	VH(i)		H(d)	H(i)
Methylene chloride	VH(o)	VH(o)			H(o)		H(i)	H(d)
Naphthalene	VH(o)	H(o)	**M(i)** **H(o)** H(i)	H(o)	H(o,i)		H(i)	H(o,i)
Nickel	VH(o)	H(i) H(o)	H(i)		H(i)		H(i)	H(o)
PAHs	VH(i,o)	**VH(o)** M(o) **H(d)**	L9I0	**H(o)**	**VH(o)**	L(i) **H(d)**	H(i) H(d)	
PCBs		**H(d)**	H(i)	**H(o)**			**H(d)** H(d) H(d,i) **H(o,d)**	H(i)
Tetrachloroethylene	VH(i)	**H(o,i)**		**VH(o)**	VH(o)		H(i)	H(i)
Toluene	VH(i)	H(i)		**H(i)**	H(i)			L(i)
1,1,1-Trichloroethane	VH(i)	VH(o) **VH(o)** H(o)	H(i)	H(i) L(i) H(i),L(i) **VH(o),H(i)** H(i)	VH(i)			H(i) VH(o)
Trichloroethylene	VH(i)	**H(o)**		H(o)	VH(o)	H(d)	VH(d)	
Vinyl chloride	VH(i)	H(i)	H(i) **VH(o)**	H(i)	H(o)			
Zinc				H(i) **VH(o)**	H(o)	H(i) H(i)		H(o)

KEY: VH=Very High exposure or dose; H=High exposure or dose; M=Medium exposure or dose; L=Low exposure or dose; i=inhalation; o=oral; d=dermal; it=intratracheal; Example: H(o,i)=High exposure/dose by oral OR inhalation) exposure. Highlighted toxicities are from laboratory animal studies.

different routes of exposure and viewing such exposures in light of overall body burdens of toxicants is important.

A review of Table 4.4 permits some general observations about the noncarcinogenic toxicities of the 30 CEPPS:

- No CEPPS is without any noncarcinogenic toxicity if exposure levels are very high. This suggests that these 30 CEPPS should, in a precautionary sense, be considered potentially harmful to populations that live near sites with completed exposure pathways unless evidence to the contrary is presented.
- About two-thirds of the CEPPS have toxicity information for all 10 toxicity endpoints listed in Table 4.4, which indicates a considerable toxic threat to communities exposed to the CEPPS.
- Relatively speaking, immunotoxicity data are lacking for the greatest number of CEPPS. This is a serious deficiency in the knowledge that health and risk assessors need because the immune system plays an essential role in protecting one's health.
- Human toxicity data exist for 28 of the 30 CEPPS. However, these data are primarily from occupational health studies and generally represent much higher exposure levels than those typically found in populations living near hazardous waste sites.
- Toxicity data in Table 4.4 generally come from exposures that are "very high" or "high" and as such are probably not representative of general environmental exposures.
- The route of exposure (oral, inhalation, dermal) can affect a CEPPS's toxicity because, in part, of differences in absorption, distribution, and elimination of substances.

Populations at risk of adverse health effects from exposure to CEPPS surely include persons of reproductive age, the elderly, fetuses, infants, and children. Toxicity data like those in Table 4.4 should be viewed as the basis for concern when such populations are exposed to releases from hazardous waste sites. When many hazardous waste sites are aggregated for analysis (described in Chapter 5), the several epidemiologic studies associate residential proximity to uncontrolled hazardous waste sites with noncarcinogenic toxicities such as adverse reproductive effects, immunologic effects, neurologic deficits, and liver disease. Limitations of these human health studies are also described in Chapter 5.

Minimal Risk Levels

Health assessors and risk managers use numerical statements for the toxicities of hazardous substances. Risk assessment methods are most often used for carcinogenic substances (Chapter 2). This results in point estimates or ranges of health risk that are based on various exposure scenarios. The ATSDR uses the numerical Minimal Risk Levels (MRLs) for noncarcinogenic toxicities of substances.

An MRL is an estimate of what daily human exposure to a hazardous substance is likely to be without appreciable risk of adverse noncancer health effects over a specified exposure duration. These substance-specific estimates are intended to serve as screening levels (Risher and DeRosa, 1997). Public health assessors use MRLs to identify contaminants and potential health effects that may be of concern at hazardous waste sites. It is important to note that MRLs are not intended to define cleanup or action levels for the EPA or other agencies.

The ATSDR develops MRLs during the preparation of Toxicological Profiles. They are derived when the ATSDR determines that reliable and sufficient data exist to identify the target organ(s) of effect or the most sensitive health effect(s) for a specific duration of exposure. MRLs are based only on noncancer health effects and not on a consideration of cancer effects. Inhalation MRLs are exposure concentrations expressed in parts per million (ppm) for gases and volatiles, or milligrams per cubic meter (mg/m^3) for particles. Oral MRLs are expressed as daily human doses of milligrams per kilogram per day (mg/kg/day).

The no-observed–adverse-effect approach, incorporating uncertainty factors, is used by the ATSDR to derive MRLs for individual hazardous substances. MRLs are set at exposure levels below what might cause adverse health effects in the persons most sensitive to such substance-induced effects. MRLs are derived for oral and inhalation exposure for acute (1–14 days), intermediate (15–364 days), and chronic (365 days and longer) durations. The ATSDR has not developed MRLs for the dermal route of exposure because dermal exposure data are usually lacking. MRLs are generally based on the most sensitive substance-induced toxic endpoint considered relevant to human health. The ATSDR does not use serious health effects (e.g., irreversible liver damage) as a basis for establishing MRLs. Moreover, exposure to a level above the MRL does not mean that adverse health effects will occur. Health assessors must judge on a site-specific basis.

The ATSDR intends MRLs as screening tools to help health professionals decide when to investigate further. The public health assessment example contained in Chapter 2 used MRLs to screen contaminant levels at the Escambia Wood–Pensacola NPL site. MRLs can be viewed as a mechanism to help identify those hazardous waste sites that are not expected to cause adverse health effects to community residents. Most MRLs contain some degree of uncertainty because precise toxicologic information is lacking on persons who might be most sensitive (e.g., infants) to the effects of hazardous substances. The ATSDR uses a conservative (i.e., protective) approach to address these uncertainties consistent with the public health principle of disease prevention. Although human data are preferred, MRLs must often be based on animal toxicology studies because relevant human studies are lacking. In the absence of information to the contrary, ATSDR assumes that humans are more sensitive than laboratory animals are to the effects of hazardous substances and that certain persons may be particularly sensitive. Therefore, the resulting MRL may be as much as a 100-fold below levels shown to be nontoxic in laboratory animals.

MRLs are available for 22 of the 30 CEPPS. Table 4.5 contains the oral and inhalation MRLs for acute, intermediate, and chronic durations of exposure. Uncertainty factors and toxic endpoints applicable to each MRL are listed in the table. For example, the MRL for benzene for acute inhalation exposure is 0.05 ppm, with an uncertainty factor of 300 used to derive the MRL. The toxic endpoint for this MRL is benzene's effect on immune function.

Toxicologic Data Needs

CERCLA directs the ATSDR to initiate research to fill key data gaps for each priority hazardous substance (Table 4.2). In 1997, 194 key data gaps were being filled for the 50 top-ranked CERCLA priority hazardous substances (ATSDR, 1996d; ATSDR 1997a). The data gaps include both gaps in knowledge about the toxicity of individual substances and also gaps in human exposure characterization. Data gaps for priority substances are filled through an applied research program that consists of government-funded projects at universities and federal government laboratories, research through enforce-

Table 4.5. Completed Exposure Pathways Priority Substance (CEPPS) for Which Minimal Risk Levels (MRLs) Are Available.

Substance	Route	Duration	MRL	Uncertainty Factors	Endpoint
Arsenic, inorganic	Oral	Chronic	0.0003 mg/kg/day	3	Skin
Benzene	Inhalation	Acute	0.05 ppm	300	Immunotoxicity
Cadmium	Inhalation	Chronic	0.0002 mg/m³	10	Renal
	Oral	Chronic	0.0007 mg/kg/day	3	Renal
Carbon tetrachloride	Inhalation	Acute	0.2 ppm	300	Hepatic
		Intermediate	0.05 ppm	100	Hepatic
	Oral	Acute	0.02 mg/kg/day	300	Hepatic
		Intermediate	0.007 mg/kg/day	100	Hepatic
Chloroform	Inhalation	Acute	1 ppm	30	Hepatic
		Intermediate	0.05 ppm	100	Hepatic
		Chronic	0.02 ppm	100	Hepatic
	Oral	Acute	0.3 mg/kg/day	100	Hepatic
		Intermediate	0.1 mg/kg/day	100	Hepatic
		Chronic	0.01 mg/kg/day	1000	Hepatic
Chromium, hexavalent	Inhalation	Intermediate	0.00002 mg/m³	10	Respiratory
		Chronic	0.00002 mg/m³	10	Respiratory
DEHP	Oral	Acute	1 mg/kg/day	100	Reproductive
		Intermediate	0.4 mg/kg/day	100	Developmental
1,2-Dichloroethane	Inhalation	Acute	0.2 ppm	10	Immunotoxic
1,1-Dichloroethene	Oral	Chronic	0.2 ppm	300	Hepatic
	Inhalation	Intermediate	0.2 mg/kg/day	300	Renal
	Oral	Intermediate	0.02 ppm	100	Hepatic
1,2-Dichloroethene, *trans*	Inhalation	Chronic	0.009 mg/kg/day	1000	Hepatic
	Oral	Acute	0.2 ppm	1000	Hepatic
		Intermediate	0.2 ppm	1000	Hepatic
		Intermediate	0.2 mg/kg/day	100	Hepatic

Manganese	Inhalation	Chronic	0.0003 mg/m³	100	Neurologic
Mercury	Inhalation	Acute	0.00002 mg/m³	100	Developmental
	Inhalation	Chronic	0.000014 mg/m³	100	Neurologic
Mercury, inorganic	Oral	Acute	0.007 mg/kg/day	100	Renal
	Oral	Intermediate	0.002 mg/kg/day	100	Renal
Methylene chloride	Inhalation	Acute	0.4 ppm	100	Neurologic
	Inhalation	Intermediate	0.03 ppm	1000	Hepatic
Naphthalene	Inhalation	Chronic	0.06 mg/m³	100	Hepatic
	Inhalation	Chronic	0.002 ppm	1000	Respiratory
	Oral	Acute	0.05 mg/kg/day	1000	Neurologic
	Oral	Intermediate	0.02 mg/kg/day	300	Hepatic
Nickel	Inhalation	Intermediate	0.0001 mg/m³	30	Respiratory
PCBs	Oral	Chronic	0.00002 mg/kg/day	300	Immunotoxicity
Tetrachloroethylene	Inhalation	Acute	0.2 ppm	10	Neurologic
	Inhalation	Chronic	0.04 ppm	100	Neurologic
	Oral	Acute	0.05 mg/kg/day	100	Development
Toluene	Inhalation	Acute	3 ppm	30	Neurologic
	Inhalation	Chronic	1 ppm	30	Neurologic
	Oral	Acute	0.8 mg/kg/day	300	Neurologic
	Oral	Intermediate	0.02 mg/kg/day	300	Neurologic
1,1,1-Trichloroethane	Inhalation	Acute	2 ppm	100	Neurologic
	Inhalation	Intermediate	0.7 ppm	100	Neurologic
Trichloroethylene	Inhalation	Acute	2 ppm	30	Neurologic
	Inhalation	Intermediate	0.1 ppm	300	Neurologic
Vinyl chloride	Oral	Acute	0.5 mg/kg/day	100	Developmental
	Oral	Intermediate	0.002 mg/kg/day	100	Developmental
	Inhalation	Acute	0.5 ppm	100	Developmental
	Inhalation	Intermediate	0.03 ppm	300	Hepatic
	Oral	Chronic	0.00002 mg/kg/day	1000	Hepatic
Zinc	Oral	Intermediate	0.3 mg/kg/day	3	Hemato
	Oral	Chronic	0.3 mg/kg/day	3	Hemato

able consent actions between the EPA and industry through authorities in the Toxic Substances Control Act (Auer, 1994), and research voluntarily conducted by private industry. All these mechanisms will ultimately improve the quality of risk and public health assessments through better toxicologic science and data.

Key data needs in exposure characterization for the CEPPS were listed and discussed in Chapter 3. In a similar vein, key toxicologic data needs are listed in Table 4.6 for the 30 CEPPS. The ATSDR and the EPA have yet to establish whether any priority data needs exist for eight CEPPS (1,1,1-trichloroethane, 1,1-dichloroethene, copper, barium, *trans*-1,2-dichloroethene, antimony, benzo(a)pyrene, and naphthalene). As an illustration, Table 4.6 shows that four toxicologic data needs have been established for benzene: (1) dose-response data in animals for acute- and intermediate-duration oral exposure, including extended reproductive organ histopathology, (2) 2-species developmental toxicity study via oral exposure, (3) neurotoxicology battery of tests via oral exposure, and (4) epidemiologic studies on the health effects of benzene, with special emphasis on immunotoxicity. Note that benzene data needs for the 2-species developmental toxicity study and neurotoxicity testing are currently shown in bold, indicating that these data gaps are currently being filled (ATSDR, 1996d; ATSDR, 1997a).

A perusal of the toxicologic data needs listed in Table 4.6 reveals a mix of needed research. Of the 64 toxicologic data needs listed in the table, 36 (56%) are currently being filled through research. This indicates a substantial body of toxicologic research is under way, but with an equally substantial work awaiting. More than half (39) of the 64 toxicologic data needs fall into two categories: dose-response studies (22) and reproductive toxicity studies (17). Given the historical human health importance of the CEPPS, it is surprising that 22 data needs are specific to acquiring very fundamental information, dose-response data.[2] It is noteworthy that the need for reproductive toxicologic data is getting emphasis. On the other hand, the lack of identified data needs in immunotoxicology is puzzling, given the importance to health of the immune system. This may reflect uncertainty about how to evaluate substances for toxicity to immune function. Seven of the 64 toxicologic data needs are for new or additional carcinogenesis data (TCE, PCBs, chloroform, zinc, manganese, PAHs, and DEHP). This relatively small number reflects the emphasis toxicologists have put on assessing the carcinogenicity of substances. This has produced information on the carcinogenic potential of many substances, but at the expense of data for the noncarcinogenic toxicities of the same substances.

TOXICOLOGY OF CHEMICAL MIXTURES

Uncontrolled hazardous waste sites typically contain mixtures of hazardous substances. Data in HazDat confirm that a hundred or more different substances can be found in a single waste site. They are found in widely varying combinations in water, soil, and air. Some combinations of substances can be more hazardous than any single substance released alone. Information about chemical mixtures is therefore useful in conducting site-specific risk and public health assessments. Knowledge of chemical mixtures is also useful for designing human epidemiologic investigations and toxicologic studies. However, the toxicologic study of mixtures of hazardous substances is a challenge (e.g., Yang, 1996). What specific mixtures or combinations of substances should

---

[2] Ten of these 22 studies also include acquiring reproductive effects data.

Table 4.6. Key Toxicologic Data Needs for 30 Completed Exposure Pathways Priority Substances (excerpted from ATSDR, 1996d; ATSDR, 1997a).

Substance	Toxicologic Data Gaps
Antimony	To be determined
Arsenic	Comparative toxicokinetic studies to determine whether an appropriate animal species can be identified
Barium	To be determined
Benzene	(1) **Dose-response data in animals for acute- and intermediate-duration oral exposure. The subchronic study should include extended reproductive organ histopathology, (2) 2-species developmental toxicity study via oral exposure, (3) neurotoxicity battery of tests via oral exposure,** (4) epidemiologic studies on the health effects of benzene (special emphasis: immunotoxicity)
Benzo(a)pyrene	To be determined
Beryllium	(1) **Dose-response data in humans for acute- and intermediate-duration inhalation exposures. The subchronic study should include extended reproductive organ histopathology, (2) 2-species developmental toxicity study via inhalation exposure, (3) immunotoxicology battery of tests via oral exposure**
Cadmium	No toxicologic data gaps; exposure data gaps are listed in Table 3.3.
Carbon tetrachloride	(1) Dose-response data in animals for chronic oral exposure. The study should include extended histopathology of reproductive organs and nervous system tissue, and demeanor, **(2) immunotoxicology battery of tests via oral exposure**
Chloroform	(1) Dose-response data in animals for intermediate-duration oral exposure, (2) epidemiologic studies on the health effects of chloroform (special emphasis: cancer; neurotoxicity, reproductive and developmental toxicity, hepatotoxicity, renal toxicity)
Chromium	(1) **Dose-response data in animals for acute-duration oral exposure to chromium (VI) and (III) and for intermediate-duration oral exposure to chromium (VI), (2) multigenerational reproductive toxicity study via oral exposure to Cr(III) and Cr(VI), (3) immunotoxicology battery of tests following oral exposure to Cr(III) and Cr(VI), (4)** 2-species developmental toxicity study via oral exposure to Cr(III) and Cr(VI)
Copper	To be determined
1,1-Dichloroethane	(1) **Dose-response data in animals for acute- and intermediate-duration oral exposures. The subchronic study should include an evaluation of immune and nervous system tissues, and extended reproductive organ histopathology, (2)** dose-response data in animals for chronic inhalation exposures. The study should include an evaluation of nervous system tissues
1,2-Dichloroethane	(1) **Dose-response data in animals for acute- and intermediate-duration oral exposures. The subchronic study should include an evaluation of immune and nervous system tissues, and extended reproductive organ histopathology, (2)** dose-response data in animals for chronic inhalation exposures. The study should include an evaluation of nervous system tissues

Table 4.6. Key Toxicologic Data Needs for 30 Completed Exposure Pathways Priority Substances (continued).

Substance	Toxicologic Data GAPS
1,1-Dichloroethene	To be determined
trans-1,2-Dichloroethene	To be determined
DEHP [di(2-ethylhexyl)phthalate]	(1) Epidemiologic studies of the health effects of DEHP (special emphasis: cancer), (2) dose-response data in animals for acute- and intermediate-duration oral exposures. The subchronic study should include an extended histopathologic evaluation of the immunologic and neurologic systems. (3) multigenerational reproductive toxicity study by oral exposure, **(4) comparative toxicokinetic studies**
Lead	**(1) Mechanistic studies on the neurotoxic effects of lead**
Manganese	(1) Dose-response data for acute- and intermediate-duration oral exposures (the subchronic study should include reproductive histopathology and an evaluation of immunologic parameters including Mn effects on plaque-forming cells (SRBC), surface markers (D4:D8 ratio), and delayed hypersensitivity reactions, (2) toxicokinetic studies on animals to investigate uptake and absorption, relative uptake of differing Mn compounds, metabolism of Mn, and interaction of Mn with other substances following oral exposure, (3) epidemiologic studies on the health effects of Mn (special emphasis includes neurologic, reproductive, developmental, immunologic, and cancer)
Mercury, metallic	**(1) Multigenerational reproductive toxicity study by oral exposure, (2) dose-response data in animals for chronic-duration oral exposure, (3) immunotoxicology battery of tests by oral exposure**
Methylene chloride	**(1) Dose-response data in animals for acute- and intermediate-duration oral exposure. The subchronic study should include extended reproductive organ histopathology, neuropathology and demeanor, and immunopathology, (2) 2-species developmental toxicity study via the oral route**
Naphthalene	To be determined
Nickel	(1) Epidemiologic studies of the health effects of nickel (special emphasis: reproductive toxicity), (2) 2-species developmental toxicity study via the oral route, (3) neurotoxicology battery of tests via oral exposure, (4) dose-response data in animals for acute- and intermediate-duration oral exposures
PAHs	**(1) Dose-response data in animals for intermediate-duration oral exposures. The subchronic study should include extended reproductive organ histopathology and immunopathology, (2) dose-response data in animals for acute- and intermediate-duration inhalation exposures. The subchronic study should include extended reproductive organ histopathology and immunopathology, (3)** 2-species developmental toxicity study via inhalation or oral exposure, **(4)** mechanistic studies how mixtures of PAHs can influence the ultimate activation of PAHs, and on how PAHs affect rapidly proliferating tissues, **(5)** epidemiologic studies on the **health effects of PAHs (special emphasis includes cancer, dermal hemolymphatic, and hepatic)**

PCBs	**(1) Dose-response data in animals for acute- and intermediate-duration oral exposures, (2)** dose-response data in animals for acute- and intermediate-duration inhalation exposures, **(3) epidemiologic studies on the health effects of PCBs (special emphasis: immunotoxicity, gastrointestinal toxicity; liver, kidney, thyroid toxicities, reproductive and developmental toxicity), (4) chronic toxicity and oncogenicity via oral exposure, (5) PCB congener analysis**
Tetrachloroethylene	**(1) Dose-response data in animals for acute-duration oral exposure, including neuropathology and demeanor, and immunopathology, (2) multigenerational reproductive toxicity study via oral exposure, (3) 2-species developmental toxicity study via oral exposure, (4)** dose-response data in animals for chronic-duration oral exposure, including neuropathology and demeanor, and immunopathology
Toluene	**(1) Dose-response data in animals for acute- and intermediate-duration oral exposures. The** subchronic study should include an extended histopathologic evaluation of the **immune system, (2) neurotoxicology battery of tests via oral exposure, (3)** mechanism of toluene-induced neurotoxicity, **(4) comparative toxicokinetic studies**
1,1,1-Trichloroethane	To be determined
Trichloroethylene	**(1) Neurotoxicity battery of tests via the oral route, (2) immunotoxicity battery of tests via the oral route, (3)** epidemiologic studies on the health effects of TCE (special emphasis: cancer, hepatotoxicity, renal toxicity, developmental toxicity, neurotoxicity)
Vinyl chloride	**(1)** Dose-response data in animals for chronic-duration inhalation exposure, **(2) multigenerational reproductive toxicity study by inhalation exposure, (3)** mitigation of vinyl chloride-induced toxicity, **(4) 2-species developmental toxicity study via inhalation**
Zinc	**(1) Dose-response data in animals for acute- and intermediate-duration oral exposures. The** subchronic study should include an extended histopathologic evaluation of the **immunologic and neurologic systems, (2)** multigenerational reproductive toxicity study via the oral route, (3) carcinogenicity testing (2-year bioassay) via oral exposure

Notes: (1) **Data gaps shown in bold print are currently being filled through research (ATSDR, 1996d for details).** (2) Dose-response data needs are also listed in Table 3.3 because both exposure and toxicity endpoints are investigated.

be tested for toxicity? At what exposure levels? This section describes a strategy for assessing the toxicity of chemical mixtures.

For hazardous waste sites, exposure is more likely to multiple hazardous substances than to a single substance. Extrapolation from well-controlled toxicologic studies on single chemicals (when exposures are often treated as time-weighted averages) to exposure scenarios typical of the ambient environment is a challenge. Typical exposures not only entail multiple chemicals, but also have dramatically different patterns of exposure (e.g., combined routes of exposure) from those often used in toxicologic research (e.g., only one route of exposure).

In addition to the substance-specific applied research CERCLA mandates, CERCLA also mandates the ATSDR to develop methods to determine the health effects of hazardous substances commonly found in combination with other substances. To address this CERCLA mandate, the ATSDR is pursuing a strategy that consists of three elements: (1) a trend analysis to identify commonly recurring combinations of chemicals, (2) an assessment of available information on potential interactions of substances, and (3) an applied research program on chemical mixtures (Johnson and DeRosa, 1995). A brief description of these three elements follows. Quantitative structure-activity relationships and physiologically-based pharmacokinetics models are used in support of the three elements.

Trend Analysis

What are trends evident in the combinations of hazardous substances released from hazardous waste sites that have completed exposure pathways? The ATSDR used the HazDat database to identify binary pairs of hazardous substances most often reported in public health assessments (Fay, 1997). It identified 462 NPL sites that had one or more completed exposure pathways and at least one binary pair of hazardous substances. This led to a list of 41,204 binary pairs of substances; 47% occurred in soil, 31% in water (ground or surface), 6% in air, 3.6% in biologic media, and 12% in miscellaneous sources. Table 4.7 contains the 25 most frequently found binary pairs, with carcinogenic substances (as categorized in Table 4.3) shown in bold. For example, trichloroethylene and tetrachloroethylene occur most often together at sites that have completed exposure pathways. All 16 hazardous substances listed in Table 4.7 are CEPPS. Trichloroethylene and lead appear most often as members of binary pairs found in completed exposure pathways.

The carcinogenic potential of the binary pairs of hazardous substances can be seen by noting the substances in Table 4.7 shown in bold. In 10 binary pairs both members are carcinogens, and nine additional pairs contain one carcinogen each. Thus, 19 of the 25 most-often-released binary pairs are potentially carcinogenic. This raises the question of whether releases from uncontrolled hazardous waste sites are contributing to elevated rates of human cancer and other adverse health effects.

Such trend analysis can also identify some "indicator" or "typical" mixtures that might be characteristic of uncontrolled hazardous waste sites. Indicator chemicals could then be tested in toxicologic research and tests that screen mixtures for toxicity.

Assessment of Interactions

The second element of the ATSDR's strategy on chemical mixtures is assessing interactions among the chemicals that constitute a given mixture. The assessment must incorporate insights into how mechanisms of action for individual chemicals might

Table 4.7. Top 25 Binary Combinations of Contaminants Released from 462 NPL Sites That Have Completed Exposure Pathways (Fay, 1997).

Sites	First Substance	Second Substance
135	**Trichloroethylene**	**Tetrachloroethylene**
95	Lead	**Arsenic**
87	Lead	**Cadmium**
77	1,1,1-Trichloroethane	**Trichloroethylene**
77	Lead	**Chromium**
74	**Benzene**	**Trichloroethylene**
68	**Arsenic**	**Cadmium**
66	**1,1-Dichloroethene**	**Trichloroethylene**
64	**Arsenic**	**Chromium**
63	1,1,1-Trichloroethane	**Tetrachloroethylene**
63	**Benzene**	**Tetrachloroethylene**
62	**Trichloroethylene**	Lead
60	Lead	Zinc
56	1,1-Dichloroethane	**Trichloroethylene**
55	**1,1-Dichloroethene**	**Tetrachloroethylene**
54	**Cadmium**	**Chromium**
52	**Chloroform**	**Trichloroethylene**
51	Lead	Manganese
49	**Cadmium**	Zinc
49	Copper	Zinc
49	Lead	Copper
48	**Trichloroethylene**	**1,2-Dichloroethane**
47	1,1,1-Trichloroethane	1,1-Dichloroethane
47	1,1,1-Trichloroethane	**1,1-Dichloroethene**
47	Lead	Mercury

relate to mechanisms of interaction. Systematic evaluation of available information for specific chemicals and chemical mixtures could identify the research needed to set rules that can be generalized to other situations. The weight-of-evidence scheme shown in Table 4.8 illustrates one such systemic assessment of information.

Applying the scheme described in Table 4.8, literature on mechanisms of toxicologic action and interaction has been critically evaluated for more than 200 of the 275 CERCLA priority hazardous substances (ATSDR, 1992c, 1994; ATSDR, 1996f). Using this information, the weight of evidence for interactions can be characterized as part of an interaction profile predicated on mechanistic insights.

To convey in summary form the weight of evidence for interaction, available information can be placed in an alphanumeric format, as summarized in Table 4.8. Data can be classified on the basis of mechanistic insights such as: direct and unambiguous mechanistic data; mechanistic data inferred from related compounds; or mechanistic data inferred considered inadequate or ambiguous. Similarly, toxicologic importance can be classified as: demonstrated directly; indirectly inferred from the study of related compounds; or unclear in terms of toxicologic significance (Mumtaz et al., 1994). Shorthand descriptors are an adjunct to, not a surrogate for, an overall narrative that describes the weight of evidence for a chemical mixture's interactions. Because chemical interactions

Table 4.8. Weight-of-Evidence Scheme for the Qualitative Assessment of Chemical Interactions (Mumtaz et al., 1994).

Determine whether the interaction of the mixture is additive (=), greater than additive (>), or less than additive (<).

Classification of mechanistic understanding:

I. Direct and unambiguous mechanistic data: The mechanism(s) by which the interactions could occur has been well characterized and leads to an unambiguous interpretation of the direction of the interaction.

II. Mechanistic data on related compounds: The mechanism(s) by which the interactions could occur is not well characterized for the compound of concern but a structure-activity relationship, either quantitative or qualitative, can be used to infer the likely mechanisms and the direction of the interaction.

III. Inadequate or ambiguous mechanistic data: The mechanism(s) by which the interactions could occur has not been well characterized or information on the mechanism(s) does not clearly indicate the direction the interaction will have.

Classification of toxicologic significance:

A. The toxicologic significance of the interaction has been demonstrated directly.

B. The toxicologic significance of the interaction can be inferred or has been demonstrated in related compounds.

C. The toxicologic significance of the interaction is unclear.

Modifiers:

1. Anticipated exposure duration or sequence
2. Different exposure duration or sequence
   a. *In vivo* data
   b. *In vitro* data
      i. Anticipated route of exposure
      ii. Different route of exposure

---

typically involve saturable biologic phenomena, the interactions narrative should address the degree to which interactions at high doses can be extrapolated to lower doses.

To illustrate the weight-of-evidence scheme outlined in Table 4.8, a "less than (<)" or "greater than (>)" sign indicates whether the interaction data suggest antagonism or some sort of potentiation; for example, additivity or synergism. A Roman numeral denotes whether the mechanistic data were on the compounds of interest (I) or were based on related compounds (II). Similarly, the uppercase letters indicate whether the data are of direct toxicologic significance (A), had been inferred from related compounds (B), or were unclear in their toxicologic significance (C). Additional designations, such as Arabic numerals or small letters, are used to designate the appropriateness of the route of exposure or whether the data were derived from *in vivo* or *in vitro* assays. An application of the scheme outlined in Table 4.8 can be found elsewhere (ATSDR, 1997c).

Applied Research on Chemical Mixtures

The use of information databases to derive the degree of potential interactions of a complex mixture is limited. A direct approach is to use functional toxicology screens to evaluate the toxicity of mixtures (McLachlan, 1993). These screens consist of cell populations, called host cells, derived from experimental animals. The cells are maintained in test tubes or cell cultures. Using genetic engineering techniques, genetic material to control a particular function (e.g., carcinogenic response) is added to genetic material of the host cells.

In addition, a "reporter" sequence of genetic material is added to the host cell. This permits the activity of the added genetic material to be "reported" in response to the presence of a hazardous substance, using photoluminescence as an index of toxic response.

These *in vitro* screens allow an assessment of the degree to which a single compound or multiple compounds might exhibit "dioxin-like" or estrogenic activity or some other functional activity. Results obtained with single chemicals and with equivalent doses of chemical mixtures can be compared to indicate the degree to which the potential joint toxic action is additive, less than additive (antagonistic), or more than additive (synergistic). This could also be assessed with a similar dose-adjustment approach using other (*in vivo* or *in vitro*) traditional toxicologic tests.

Another approach for assessing toxicities of chemical mixtures is the further development of mathematical models and computational approaches, including physiologically based kinetic models (e.g., Yang et al., 1995). These models can be used to define levels of chemicals or chemical mixtures to be assessed *in vitro*, such as in functional toxicology screens.

Similarly, structure-activity relationships can be used to extend the range of plausible inference and extrapolation within and across chemical classes (e.g., el-Masri et al., 1997; Verhaar et al., 1997). This approach compares the known toxicity of a substance that has a known molecular structure with compounds of similar molecular structure for which experimental toxicologic data is lacking. Although no single assay screen or method is likely to provide all necessary information, such approaches, in aggregate, can provide important insights about the overall public health implications of exposure to chemical mixtures.

The comments in this section outline one agency's approach to assessing chemical mixtures. This should not be interpreted as indicating that no toxicologic testing of complex mixtures has occurred. Quite the contrary, several toxicologists have reported findings from studies of complex mixtures that are relevant to hazardous waste sites. Because this body of research is nascent, a comprehensive review of findings would be premature. However, three studies will suffice to describe how these kinds of studies are conducted.

Simmons et al. (1988) evaluated the lethality and hepatotoxicity of a representative set of complex waste mixtures. Male rats, exposed by gavage to samples of 10 sets of complex mixtures, were evaluated 24 hours later. The mixtures were obtained from the input stream of six hazardous waste incinerators and were analyzed for concentrations of organic chemicals on an EPA list of priority chemicals. Seven of the 10 samples caused death at doses ranging from 1 to 5 mL/kg body weight; 8 of the 10 samples were hepatotoxic based on histopathologic evaluation. The investigators concluded, "In general, the hepatotoxicity of the waste samples did not appear to be readily predicted from (partial) chemical characterization data." This outcome must be tempered, however, because only partial chemical characterization of the complex mixtures was done. This study is nevertheless important in illustrating some challenges faced by investigators who conduct studies of complex mixtures.

Germolec et al. (1989) studied a complex mixture of 25 contaminants commonly found in groundwater at hazardous waste sites. The mixture was administered in drinking water to female mice. The dosing lasted 14 or 90 days, and immune function was assessed for both durations of exposure. No overt signs of toxicity were observed throughout the study, but immune-function changes were notable and included suppression of hematopoietic stem cells and antigen-induced antibody-forming cells. No effects were seen on T lymphocyte function or numbers of T and B lymphocytes. Moreover, mice treated with the highest concentration of the chemical mixture exhibited lower resistance to an infectious agent, indicating an impaired immune-system response.

In another study, male rats were administered drinking water containing a mixture of seven substances (and submixtures of these substances) selected for their presence in groundwater at hazardous waste sites (Constan et al., 1995). The seven substances were arsenic, benzene, chloroform, chromium, lead, phenol, and TCE. Male F344 rats were administered the mixture for six months. The study included a time-course experiment using toxicologic testing at three and ten days and one, three, and six months. No differences in weight gain, body weight, liver weight ratios, and liver-associated plasma enzymes were found throughout the study. Primary focus of the study was to measure the proliferation of hepatocytes using a technique that labeled the nuclei of proliferating cells. Significant increases in hepatocellular labeling were observed at three days, ten days, and one month after treatment with the full mixture, as well as with organic or inorganic submixtures. The proliferation of hepatocytes occurred in the absence of other signs of toxicity.

The finding of increased hepatocellular proliferation led the same investigators to evaluate apoptosis in the livers of male rats exposed to the same mixture of substances in drinking water used in their prior research (Constan et al., 1996). Apoptosis, the process by which cells undergo programmed death, occurs when cells are irretrievably damaged, whether cancerous or not. The schedules for administering the complex mixture and testing were the same as in the investigators' prior study (Constan et al., 1995). Immuno-histochemical staining for apoptosis was the primary test method. Apoptosis activity was found to be maximum after one month of exposure, whereas hepatocellular proliferation reached a maximum rate after ten days of exposure. The investigators concluded that apoptosis correlated with changes in cell proliferation. Moreover, the effects on cell proliferation and apoptosis occurred from a relatively low dose of a chemical mixture.

Because of the multiple combinations of test mixtures used, toxicologic studies of complex mixtures, like those illustrated here, challenge investigators, especially if the studies involve large numbers of laboratory animals. Chemical mixtures must be carefully characterized and doses precisely administered to animals, usually over long periods of time. In some toxicologic studies, mixtures have been selected as representative of those found in groundwater at hazardous waste sites. The question arises whether the results can be extrapolated to other mixtures. The weight-of-evidence and associated alphanumeric approaches described here are intended to convey a confidence level when addressing this question. These methods are currently the subject of applied research to further assess their utility. Nonetheless, toxicologic studies of complex mixtures are vital if the human health impact of real community exposures and associated health concerns is to be interpreted.

SYNTHESIS

The dose makes the poison. This is the first and most fundamental principle of toxicology. This basic principle is often misunderstood or misused in relation to hazardous waste issues. For example, because concentrations of hazardous substances found in environmental media are often "low," some scientists and special interest groups have averred that no health effects would occur. They assert that health effects occur only at very high levels of exposure, citing occupational health studies of workers who did not manifest adverse health effects at even relatively high levels of exposure in the workplace. Other groups assert that these "low" levels *must* be associated with the health problems they have observed in neighbors or family members because they think substances are toxic at any level of exposure. Where is the truth? Understanding the difference between exposure and dose and gaining a knowledge of the basic elements of toxicology are fundamental to getting at the "truth."

Any diverse population exposed to a hazardous substance under nonlethal exposure conditions will express a range of toxic reactions. Some persons within the population will be more susceptible to the substance's toxicity than will others. Susceptible groups include children, elderly persons, fetuses, persons with certain genetic disorders, and persons with infirmities.

Knowing the toxicity of substances at, and released from, hazardous waste sites is fundamental for assessing a site's hazard to human health. An examination of the HazDat database reveals that more than 2,000 substances have been identified in samples obtained by the EPA in their remedial investigations of NPL sites. The primary chemical classes are: volatile organic compounds (VOCs) (at 72% of sites), inorganic substances (71%), halogenated pesticides (38%), polyaromatic hydrocarbons (32%), phenols/phenoxy acids (23%), phthalates (22%), nitrosamines/ethers/alcohols (16%), and organophosphates (5%). Within these chemical classes are substances that differ widely in toxicologic and human-exposure properties.

All substances released from hazardous waste sites are not equally hazardous to human health. The toxicology of those substances found in completed exposure pathways is particularly important because these substances have the greatest potential for human exposure. There are 30 substances found at 6% or more of NPL sites that have completed exposure pathways. In this text, these are referred to as Completed Exposure Pathways Priority Substances (CEPPS). Of the 30 substances, four are known human carcinogens (arsenic, benzene, chromium, vinyl chloride), and 14 are reasonably anticipated to be carcinogens (benzo[a]pyrene, beryllium, cadmium, carbon tetrachloride, chloroform, 1,2-dichloroethane, 1,1-dichloroethene, di[2-ethylhexyl]phthalate, methylene chloride, nickel, PAHs, PCBs, tetrachloroethylene, TCE). Thus, at least 18 of the 30 CEPPS represent carcinogenic hazards to community populations exposed to them.

An examination of the carcinogenic potential of binary pairs of hazardous substances most often released from sites with completed exposure pathways shows 19 of the 25 most common binary pairs contain one or more carcinogen. This again raises the question of whether releases from uncontrolled hazardous waste sites are contributing to elevated rates of human cancer.

In addition to the carcinogenic potential that some CEPPS possess, health officials must be concerned about other toxicity endpoints. The term noncarcinogenic toxicity is used here to refer to toxic endpoints other than carcinogenicity. Some general observations can be made about the noncarcinogenic toxicities of the 30 CEPPS:

- No CEPPS is without any noncarcinogenic toxicity if exposure levels are very high. This suggests that, until evidence to the contrary is available, these 30 CEPPS should, in a precautionary sense, be considered potentially harmful to populations that live near sites with completed exposure pathways.

- Toxicity information available on about two-thirds of the CEPPS for all 10 toxicity endpoints listed in Table 4.4 indicates a considerable toxic threat to communities exposed to the CEPPS.
- Immunotoxicity is the endpoint most often lacking for the CEPPS. This is a serious deficiency in knowledge needed by health and risk assessors because the immune system plays an essential role in protecting one's health.
- Human toxicity data exist for 28 of the 30 CEPPS. However, these data are primarily from occupational health studies and generally represent much higher exposure levels than those typically found for populations living near hazardous waste sites.
- Toxicity data in Table 4.4 generally come from exposures that are "Very High" or "High" and thus are probably not representative of general environmental exposures.
- The route of exposure (oral, inhalation, dermal) can affect the toxicity because in part of differences in absorption, distribution, and elimination of substances.

It bears repeating that the carcinogenic and noncarcinogenic toxicities associated with specific hazardous substances found in completed exposure pathways do not necessarily translate into actual adverse human health effects. One must ascertain the extent of actual exposure experienced by populations at risk and the susceptibility of members of those populations. The extent of hazard in communities impacted by hazardous waste sites must be determined on a case-by-case, weight-of-evidence approach, preferably through public health assessments.

Chapter Five

Human Health Studies

What evidence associates adverse human health effects with hazardous waste sites and other sources of uncontrolled releases of hazardous substances? How extensive and robust is the human health effects database? When should a health study be conducted? This chapter summarizes basic concepts in epidemiology and presents the key literature on associations between human exposure to hazardous substances and impacts on human health. The chapter is organized around ATSDR's Seven Priority Health Conditions on the health effects of substances released from waste sites.

Scientific literature on human exposure to hazardous substances indicates that sufficient exposure to some individual substances, or combinations of substances, will cause adverse health effects.[1] The literature base supporting this assertion comes from studies of workers, laboratory animals, and, to a lesser extent, community residents.

Concerning exposure to hazardous substances released from waste sites or from emergency events like chemical spills, two broad groups of populations are of primary concern. One group is persons whose places of residence may lead to exposure to substances released from hazardous waste sites. Within this group are persons of special concern because of vulnerability to hazardous substances. In particular, young children are known to be at elevated health risk because they absorb through the gut greater doses of hazardous substances than do adults for comparable doses. The other broad group is waste site cleanup workers, workers at waste disposal facilities, and persons who respond to chemical spills and similar events. Chapter 6 summarizes the safety and health risks that hazardous waste workers and emergency responders have from exposure to uncontrolled releases of hazardous substances.

This chapter summarizes key health studies of persons living near hazardous waste sites. Most of these studies have used the principles and practices of epidemiology. A summary of the basic principles, practices, strengths, and limitations of epidemiologic investigations is included as prologue to describing findings from specific epidemiologic studies. This prologue will help clarify the nature, strength, and limitations of the described studies.

---

[1] Use of the term "sufficient exposure" implies that a threshold exposure must be achieved for a toxicant to be harmful to an individual's health. For some toxicants the threshold dose can be quite small.

Before proceeding, a definition of epidemiology is in order. According to one source, an international panel has provided the following definition (Tyler and Last, 1992):

Epidemiology is the study of the distribution and determinants of health-related states and events in specified populations and the application of this study to the control of health problems.

Thus, epidemiology deals with health, events, specified populations, and the application of findings from epidemiologic investigations to control health problems. Within the field of epidemiology are specialty areas. One specialty area, environmental epidemiology, will be discussed in a subsequent section.

Since the early 1980s, a considerable number of epidemiologic studies have investigated the health of persons who live near hazardous waste sites. Some studies report adverse health effects in communities or populations exposed to hazardous substances released from waste sites or from uncontrolled chemical releases. These studies have examined many health outcomes, including cancer, adverse reproductive outcomes, and toxicity to organ systems (e.g., Buffler et al., 1985). The sections that follow provide summary descriptions of findings from various kinds of epidemiologic investigations of persons in communities potentially impacted by uncontrolled releases of hazardous substances.

Other epidemiologic studies of community residents putatively exposed to substances from hazardous waste sites are "negative" in the sense that no statistically significant increases in adverse health effects were found (Grisham, 1986; Upton et al., 1989). Although under the conditions of the studies, such findings may represent the true health status of the communities, negative or equivocal findings from epidemiologic investigations can occur from limitations in how the studies are conducted.

Epidemiologists have advised on some fundamental issues that can affect the outcome of a health study. Heath observed that the epidemiologist must be aware of three particular difficulties that can severely limit the power of an epidemiologic investigation (Heath, 1983). First, consideration must be given to the size of the study population needed to demonstrate a particular health effect. Generally, if the health outcome to be assessed is rare, the population to be studied will need to be relatively large. Second, health effects can have long or variable latencies. Some diseases (e.g., cancer) develop only years or decades after the start of exposure to an environmental hazard. If a health study is conducted before the latent time of the health effect, the outcome will be "negative." Third, the biologic effect under study can be clinically nonspecific. For example, exposure to low levels of solvents can produce a pattern of symptoms that includes headache, low-grade nausea, mild dizziness, and fatigue.

Some of these limitations and confounders have the potential for being palliated through better assessment of individuals' exposures to hazardous substances and by combining sites for study. It is possible to combine sites by matching such factors as common environmental characteristics or by matching sites that have common features. A review panel consisting of experts in environmental epidemiology and occupational and community medicine identified several advantages in conducting multisite studies (ATSDR, 1992d). The panel noted that the primary advantages of a multisite study design are (1) increased sample size and greater statistical power, both of which enhance the validity of findings from a study, and (2) added diversity in the sample, thereby improving the likelihood that findings are real and not due to random variations. As a caution, the panel noted that it is inappropriate to pool data from multiple sources without the necessary advance planning on how to coordinate data collection and ensure its quality. An example of a multisite lead and cadmium exposure investigation (ATSDR, 1995a) that conforms with the panel's advice on multisite studies is found in Chapter 3 and elsewhere in Chapter 5.

As a prelude to describing findings from hazard waste investigations, and to appreciate the nature and consequences of health studies of communities impacted by hazardous waste, the following section summarizes the Love Canal, New York, episode, an event that had a major impact on the enactment of CERCLA by Congress. This example provides a sense of a community's health concerns pertaining to hazardous substances and introduces terms that will be defined later in the chapter.

LOVE CANAL, NEW YORK, EPISODE

Love Canal, near Niagara Falls, New York, was the first large residential area known to be affected by buried hazardous waste. The community had been built over an abandoned chemical waste dump, which resulted in the presence of hazardous substances in households and raised concern for the impact on residents' health. Events at Love Canal had a signal effect on shaping concern for the human health consequences of hazardous waste. The enactment of CERCLA by the U.S. Congress was strongly influenced by the Love Canal episode. Public attention directed to Love Canal and the ensuing legislation were both based on the potential for "hidden" toxic waste to harm human health.

The Love Canal episode helped shape the American public's perception of hazardous waste sites as a major environmental hazard. Because of Love Canal, the American public perceived hazardous waste as a major environmental hazard. As discussed in Chapter 8, dread is a key factor in shaping the public's perception of the risk of technologic hazards. Surely contributing to the sense of dread in Love Canal residents was the perception that one's children and home were unsafe.

Some background about Love Canal is instructive. As recounted by Paigen et al. (1985, 1987), Love Canal had a long history. In 1896 William T. Love tried to build a navigable canal between the upper and lower Niagara River. His attempt was unsuccessful; a 3,000 meter channel was dug, but fell short of connecting with the Niagara River when money ran out. Waste burial in the canal started in the 1920s and ended about 1953 (Janerich et al., 1981). The Hooker Electrochemical Company, which later became the Occidental Chemical Corporation, purchased the canal in 1942 for depositing its chemical waste. By 1953 an estimated 19,000 metric tons of liquid and solid waste had been placed in the canal and covered by soil.

The property was later purchased by the Niagara Falls Board of Education. A school and playground were constructed over the central part of the canal site; residential housing was built along the northern and southern banks of the canal. An environmental sampling survey in 1977 found that hazardous substances had migrated from the waste dump into the basements of nearby homes and storm sewers. More than 200 hazardous substances were found in the Love Canal dump site, including benzene and lindane (Paigen et al., 1985).

In August 1978, the New York State Department of Health and President Jimmy Carter declared a health emergency existed in the Love Canal area (Hernan, 1994). This led to closing the area's school and evacuating 235 families living within 120 meters of the canal (Paigen et al., 1985). Other evacuations occurred in 1979 of families with

"While the different government agencies argued over who was responsible, the residents were left waiting and wondering what their health risks were, whether they should be getting health care, or whether they ought to flee from their homes to protect themselves, and especially their children, even if it meant financial ruin for the family."

Lois Marie Gibbs (1981), President, Love Canal Homeowners' Association, Inc.

pregnant women or children under the age of two years living in the southern half of the Love Canal community. Approximately 500 additional families were relocated in 1980, and their homes were bought by the state (Hernan, 1994). In May 1980, the federal government offered relocation to all residents of Love Canal. In 1995 the Love Canal area was declared safe for rehabilitation following remediation of the site.

Love Canal was the first residential community known to be impacted by hazardous waste and, therefore, the first community to be the subject of response actions by government. The community's frustration with how government responded to the Love Canal crisis is illustrated by the comment from Lois Marie Gibbs shown in the inset box. Personal experience suggests Gibbs's comment still characterizes the feelings of many communities impacted by hazardous waste sites.

Several health investigations of Love Canal residents have been conducted. The first attempt to assess the effect of hazardous chemicals on Love Canal residents' health was conducted as a pilot study by the EPA, which facilitated the collection of blood samples from 36 self-selected Love Canal residents (as cited in Heath et al., 1984). Abnormal cytogenetic findings consisting of large "acentric" chromosome fragments were reported, although no differences in overall chromosome aberration frequencies were found in comparison to laboratory historical data (Picciano, 1980). The conducting laboratory was not blind to the origin of the samples, and no comparison group from Niagara Falls was used. Nonetheless, the findings of "abnormal chromosomes" fueled speculation that hazardous substances were causing adverse health effects in Love Canal residents. Moreover, the study gave the imprimatur of science to the residents' health concerns, and contributed to a wider news media coverage than might have occurred in the absence of "scientific data."

Subsequent cytogenetic analyses did not confirm the 1980 findings. Investigators from the Centers for Disease Control performed cytogenetic analyses on cells in peripheral blood drawn between December 1981 and February 1982 from 46 current or past residents of the Love Canal area (Bender and Preston, 1983; Heath et al., 1984). The study's participants included 17 persons for whom cytogenetic analyses had been performed in 1980 (i.e., Picciano, 1980) and 29 persons who in 1978 lived in seven homes that abutted the canal and in which environmental surveys showed the presence of hazardous substances from the canal waste site. (The mix of current and past residents of Love Canal is unstated.) The comparison group comprised 25 persons from another section of Niagara Falls. Cytogenetic samples were coded so that laboratorians were unaware of the exposure status of each person's sample. Results showed no increase in frequencies of either chromosomal aberrations or sister chromatid exchange, placing in question the accuracy of the Picciano findings.

A series of health investigations began in 1980. The prevalence of health problems in children living in the Love Canal community was investigated under sponsorship of the Environmental Defense Fund (Paigen et al., 1985). The study population consisted of 523 Love Canal and 440 comparison children. The comparison group was selected from the Niagara Falls area. The mean ages were 117.6 months for Love Canal children and 98 months for comparison children. A questionnaire was administered to the study groups in May 1980. Interviewers questioned parents and children about children's health problems as diagnosed by physicians. Two surrogate measures of exposure were constructed: (1) distance of residence from the canal, grouping homes into concentric 120-meter-wide bands, and (2) the proximity of homes to "wet" areas and areas filled with rubble, which created porous soil, as contrasted with the generally low-porosity clay of the Love Canal residential area.

The investigators found increased prevalences of seven health problems among Love Canal children compared with control children: seizures, learning problems, hyperactivity, eye irritation, skin rashes, abdominal pain, and incontinence. Each health problem showed evidence of a dose-response for both surrogate measures of exposure. These findings are tempered by possible recall bias of study participants, a higher level of motivation by Love Canal residents to participate in the study, and surrogate exposure indices rather than actual exposure data.

Two studies were conducted of Love Canal residents' reproductive outcomes and infants' development. One study compared growth of Love Canal children with a comparison group (Paigen et al., 1987). Anthropometric measures of children, stature of parents, demographics, and health information were obtained for 493 children living near Love Canal and 428 children from census tracts matched to the Love Canal community. Children born and spending at least 75% of their life in the Love Canal area (n = 172) had significantly (p = 0.004) shorter stature for age percentile (46.6 $\pm$ SE 2.2) than did 404 comparison children (53.3 $\pm$ SE 1.4). This difference could not be accounted for by differences in parents' stature, socioeconomic status, nutrition, birth weight, or chronic illness.

In another study, the incidence of low birth weight among white, live-born infants from 1940 through 1978 was studied in various sections of the Love Canal area (Vianna and Polan, 1984). Infants who weighed 2,500 grams or less at birth were considered low birth weight. The investigators gave particular attention to the swale areas of Love Canal because they served as drainage areas and, therefore, may have had concentrated hazardous substances in the soil.

The most compelling finding from the Vianna and Polan study is shown in Figure 5.1,[2] which gives the five-year moving average for percentages of low birth weights among infants born to parents living in the swale area for the period 1940 through 1975. A significant excess of low-birth-weight births was found in the historic swale area from 1940 through 1953, the period when various chemicals were dumped in the waste site. The investigators state, "For the period of active dumping (i.e., prior to 1954), the swale

---

[2] Reprinted with permission from Vianna and Polan (1984). Incidence of low birth weight among Love Canal residents. *Science,* 226:1217–9. Copyright 1984, American Association for the Advancement of Science.

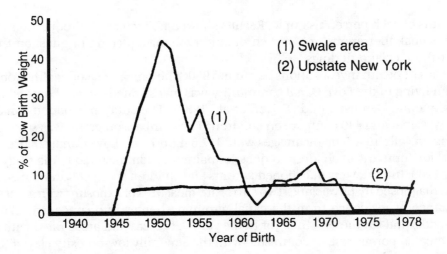

Figure 5.1. Five-year moving averages for percentages of low birth weights among Love Canal's swale area infants (Vianna and Polan, 1984).

area's percentage of low weight births was higher than in upstate New York (z test, $p < 0.0001$) and the rest of the canal (z test, $p < 0.012$)." These studies of the stature and birth weight of infants born to Love Canal residents were important for validating the health concerns of the area's residents and for providing early evidence of the need to interdict releases from hazardous waste sites.

Cancer incidence in the Love Canal area was investigated by the New York State Department of Health (Janerich et al., 1981). Cancer data (i.e., number of cancer cases) from the New York Cancer Registry were examined for relation to persons' proximity to Love Canal. Standardized incidence ratios (SIRs), based on registry data and census tract population estimates, were calculated for each cancer site (e.g., liver cancer) by dividing the observed number of cancer cases in the Love Canal tract by the number expected for that census tract. Rates for all anatomical sites of cancer for the years 1966 to 1977 were ascertained, as were rates for the cancers of special interest (liver cancer, lymphoma, leukemia) for the period 1955 to 1965. Age- and sex-specific cancer incidence rates were calculated for 10 major cancer sites and five age groups for each of the 25 census tracts in the city of Niagara Falls; one tract included the Love Canal area. The analysis showed no cancer rate to be consistently elevated in comparison with statewide cancer rates. The investigators noted that the effects of socioeconomic factors, air pollution levels, and cigarette smoking could not be assessed in their study design. They observe that uncertain latency periods for the cancers investigated could have obscured the study's findings.

In June 1994, the State of New York and the Occidental Chemical Corporation settled legal claims. The chemical company agreed to pay the state $98 million and to assume the continuing operation and maintenance of remediation activities at the Love Canal site (Hernan, 1994). Remedial actions at Love Canal have cost state and federal governments more than $150 million (Hernan, 1994). A retrospective health investigation of former Love Canal residents is scheduled to begin in 1997 as part of the final settlement between the waste site's Potentially Responsible Parties and the federal and New York state governments.

BASICS OF EPIDEMIOLOGY

Having discussed how various studies addressed the health of Love Canal residents and the impact of the studies, it is useful now to describe some key concepts in epidemiology. One key concept is the design by the epidemiologist of the health study. Several kinds of study designs are used by epidemiologists; the choice for a particular study will be determined by the purpose of the study and conditions attending the study population. Each type of study design has important implications for assessing the strength between exposure to an environmental agent and any resultant health effects in the population being studied.

The study design chosen by the epidemiologist to examine the relation between exposure to hazardous substances and an adverse health effect (e.g., cancer) will be determined by both the nature of the biologic effect expected from exposure and the time course of the postulated health effect. Environmental epidemiologists, unlike toxicologists, cannot manipulate the environmental conditions of interest to them. Because of ethical concerns, exposure to hazardous substances cannot be imposed on a human study population in a community and controlled in ways that permit observations and measurements of adverse health effects.[3] Epidemiologists, however, have a variety of study designs to apply to particular health investigations that control for some confounders (e.g., age of subjects).

In general, epidemiologists conduct two major kinds of studies, *experimental* and *observational*, to assess relationships between suspected risk factors (e.g., exposure to hazardous waste) and disease (e.g., birth defects) (Beaglehole et al., 1993). Each kind can be further divided according to study design and purpose of study. Table 5.1 contains the major groups of classical epidemiologic study designs, together with strengths and weaknesses of the designs.

Experimental Studies

In experimental (or intervention) studies, the investigator attempts to modify or influence conditions in the study population. The major experimental study design is the randomized controlled trial in which patients serve as the study group; for example, studies that test the efficacy of new drugs when administered to terminally ill persons. Because experimental epidemiologic studies involve human beings as subjects of research, they must conform with practices that ensure the ethical conduct of research. No experimental epidemiologic studies of populations exposed to hazardous waste have been conducted.

Observational Studies

Observational studies can be classified broadly as either **descriptive** or **analytic**. "Observational studies allow nature to take its course: the investigator measures but does not intervene" (Beaglehole et al., 1993). The investigator must estimate or measure a study group's level of exposure to a hazardous substance or some other environmental agent, rather than manipulate levels of exposure to the toxicant of interest.

---

[3] Experimental research with human subjects is conducted under closely controlled conditions as laboratory studies or clinical trials, as discussed later in the chapter.

Table 5.1. Most Common Types of Epidemiologic Studies (WHO, 1983; Beaglehole et al., 1993; NRC, 1997).

Type of Study	Alternative Name	Population	Exposure	Health Effect	Weaknesses	Strengths
A. Experimental						
1. Randomized controlled trials	Intervention Clinical trials	Patients	Controlled and known	Measured during the study	Expensive; ethical considerations	Strong evidence for causality
2. Field trials		Healthy individuals	Controlled and known	Measured during the study	Can be expensive; ethics	Good evidence for causality
3. Community trials	Community intervention studies	Communities	Controlled and known	Measured during the study	Can be expensive; ethics	Good evidence for causality
B. Observational						
1. Descriptive						
a. Case reports		Patients	Current or past	Clinical findings	Difficult to relate exposure to effects	Useful to formulate hypotheses
b. Surveillance systems		Community or special groups	Current	Morbidity and mortality	Difficult to relate exposure to effects	Inexpensive when using existing monitoring and surveillance data
c. Ecologic	Correlational	Grouped populations	Current or past	Morbidity and mortality	Confounders are difficult to assess	Simple to conduct, drawing upon existing databases
d. Cluster studies		Groups or areas within communities	Current or past	Morbidity and mortality	Study may have low statistical power; low number of subjects	Responds to community concerns

2. Analytic

			Exposure	Health effects	Disadvantages	Advantages
a. Cross-sectional	Prevalence	Community or special groups; exposed vs. nonexposed groups	Current	Current health effects	Hard to establish cause-relationship; current exposure may be irrelevant to current disease	Can be done quickly, can use large populations; can estimate extent of problem (prevalence)
b. Case-control	Case-referent	Usually small groups; diseased (cases) vs. nondiseased (controls)	Occurred in past and determined by records or interview	Known at start of study	Difficult to generalize because of small study group; some incorporated bias	Relatively inexpensive; useful for studying rare diseases
c. Cohort	Follow-up					
1. Retrospective		Special groups, patients	Exposure occurred in the past; need records of past exposure	Health effects occurred in the past; need health records	Exposure and health records may be unavailable or inaccurate	Less expensive and quicker to conduct than a prospective cohort study
2. Prospective		Special groups, patients	Current exposure; requires measurement	Current health effects data	May take years to conduct to exceed latency of disease outcome studied	Can provide strong information on disease causation

Descriptive Studies

Descriptive epidemiology addresses questions about how much (e.g., how many cases of cancer occurred), when (i.e., time), where (i.e., place), and to whom (i.e., person) (Goodman and Peavy, 1996). It includes time-series analyses and prevalence studies that analyze large sets of data. Descriptive studies often help in generating hypotheses and are usually limited to describing the occurrence of a disease in a target population; such studies are often the first step in any epidemiologic investigation of a putative disease outbreak. Descriptive studies include case reports, surveillance systems, ecologic studies, and cluster studies.

Case Reports: Unusual findings that relate a disease to a new cause or etiology of the disease are often given in medical literature as case reports. These reports typically describe the disease, the medical intervention, and putative cause of the disease. Case reports can consist of a relatively small number of patients, as few as one, and, therefore, do not represent a "study" in the sense that a hypothesis was being tested. Case reports are valuable for forming hypotheses that can be evaluated in epidemiologic follow-up studies.

Surveillance Systems: Public health surveillance is defined as "...the ongoing, systematic collection, analysis, and interpretation of health data essential to the planning, implementation, and evaluation of public health practice, closely integrated with the timely dissemination of these data to those who need to know..." (Thacker and Berkelman, 1992). As used by public health practitioners, a surveillance system refers to a data-collecting system that monitors the occurrence of a specific disease (disease surveillance) or the distribution of exposure to potential hazards (hazard surveillance). An example of a disease surveillance system is the reporting to some state health departments by physicians and hospitals of cases of cancer or birth defects in newborn infants. Surveillance systems provide an early warning of situations in which epidemiologic investigations or other public health actions should be taken.

Ecologic Studies: Ecologic (or correlational) studies explore the statistical connection between disease and estimated exposures in population groups rather than in individuals (NRC, 1997). Data from vital records, hospital discharges, or disease registries (e.g., cancer mortality registry) are combined with grouped data or estimates of exposure to specific environmental factors. An example of an ecologic study is the examination of birth defects data from a state congenital malformations registry, grouped according to parental proximity to hazardous waste sites. Ecologic studies can be difficult to interpret because the investigator is seldom able to examine directly the various potential explanations for findings (Beaglehole et al., 1993). One significant problem is the lack of individual exposure data. Ecologic studies are appealing to investigators as a first step in assessing adverse health patterns, often in large populations, because they are simple to conduct in comparison with other epidemiologic study designs.

Cluster Studies: A cluster study is a descriptive study of the population in a geographic area, occupational setting, or other small group in which the rate of a specific adverse health effect is thought to be, or known to be, much higher than expected (NRC, 1997). Clusters of disease or illness are frequently the basis of concern in communities located near hazardous waste sites. Typically, persons in the community relate some perceived excess of serious disease, like cancer, with the presence of the waste site.

Analytic Studies

Analytic studies include cross-sectional, case-control (retrospective) and cohort (follow-up) studies and typically test hypotheses of interest to the epidemiologist. However, both descriptive and analytic studies can generate and test hypotheses (NRC, 1991b).

Cross-Sectional Studies: In a cross-sectional study, measures of exposure to a toxicant are correlated with levels of absorption of the suspected toxicant or rates of occurrence of morbidity. Such an *exposure-based, cross-sectional study* is structurally similar to the cohort study (described later) in that the study groups are initially defined by the presence or absence of the exposure of interest, and then the disease status of individuals within the exposed group is ascertained (Johnson et al., 1987). The cross-sectional study gives only a 'snapshot' in a particular time period of the differences, if any, between an exposed group and a comparison group that has no exposure to the toxicant of concern. In addition to providing information only for the time period of observation, a cross-sectional design may not be able to account for confounding variables other than exposure classification. The inability to assess pre-exposure that functions in a cross-sectional design is a very real problem in interpreting cross-sectional studies.

Case-Control Studies: In a case-control (case-referent) study, persons with disease or organ system impairment ("cases") are compared with individuals who lack the condition ("controls," "referents," "comparisons") with respect to their rate of exposure to the agent(s) of interest (Johnson et al., 1987). In a case-control study, subjects are selected without regard to toxicant exposure. Previous exposure is assessed only after the subject has been identified as meeting the criteria for inclusion in the study (as a case or control subject). Exposure is then assessed 'blind' (i.e., without the investigator knowing whether the person is a case or a control subject). The third stage is to compare the rate of exposure between cases and controls (*ibid.*). A case-control approach requires that the diagnosis of a subject as a "case" be unbiased by the investigator's perception of any link between the person's exposure and the disease of concern. These same concerns of potential bias apply to a retrospective cohort study.

Cohort Studies: A cohort is a defined population group followed in an epidemiologic study. In cohort studies, comparable groups of exposed and unexposed persons are followed to ascertain the incidence of disease or mortality caused by disease in association with the exposure (NRC, 1991b). Cohort studies are of two kinds. The *retrospective cohort study* relates a complete set of outcomes already observed in a defined population to exposures that occurred earlier. The *prospective cohort study* follows a group of persons, and the changes in their health condition are related to changes in exposure history during the follow-up period.

In rare instances, conducting such a study on a cohort of persons without prior exposure may be possible, administering to them a set of biomedical or other tests, both before exposure begins and at one or more times after exposure. The prospective cohort study design allows an evaluation of the relationship between exposure and the extent to which a health condition deteriorates beyond what is expected through normal aging of the cohort being studied.

When existing cases are evaluated to establish their relation to exposure, the choice will lie between a case-control study, in which cases and controls are compared with respect to their exposure experience, and a retrospective cohort study, in which the exposure history of all subjects from a given time period is ascertained and the proportion of cases at different levels of exposure is assessed.

When "cases" cannot be identified in advance, or when the designation of a subject as a "case" is known to depend in some systematic way on the person's previous exposure history (such as cases assessed for medical purposes), special studies may have to be designed to evaluate the extent of the health effect. Many studies of such health effects have been cross-sectional in design. In these studies, the performance of exposed persons on a series of biomedical or other tests is compared with that of nonexposed persons. Test results of the latter provide an estimate of the functional level exposed persons would have had were they not exposed to the toxicant of interest.

The reversibility of an adverse health effect also may be measured by assessing the functioning of a person on two or more occasions, first under conditions of high exposure and again when exposure has either been reduced following improved environmental measures or eliminated by complete removal from exposure. Typically, such "follow-up" studies occur over a period of months or even years.

A special kind of case-control (case-referent) study has received attention as a means of identifying specific exposures responsible for an excess of a particular disease which was initially observed in a larger cohort study. In such a *nested case-control study* in a community, cases of disease are identified among persons in one community, and controls without the disease are identified from the same community. Detailed individual histories with specific exposure estimates are compiled for cases and controls and frequencies of exposure to specific substances are compared.

Table 5.1 shows the most common types of epidemiologic studies and various details about each kind of study design. Particularly important is the information on strengths and weaknesses known to be associated with each kind of study design. This information is useful for interpreting the findings from studies of populations exposed to hazardous substances from waste sites, as described later in the chapter. For example, Table 5.1 indicates that cross-sectional studies of communities exposed to releases from hazardous waste sites can be done quickly and can examine large populations but are limited in their strength to establish cause-effect relationships.

Study Design Considerations

An epidemiologic study involves comparing health data between groups of people. The data in hand may be 'cases' or health records, but ultimately some kind of data comparison is made between the group of interest (e.g., a group exposed to hazardous substances) and a comparison group. The number of persons constituting each group, or each group's dataset, is a key consideration in designing any epidemiologic study. The determination of sample size and attendant power calculation is an essential component of epidemiology. This is sometimes misunderstood by community groups, who may expect any health study to be able to answer their health concerns even when group sizes are small. Other key definitions and concepts also need introduction at this point.

Two important definitions that relate to the epidemiologic studies to be summarized in subsequent sections are *population* and *sample*. *Population* is a statistical term denoting all the objects, events, or subjects in a particular class (Hensyl, 1987). The application of this definition requires a determination of class and all members of the class. For example, a population could be represented by all persons in a community exposed to solvents released from a particular hazardous waste site. A population could also be all infants with birth defects born in a defined geographic area; for example, a state. Often it is impractical or unnecessary to investigate all objects, events, or subjects in a population, and a sample of the population is taken. A *sample* is a specified portion of a population selected for study with the hope that it is representative of the entire population (Peavy, 1996). The observations, measurements, and analyses of the sample are subjected to statistical methods to draw conclusions from the sample about the population's characteristics.

Prevalence is defined as the proportion of ill persons in a population at a given time and is expressed as a simple percentage. *Incidence* is the rate of development of disease in a population and can either be expressed as incidence density or cumulative incidence. As Dicker (1996) noted, prevalence refers to existing cases of a health condition in a population, and incidence refers to new cases. An estimate of the sample size needed in a study requires some knowledge of the distribution of the outcome variable in the referent population (or in a case-control study, the frequency of exposure) and a decision on the importance of avoiding each of two kinds of error: *type I or "alpha" error*, which is the error of claiming that the exposure is associated with the outcome when, in fact, it is not and *type II or "beta" error*, which is reporting that exposure is not associated with the outcome when, in fact, it is. As a general rule, the probability for a type I error should normally not be greater than 5%, and for a type II error, not greater than 20%.

Power is a statistical measure of the potential of the study to find an association. Power, in the statistical sense, is defined as one minus the probability of type II error, or simply 1–beta. In addition, the aggregate measure of effect to be detected between the study groups (e.g., rate ratio or rate difference) must be specified for the power calculation. With such guidelines and some informed guesswork drawn from previous studies using the same outcome variable, the sample size can be computed using the formula appropriate for each design (Johnson et al., 1987).

Marsh and Caplan said (1987),[4] "In the context of hazardous waste studies, statistical power can be defined as the probability that an adverse health effect of a specific size will be detected when it is present in the target population from which the sample was drawn. Power is an extremely important consideration since it helps to determine study design and provides an objective basis from which to meaningfully interpret study results." Statistical power, they indicate, is a function of five study parameters, listed in Table 5.2. Marsh and Caplan provide the following elaboration on each of these parameters:

- *The size of the study and control groups.* In general, statistical power increases as sample size increases.
- *The variability of the health outcome under study.* In general, power is inversely related to the variability of the health outcome in the population under study.
- *The predetermined statistical significance level or type I error that will be accepted as confirmation of an association between exposure and health outcome.* Power is directly related to the significance level, with all other parameters fixed.
- *The magnitude of the expected association between exposure and outcome.* With all other parameters fixed, power is directly related to the magnitude of this association.
- *The design of the study and statistical techniques used for analysis.* Various special design and analytic techniques can be used to enhance power. These include refining the history of exposure to avoid misclassification bias, refining the response variable to conform with an anticipated biologically coherent health outcome, increasing sample

---

[4] Excerpted with permission from *Health Effects from Hazardous Waste Sites*, Andelman, J.B. and D.W. Underhill, Eds., Lewis Publishers, 1987. Copyright CRC Press, Boca Raton, Florida.

Table 5.2. Factors that Determine Statistical Power in Epidemiologic Studies (Marsh and Caplan, 1987).

- The size of the study and control groups.
- The variability of the health outcome under study.
- The predetermined statistical significance level or type I error that will be accepted.
- The magnitude of the expected association between exposure and outcome.
- The design of the study and statistical techniques used for analysis.

Table 5.3. Sample Size Requirement for a Study of Nerve Conduction Velocity (Johnson et al., 1987).

Difference in Means (m/s)[a]	Sample Size Required[b]
2	437
4	87
8	22
10	14

[a] Difference between the two study groups in meters per second.
[b] Assuming: alpha = 0.05, beta = 0.10, Standard Deviation = 9 m/s.

size through an intensified case finding, use of continuous rather than discrete health outcome variables, and other techniques (Marsh and Caplan, 1987).

As an example of sample sizes required to detect a difference in means between an exposed and a comparison group (e.g., a cross-sectional design), consider a study in which the investigator's concern is the neurotoxicity of lead. The measure of choice is the functional integrity of the median nerve determined by the velocity at which the nerve conducts an impulse in response to an electrical stimulus (Johnson et al., 1987). Table 5.3 illustrates the impact that differences in the mean values have on sample size. Assume that the standard deviation is 9 meters per second (m/s) in both groups and that the investigator is willing to accept a 5% probability of a type I error (i.e., alpha error) and a 10% probability of a type II error (i.e., beta error). This is a study with 90% power (i.e., 1–beta = 1.0 – 0.1 = 0.9, or 90%).

Table 5.3 shows that the sample size needed for the exposed and nonexposed groups varies according to the size of the difference between means under consideration. Sample size and study power are influenced by the variability of the outcome measures being used. As data in the table illustrate, the smaller the difference in means measured between the two groups, the larger must be the sample size to maintain the same power. For example, at least 437 subjects will be required for a study with 90% power if the difference in mean values of nerve conduction velocity between the exposed group and the comparison group is 2 m/s.

When the sample size is fixed by external constraints (e.g., a fixed number of persons exposed to a toxicant of concern), the power of the study to detect a specified difference in an outcome variable can be calculated in a manner analogous to the sample size estimation. Given a specified sample size, alpha level, and prevalence in the unexposed population, a study becomes less powerful as smaller differences in prevalence between groups are sought. In a study of health symptoms reported by two groups of 30 individu-

Table 5.4. Power Estimates with Varying Prevalence Differences (Johnson et al., 1987).

Prevalence Difference[a] Exposed–Unexposed	Power
10	0.23
15	0.39
20	0.56
25	0.71
30	0.83
35	0.91
40	0.96

[a] Assuming alpha = 0.05, p (prevalence in unexposed) = 0.06.

als each, the power changes with the difference in prevalence that the study seeks to detect in the two groups (Table 5.4). If the investigator wishes to achieve a 90% or greater statistical power, a difference in prevalence of 35 will be required under the conditions shown in the table. As the prevalence difference between exposed and referent groups decreases, the power decreases as well. The difference in prevalence sought should be influenced primarily by the biologic significance of the specified change in the health parameter under study.

In summary, power and sample size are directly related. At fixed levels of alpha, beta, and the difference to be detected between groups, the larger the sample size, the greater the study's power. Increasing statistical power can also be accomplished by specifying that a larger difference must be detected between the two groups or by selecting an outcome measure (i.e., test) with less variability (i.e., low standard deviation in the reference group). A detailed discussion of sample size determination, study power, choice of statistical test, and other details can be found elsewhere (e.g., Lwanga and Lemeshow, 1991).

MEASURES OF ASSOCIATION

As Dicker (1996) discussed, exposure and the health event being studied can be described in many epidemiologic studies as a binary variable (i.e., yes or no). When this is the case, the relationship between exposure and health event can be organized into a 2×2 table. Such a 2×2 table is shown in Table 5.5. Disease status (e.g., well or ill) is placed along the rows of the 2×2 table, and exposure status (e.g., exposed or unexposed) is placed along the columns of the table. The intersection of a row and column is called a cell. The letters a, b, c, and d within the four cells of the 2×2 table denote the number of persons with the health condition of concern according to their exposure status. For example, cell b contains the number of persons who are well and exposed to the condition of concern. Table 5.5 illustrates the concept of proportional risk, or attack rate, for the health event according to exposure status. Proportional risk (or attack rate) is the fraction of a group of persons who develop disease or illness during a specified period of time. For example, the proportional risk (or attack rate) in the unexposed group is the number of persons ill (i.e., cell c) divided by the total number of unexposed persons (i.e., h_0).

A measure of association quantifies the strength or magnitude of the statistical association between the exposure and the health event of interest. Measures of association

Table 5.5. Data Layout and Notation for the Standard 2×2 Table (Dicker, 1996).

	Ill	Well	Total	Risk (Attack Rate)
Exposed	a	b	$h_1 = a + b$	a/h_1
Unexposed	c	d	$h_0 = c + d$	c/h_0
Total	v_1	v_0	$t = v_1 + v_0$	v_1/t

are sometimes called measures of effect because they quantify the effect of being exposed on the incidence of disease, if the exposure is causally related to the disease (Dicker, 1996). The following measures of association are commonly used.

Relative Risk: The relative risk (RR) or risk ratio is the risk in the exposed group divided by the risk in the unexposed group, and is expressed mathematically as:

$$\text{Relative risk (RR)} = \text{risk}_{\text{exposed}}/\text{risk}_{\text{unexposed}} = (a/h_1)/(c/h_0)$$

The relative risk represents the excess risk in the exposed group compared with the unexposed group (Dicker, 1996). The relative risk will be greater than 1.0 when the risk is greater in the exposed group than in the unexposed group, and less than 1.0 when the risk in the exposed group is less than the risk in the unexposed group.

Odds Ratio: The association of an exposure and a disease is measured in a case-control study by calculation of the odds ratio (OR) (Beaglehole et al., 1993), because in most case-control studies, the true size of the exposed and unexposed groups is unknown, which leads to the inability to determine the denominator on which to calculate risk. The OR is the ratio of the odds (i.e., chance) of exposure among the cases to the odds in favor of exposure among the controls. The odds ratio is calculated as:

$$\text{Odds Ratio (OR)} = ad/bc$$

In cohort studies, the measure of association most commonly used is the relative risk (RR). In case-control studies, the odds ratio (OR) is the most commonly used measure of association. In cross-sectional studies, either a prevalence ratio or a prevalence odds ratio may be calculated (Dicker, 1996).

Prevalence Ratio and Prevalence Odds Ratio: As Dicker noted, cross-sectional studies or surveys generally measure the prevalence (existing cases) of a health condition in a population rather than the incidence (new cases). "Prevalence is a function of both incidence (risk) and duration of illness, so measures of association based on prevalent cases reflect both the exposure's effect on incidence and its effect on duration or survival" (Dicker, 1996). The prevalence measures of association analogous to the relative risk (RR) and the odds ratio (OR) are the prevalence ratio and prevalence odds ratio, respectively.

Standardized Mortality Ratio: The standardized mortality ratio (SMR) is a special type of risk ratio (Beaglehole et al., 1993). The SMR is derived when the observed mortality pattern is compared with what would have been expected if the age-specific mortality rates had been the same as in a specified reference population. This procedure adjusts for differences in age distribution between the study and reference populations.

Confidence Intervals: Once a sample has been drawn from a population, it can be used to estimate characteristics of the underlying population (Beaglehole et al., 1993). Because estimates (e.g., the number of cases of cancer) vary from sample to sample in the same population, it is important to know how close the estimate derived from one sample is likely to be to the underlying population value. One way to find this out is to construct a confidence interval around the estimate; that is, to construct a range of values surrounding the estimate that have a specified probability (e.g., 95%) of including the true population values. The specified probability is called the confidence level (e.g., 95%) and the endpoints of the confidence interval are called the confidence limits (Beaglehole et al., 1993). The derivation of confidence limits is beyond the scope of this chapter; standard reference works should be consulted (e.g., Gregg et al., 1996).

Case Example of 2×2 Table: Forest Glen Mobile Home Park, New York: Data from the ATSDR and New York State Department of Health survey of health conditions for persons living in the Forest Glen mobile home park, Niagara Falls, New York provide an example of a 2×2 table (ATSDR, 1989b). Complaints from residents indicated the presence of drums of hazardous substances appearing in the park's soil when soil erosion occurred. The EPA conducted soil sampling that confirmed the presence of several hazardous substances in the soil, including PAHs, aniline, phenothiazine, and other substances. The primary area of soil contamination was in the north part of the mobile home park. The health survey found signs and symptoms of dermal contact with the substances, especially in young children who played in the contaminated area (NYDOH, 1989). This led to issuance of a public health advisory (ATSDR, 1989b), which in turn led to the EPA's relocation of the park's residents.

Case Example: 2×2 Table for Forest Glen Mobile Home Park Households Reporting Illness Symptoms

	Ill	Well	Risk
North	11	15	0.42
South	3	16	0.16

Relative risk = (11/26)/(3/19) = 2.68

The case example in the inset box illustrates how a 2×2 table could be constructed. The "ill" designation refers to persons reporting adverse health symptoms; the "north" and "south" designations refer to areas of the mobile home park associated with persons reporting symptoms. For example, 11 persons in the north area of the park reported some kind of adverse health symptom. The relative risk can be calculated as 2.68, which means a person living in the north area was approximately three times more likely to report a health complaint than was a person living in the south area. Because soil contamination was known to be higher in the north area, this relative risk gives support for associating area of residence with adverse health outcomes.

Causation in Epidemiology

Causal inference is the term used for the process of determining whether observed associations are likely to be causal; the use of guidelines and professional judgment is required. As the National Research Council noted, "The world of epidemiology, as that of any human science, seldom permits elegant inferences to be drawn about **causation**" (NRC, 1991b). A fundamental dilemma for environmental epidemiologists derives from the fact that the statistical correlation of variables (e.g., proximity to waste sites and elevated risk of birth defects) does not necessarily indicate any causal relationship among the variables, even where tests of statistical significant may be met. As the NRC observed, "Mere coincident occurrence of variables says nothing about their essential connection" (*ibid.*). Professional assessment must be made of a study's findings and inferences, and ultimately causation is inferred only after a consideration of all relevant science and epidemiology.

A reasonable inference of causation in environmental epidemiology requires asking: (1) Could the observed association be due to bias in how subjects were selected for study or in how data were collected? (2) Could the association be due to confounding of variables? (3) Could the association be a result of chance? If the answers to these three questions are no, no, and probably not, respectively, then guidelines recommended first by Hill should be applied to the association. In particular, eight basic characteristics of a study's findings should be considered. These eight considerations flow from the work of the medical statistician, Sir Austin Bradford Hill (Hill, 1965). Hill's nine recommendations have been refined by the World Health Organization and reduced to eight, which are listed in Table 5.6. A discussion of each WHO guideline follows.

Temporal Relationship: The relationship in time between an outcome and a potential causal factor must be examined closely. The cause must precede the event attributable to the toxicant. For instance, exposure to a toxicant must precede any adverse health event attributable to the toxicant. Repeated measurement of exposure at more than one point in time and in different geographic locations may strengthen the evidence for causation.

Plausibility: The plausibility of an association is strengthened if there is biologic science that supports the observation. For environmental epidemiologic studies of persons exposed to toxicants, examination of plausibility often means comparing the association with what is known about the toxicology of the substance. Consider the report of an association between birth defects in children born to mothers who, during pregnancy, drank well water found later to contain a toxicant. The investigator would ask whether this finding is plausible in view of the toxicology of the substance.

It is interesting that Hill cautioned, "It will be helpful if the causation we suspect is biologically plausible. But this is a feature I am convinced we cannot demand. What is biologically plausible depends upon the biological knowledge of the day" (Hill, 1965). Given the incomplete knowledge about the toxicologic properties of hazardous substances released from waste sites, Hill's caution should be borne in mind when assessing plausibility of an association in an epidemiologic study of a population exposed to hazardous substances.

Consistency: Consistency is demonstrated when several studies give the same result. This is particularly important when various study designs are used in different settings because the likelihood is reduced that all studies have made the same errors or misinterpretations. However, according to Beaglehole et al., a lack of consistency across studies does not exclude a causal association because different exposure levels and other conditions may reduce the impact of the causal factor in certain studies. They recommend giving extra weight to the best-designed studies when looking for consistency of association. Techniques have been developed for pooling the findings of sev-

Table 5.6. Guidelines for Causation (Beaglehole et al., 1993).

- Temporal relationship
- Plausibility
- Consistency
- Strength
- Dose response
- Reversibility
- Study design
- Judging the evidence

eral studies that have examined the same health issue. Meta-analysis is the method most often used, but the details are beyond the scope of this chapter (Beaglehole et al., 1993).

Strength: How great is the risk of disease putatively induced by a given stressor (e.g., exposure to a toxicant)? As previously noted, this is often expressed as relative risk (RR), odds ratio (OR), standardized mortality ratio (SMR), or standard fertility ratio (SFR). Each risk statistic compares the risk of disease incurred by exposed persons with that of unexposed persons (NRC, 1991b). The greater the RR, SFR, or SMR, the stronger the inferred association between observed disease and exposure. In particular, relative risks (RR) greater than two can be considered strong (Beaglehole et al., 1993). A weak association may be the consequence of confounding or bias.

Dose Response (Biologic Gradient): A dose response relationship (biologic gradient) occurs when changes in the level of a possible cause are associated with changes in the prevalence or incidence of the effect (Beaglehole et al., 1993). For environmental epidemiologic studies, a dose response relationship is suggested if increased levels of exposure to an environmental toxicant are associated with an increased rate of a specific illness in a community. The demonstration of a clear dose response relationship in unbiased studies provides strong evidence for a causal relationship between exposure or dose and adverse health effect.

Reversibility: When removal of a possible cause of association leads to a reduced disease rate or reversal of an adverse health outcome, the likelihood of the association being causal is strengthened. Several examples exist in environmental epidemiology of reversibility of adverse health effects. Cessation of cigarette smoking is associated with a reduced risk of lung cancer relative to persons who continue smoking, an observation that strengthens the causality between smoking and lung cancer. For example, as described for the Love Canal episode, the elevated rate of low-birth-weight infants abated when hazardous substance releases from the Love Canal site were reduced.

Study Design: Strength of a study design to prove causation will vary according to the type of design. Beaglehole et al. observed that the best evidence for causation comes from well-designed, competently conducted, randomized controlled trials. This kind of study is used to evaluate drug efficacy and the effectiveness of larger public health campaigns to prevent disease. For example, new drugs are often tested against placebos in human populations under very tightly controlled conditions to determine the drugs' effectiveness and any side effects. In environmental epidemiology, randomized controlled trials are not used because the investigator cannot control experimental conditions; that is, exposure of a study group to various levels of environmental toxicants. The relative ability of different study designs to infer causation is shown in Table 5.7

Table 5.7. Relative Strength of Studies for Causal Inference (modified from Beaglehole et al., 1993).

Randomized controlled trials [strong]
Cohort studies [moderate]
Case-control studies [moderate]
Cross-sectional studies [weak]
Ecologic studies [weak]

(Beaglehole et al., 1993). This table can be useful in interpreting the strength of studies summarized in other tables in this chapter.

Judging the Evidence: Judgment synthesizes one's sense of the other seven guidelines for causality. No standard guidelines exist on how to judge the evidence, but a study that combines evidence of temporal relationship, dose response outcome, plausibility, and consistency with other studies would be compelling for a causal inference.

Under CERCLA, the ATSDR is responsible for conducting epidemiologic research, including several types of studies (cluster investigations, disease and symptom prevalence studies, analytic epidemiologic studies), surveillance programs, and exposure registries. Cluster investigations and disease and symptom prevalence studies investigate the occurrence of disease in populations. Analytic epidemiology studies are conducted to evaluate the causal nature of associations between exposure to hazardous substances and disease outcomes.

Bias

The primary purpose of an environmental epidemiologic study is to describe the effect of exposure to an environmental agent on health outcome. The design and conduct of the study should mitigate against extraneous differences between the study groups causing a misinterpretation of the exposure–outcome results. Sources of variation in the study must be controlled. Two sources of variation are particularly important: variation in characteristics of the study groups and variation in quality of data collected for the study groups. These and other sources of variation represent bias in an epidemiologic study.

Marsh and Caplan (1987) note that sources of bias resulting from methodologic features of study design and analysis can be classified in several ways. They described the four primary types of bias they find likely to be encountered in epidemiologic studies of populations exposed to releases from waste sites: confounding bias, selection bias, information bias, and reporting bias.

Confounding bias refers to factors whose control will reduce or correct bias in a study. In studies of exposure to toxicants, some health outcomes (e.g., cancer) can have risk factors other than the toxicant under investigation. For example, individual characteristics such as age, race, tobacco use, diet, alcohol consumption, and occupation can confound the effects of a particular hazardous substance. Furthermore, for cohort studies, exposed and unexposed groups should be comparable in all known risk factors for the health outcome of interest. In case-control studies, case and control groups should be comparable in their *a priori* chances of exposure and in characteristics other than exposure that relate to the health outcome under study (Marsh and Caplan, 1987).

Selection bias refers to distortion in the observed health effect resulting from the method of selecting subjects from the study population. Sources of selection bias include: flaws in the choice of groups to be compared, choice of the sampling frame, loss to follow-up or nonresponse of subjects during data collection, and selective survival (Marsh and Caplan, 1987).

Information bias (misclassification bias) refers to a distortion in estimating the health effect of interest that results from measuring the exposure or health outcome in a systematically inaccurate way. Sources of error that can contribute to information bias include: a flawed measurement device (e.g., inaccurate questionnaires or interviewing of subjects), incomplete or erroneous data (e.g., environmental contamination levels), and subjects' recall of health and exposure data.

Reporting bias refers to distortions caused by estimating health effects in communities selected for study on the basis of their reporting of health problems perceived to be in excess. As noted by Marsh and Caplan (1987), "Usual tests of significance (i.e., statistical tests) may be inappropriate since one has chosen to test for an effect that has already been noted." This means that hypothesis testing, based on random sampling procedures, may not be possible in a community that is being evaluated for a health problem (e.g., cancer mortality) reported by the community as a belief.

These four kinds of bias, and others not described, can be reduced or eliminated in some study designs. Standard references on environmental epidemiology provide procedures to address potential bias in epidemiologic studies (e.g., NRC, 1991b; Gregg et al., 1996; NRC, 1997).

Clusters of Disease

A cluster study is descriptive of the population in a geographic area, occupational setting, or other small group in which the rate of a specific adverse health effect is thought to be, or known to be, much higher than expected (NRC, 1997). Clusters of disease or illness are frequently the basis of concern in communities located near hazardous waste sites. Typically, persons in the community relate some perceived excess of serious disease, like cancer, to the presence of the waste site.

Clusters often have the disadvantage of small sample size and the use of cluster data to compare with some other community or reference base. The NRC (1997) gives the example of a community that perceives a cluster of elevated rates of cardiac birth defects that occurs in a population living near a hazardous waste site. "Testing" the elevated rates by comparing measured rates of these defects in a given geographic area with rates from outside the area is a highly unreliable approach both methodologically and statistically, primarily because the sample being studied (birth defects) was not randomly selected.

When a cluster study does not confirm a community's expectations, the study's investigators will sometimes be challenged in communicating the study's findings because the "negative" results do not agree with the community's belief that excess disease is clustered in their midst. In such instances, careful attention to principles of risk communication is required (Chapter 8).

Statistical Significance

The usual purpose of an environmental epidemiologic study is to evaluate the health status of a group presumed exposed to an environmental agent of concern. Often the exposed group is compared with a group that possesses the same characteristics, but without exposure to the environmental agent of interest. Comparison of the groups leads to the use of statistical methods to estimate the likelihood that the study's findings could be attributed to chance. This comparison is called statistical significance testing.

As the NRC (1997) stated, "Statistical significance testing is used to assess the likelihood that 'positive' results of any given study represent a 'real' association." Probability values, or p-values, are measures of random uncertainty alone. They are strongly influenced by sample size, confounding, biases, and other considerations and are calculated according to the study protocol used by the epidemiologist and the conditions of the study. Values equal to or less than 0.05 are commonly accepted as indicating "statistical significance." Standard statistical references are available to use in calculating p-values (e.g., Gregg et al., 1996).

ENVIRONMENTAL EPIDEMIOLOGIC STUDIES

An investigation conducted of a disease thought to be associated with an environmental factor or hazard is referred to as environmental epidemiology. By the National Research Council's definition, *"Environmental epidemiology* is the study of the effect on human health of physical, biologic, and chemical factors in the external environment, broadly conceived. By examining specific populations or communities exposed to different ambient environments, it seeks to clarify the relationship between physical, biologic, or chemical factors and human health" (NRC, 1991b). For this chapter, environmental epidemiology is focused on the human health effects of hazardous substances released into the community environment and excludes the workplace environment.

Regarding hazardous waste and hazardous substance concerns, environmental epidemiology plays a key role in evaluating the causes of specific diseases occurring among populations exposed to hazardous substances. For example, reflecting on the Love Canal episode, the residents were very concerned that hazardous substances were causing cancer and other adverse health effects. State and federal health agencies performed epidemiologic studies in response to the community's concerns. In a different kind of study, the question of interest might be to assess the effect of different exposure levels on a documented health outcome. For example, environmental epidemiologic studies of young children exposed pre- and post-natally to lead showed blood lead levels were correlated with adverse developmental and neurologic effects at increasingly lower levels of lead exposure (e.g., Needleman, 1995).

When the rate of illness appears to be elevated in a community population, epidemiologic studies can be used to define the course of illness and to help form various intervention strategies. For example, for hazardous waste sites undergoing remediation, epidemiologic surveillance systems could be used to identify any unsuspected public health problems caused by site remediation and attendant adjustments to remediation methods.

Because epidemiology is concerned with human disease, it and biostatistics occupy a unique position in providing a basis for strategies designed to identify, control, and prevent environmental health problems. For example, public health strategies to prevent adverse health effects of lead in young children are based on findings from epidemiologic studies that identified young children, particularly minority children, as being at risk of cognitive delays, problems in hearing, and reduced IQ. The public health strategies include regulatory actions, parents' education, community education, blood lead surveillance, and deed restrictions on real estate. Reducing the impact of environmental lead on children's neurologic health and development is a very positive outcome from environmental epidemiology.

As will be discussed in Chapter 7, findings from epidemiologic studies of communities impacted by hazardous waste sites can be very useful for determining health promotion activities. These activities include training and education programs for community health care providers, community education programs, health surveillance, and medical monitoring activities. Furthermore, findings from epidemiologic studies can be very important for site-specific risk assessments and as material for establishing exposure standards for specific contaminants.

However, environmental epidemiology has drawbacks. Some key limitations stem from how epidemiologic investigations are designed and conducted. Because epidemiology deals with human illness and is performed by scientists, communities concerned about environmental hazards, like hazardous waste sites, can have an unrealistic expectation of what the health study might produce. For instance, a community concerned about a cancer cluster in its midst will expect a "health study" to identify the cause. Epidemiologists often have difficulty providing satisfactory answers because of sample size limitations, inadequate toxicologic knowledge, uncertain exposure classifications, and inadequate data on individuals' exposure levels. Epidemiologists must, therefore, explain the limitations of their work before conducting environmental epidemiologic investigations and actively involve the community in planning, conducting, and evaluating a health study.

Epidemiologic Approach to Conducting Waste Site Studies

Several epidemiologists with experience in conducting environmental health studies have offered advice on how to conduct studies of persons with exposure to hazardous substances released from waste sites (e.g., Heath, 1983; Landrigan, 1983). Landrigan notes, "The function of epidemiology in the evaluation of persons exposed to chemical dumps is to document and to define any etiologic associations which may exist between exposure and disease. Epidemiology in this context is the discipline which guides study design" (Landrigan, 1983). He outlined four basic epidemiologic tenets that must guide the evaluation of persons exposed to hazardous substances released from waste sites or similar releases: (1) the nature and extent of exposure must be documented, (2) the exposed populations must be precisely defined, (3) disease and dysfunction in the exposed populations must be diagnosed as unequivocally as possible, and (4) the relationships between exposure and disease must be evaluated with rigorous statistical methodology.

These four principles are elaborated in Table 5.8, with a fifth principle added: act on the findings of significance to the public's health. This principle should be included in any epidemiologic investigation, given that the study has at its heart a concern for the health of the population investigated.

Heath provided valuable cautionary advice on the conduct of epidemiologic studies of persons at risk of exposure to substances from hazardous waste sites (Heath, 1983).

Table 5.8. Basic Epidemiologic Principles Applied to Waste Site Studies (adapted from Landrigan, 1983; Heath, 1983).

- Document exposure
  - Inventory materials in the waste site
  - Determine environmental emissions
  - Determine routes of human exposure
  - Estimate or measure exposure
- Define exposed populations
  - Define at-risk groups
  - Select unexposed groups
- Document disease and dysfunction
  - Relate medical tests to exposure
  - Conduct focused medical exams
- Link exposure to disease
  - Compare exposed to unexposed
  - Compare exposed to reference data
  - Explore exposure-effect relationship
  - Apply Hill's postulates
- Act on findings of significance to the public's health

Regarding exposure, he observed, "Despite the presence of clearly toxic materials, human exposure may not have occurred or may be only remotely possible." He underscored the importance of exposure pathway identification and analysis. He cautioned that clinical nonspecificity will characterize the biologic effects under study of waste site community residents. This means the epidemiologist must design the study protocol to account for factors that could confound any observed biologic effect. For example, studies of birth weights must account for factors that would contribute to reduced birth weights. These confounding factors include effects of individuals' tobacco smoking, prescription drug use, alcohol consumption, occupational exposure to reproductive toxicants, nutrition, and prenatal health care.

Which organ systems and associated health effects should be the focus of epidemiologic investigations of persons exposed to releases from hazardous waste sites? Buffler et al. (1985) sought to answer this question through a review of the scientific literature on the effects of hazardous substances on human health. Occupational health studies were the predominant database used in the analysis of health effects of toxicants. Buffler et al. concluded that the specific organ systems most likely to be affected by hazardous substances released from waste sites would be heavily dependent on the route of exposure and the dose that persons receive from their exposure. Using a limited waste site database, Buffler et al. concluded that ingestion was the most likely route of exposure in waste site episodes of environmental contamination.

Assuming ingestion as the most important route of exposure for populations impacted by waste site releases and that exposure levels would be low-to-moderate, Buffler et al. concluded that the organ systems most likely affected were the hematologic system for moderate exposure levels and central nervous, liver, and reproductive systems for low-dose situations. As observed in the next section, Buffler et al.'s organ system priorities are in partial agreement with the ATSDR's priority health conditions.

FINDINGS FROM HEALTH STUDIES

A substantial number of health studies[5] have been conducted of persons potentially impacted by substances released from hazardous waste sites or from other uncontrolled releases of hazardous substances. This section summarizes key studies and findings. The summaries are organized around the ATSDR's Priority Health Conditions. The selection of studies has emphasized a positive association between a particular health outcome and exposure to substances released from the sites investigated. "Negative outcome" studies are important, but any weight-of-evidence assessment of a hazard to the public's health gives greater weight to "positive outcomes" because they may portend the need for a public health response.

Tips on Reviewing Health Studies

The following are summary descriptions of a large number of health studies of populations potentially exposed to hazardous substances released from waste sites or from waste management activities. The studies differ in many respects, including health outcomes investigated, substances putatively released from the site, epidemiologic study design, and methods of conduct of the study. Some tips are useful for comparing the strengths and weaknesses of each study:

- Sample size of the population being studied and the statistical power of the study are intertwined. In general, small sample size produces equivocal or marginal results and a difficult interpretation of exposure and health outcome associations.
- Choice of the comparison group is important. In general, the comparison group should reflect the characteristics of the exposed group but without exposure to the substance of concern. Furthermore, the comparison group should be selected without evident bias; for example, self-referred to the investigator for health study.
- Although p-values are given for many but not all studies, modern epidemiology gives more emphasis to odds ratio, relative risk, and regression coefficients as indicators of associations between exposure and health outcome. For that reason, most studies cite OR, RR, SMR, or similar statistics.
- In general, the larger the ratio (e.g., OR, RR, SMR) the stronger the association between inferred risk of exposure and disease in exposed persons.
- A wide confidence interval reduces the strength of association between exposure and inferred risk of disease for exposed individuals.
- To control bias, investigators should take precautions in the design and conduct of their study. For example, were investigators "blind" to the exposure status of individuals tested?
- Methods to measure or estimate individuals' exposure to hazardous substances should be stated by the investigator. Direct measures of exposure are generally preferable to estimates or surrogate measures.
- The study design will affect the strength of the study's findings. See Table 5.7 for further details.

---

[5] Several studies described in this chapter are cited as reports from the ATSDR or state health departments. Although they may not have been published in scientific journals, all reports have received independent scientific peer review, as required by CERCLA.

Priority Health Conditions

In 1993 the ATSDR identified seven priority health conditions after reviewing the toxicology and epidemiology literature on 70 hazardous substances identified at NPL sites (Lybarger et al., 1993). This initial screening of substances and possible health impacts was conducted by a multidisciplinary team of physicians, epidemiologists, toxicologists, and other public health specialists.

Following this initial review, the ATSDR selected the final priority health conditions on the basis of four criteria: (1) frequency of epidemiologic or toxicologic associations with the substances considered most hazardous at NPL sites, (2) severity of the adverse health condition, (3) extent of physician, public health specialist, and community concerns, and (4) ability to lessen the impact of a particular illness through medical care or illness prevention activities. The seven priority health conditions are shown in alphabetical order in Table 5.9.

These seven conditions form the basis for applied research conducted both by the ATSDR investigators and through grants to state health departments and academic researchers. The priority health conditions constitute the structural framework for this chapter. The following is a brief summary of each priority health condition, together with summaries of key health studies of persons living near hazardous waste sites. The studies are generally presented chronologically by date of publication of findings.

Birth Defects and Reproductive Disorders

Reproduction is essential for perpetuation of any species. From time of conception to the time of birth, reproduction is an elegant physiologic process, but this process does not always produce either a successful fertilization of an ovum or an offspring free from defects or disorders. Male and female reproductive systems in humans are known to be sensitive to the effects of biologic, physical, and chemical agents, as is the fetus after pregnancy occurs. The relationship between environmental toxicants and reproductive disorders is complex because exposure of either or both parents can influence the reproductive outcome of their child. Moreover, the exposure scenario of either or both parents can be complicated. For instance, health of a fetus can be affected by parental exposure to hazardous substances well before conception, just before conception, or during gestation. The wide range of adverse reproductive outcomes includes reduced fertility, spontaneous abortion, reduced birth weight, birth defects (i.e., congenital malformations), and developmental disorders in infants (Hicks et al., 1993).

Among pregnancies of less than 20 weeks duration, 15% are estimated to end in spontaneous abortion (Hicks et al., 1993). An estimated 3% of newborn infants have some kind of major developmental defect at birth. About 7% of newborns are born prematurely (i.e., before 37 weeks gestation) and 7% of infants born at full term have low birth weight (2,500 grams or less). During early childhood, other birth defects can appear and the frequency of all malformations, major and minor, increases to about 16% of live births (Hicks et al., 1993).

The significance of birth defects and low birth weight in infants is great. Birth defects cause both emotional and financial consequences. Children with severe birth defects must often bear the pain of social rejection and the cost of medical care. Health care costs can be considerable, depending on the kind of birth defect. Regarding low birth weight, one source estimates that health care, education, and child care in 1988 for the

Table 5.9. ATSDR's Priority Health Conditions (ATSDR, 1992e; Lybarger et al., 1993).

- Birth defects and reproductive disorders
- Cancer (selected anatomic sites)
- Immune function disorders
- Kidney dysfunction
- Liver dysfunction
- Lung and respiratory diseases
- Neurotoxic disorders

3.5 to 4 million children ages 0 to 15 years born with low birth weight cost $5.5 to 6 billion more than if they had had normal birth weight (Lewit et al., 1995).

Relatively little is known about effects on human reproductive health of the thousands of chemicals found in the environment. One source attributes only 7 to 11% of birth defects to known agents such as radiation, drugs, chemicals, infection, or maternal metabolic imbalance, and no cause can be identified for 55 to 70% of birth defects (Hicks et al., 1993).

The reproductive effects of hazardous substances released from hazardous waste sites are now documented in the scientific literature. Several studies have examined the effect of parental residence near hazardous waste sites and the occurrence of birth defects, low birth weight, and other adverse reproductive outcomes. These studies have relied primarily on birth defects surveillance systems maintained by some state health departments. Examples of such surveillance systems are state-based reporting by hospitals and physicians of birth defects and low birth weight. Additionally, some waste site-specific studies of residents' reproductive health have also been conducted. This paper reviews the salient scientific literature that associates parental residential proximity with adverse reproductive health.

Studies That Used Surveillance Systems

New York State: A study conducted in collaboration with the New York State Department of Health examined the association between congenital malformations in children and maternal proximity to hazardous waste sites in the state (Geschwind et al., 1992). The study populations consisted of 9,313 newborns with congenital malformations, as recorded in the New York State Congenital Malformations Registry, and 17,802 healthy comparison children. In 20 New York counties, 590 uncontrolled hazardous waste sites were selected for analysis. An "exposure risk index" was created for each respondent, incorporating the distance from and the hazard ranking score for each site within a one-mile radius of the birth residence. The investigators controlled for several possible confounding variables, including maternal age, race, education, complications during pregnancy, birth weight, length of gestation, and sex of child. Other possible confounders, such as smoking and alcohol history, maternal and paternal occupational exposures, and maternal nutritional status were not evaluated because data were lacking.

Findings indicated that maternal proximity to uncontrolled hazardous waste sites was associated with an increased risk of approximately 12% of bearing children with any kind of congenital malformation. The risk was 29% higher for nervous system malformations, 32% higher for malformations of the integument system, 16% for the musculoskeletal system, and 15% for oral cleft malformations. Higher overall malformation rates were associated with both higher exposure risk (63% increased risk) and documentation of off-site chemical leaks (17% increased risk).

New York State: As a follow-up to the Geschwind et al. study, investigators from the New York State Department of Health evaluated the risk of two kinds of birth defects (central nervous system [CNS] and musculoskeletal [MUS] defects) associated with mothers' potential exposure to solvents, metals, and pesticides through residence near hazardous waste sites (Marshall et al., 1995, 1997). Subjects were drawn from births occurring in 1983–1986 to residents of 18 urban counties in New York State. Cases were drawn from the Congenital Malformations Registry and controls were drawn from those included in the Geschwind et al. study (Geschwind et al., 1992). Potential residential exposure was based on the address at birth. Environmental data for all inactive hazardous waste sites (n = 643) were assessed. Areas within one mile of each site were classified according to the probability of exposure. The environmental ratings for 473 CNS cases, 3,305 MUS cases, and 12,436 controls were combined with data on other potential risk factors for birth defects as obtained from birth certificates, such as mother's age, race, and education level.

Infants of mothers living in areas classified as having medium or high probability of exposure to substances from uncontrolled hazardous waste sites did not show increased risk of either CNS or MUS birth defects when compared with infants whose mothers lived in low probability exposure areas (Marshall et al., 1995, 1997). The low number of persons classified in the medium and high probability of exposure categories limits the investigators' findings. Further efforts are needed to elaborate the differences between the Geschwind et al. (1992) and Marshall et al. (1995, 1997) studies.

United States: In a study of NPL sites nationally, the ATSDR investigators examined reproductive outcomes reported in the 1988 National Maternal and Infant Health Survey and compared the occurrences of adverse outcomes with mothers' residence during pregnancy, using ZIP codes as a surrogate for maternal exposure to any releases from sites (Sosniak et al., 1994). The national database contained questionnaire responses from mothers who furnished information on various factors that could affect reproductive health and the outcome of their pregnancy. The NPL database was derived from EPA records for 1990. Mothers who lived one mile or less from the nearest NPL site were defined as "exposed" to contaminants from the site; those who lived more than one mile were defined as "unexposed." Of 17,407 mothers who participated in the national survey, 1,733 were classified by the ATSDR as "exposed" to hazardous waste sites and 15,674 were classified as "unexposed." Possible confounding variables that account for adverse reproductive outcomes; for example, tobacco smoking history and quality of prenatal care, were extracted from the mothers' questionnaires. The investigators found very low birth weight, infant and fetal death, prematurity of birth, and congenital malformations were not associated with living within a one-mile radius of NPL sites during pregnancy. The lack of association between maternal proximity to NPL sites nationwide and adverse reproductive outcomes is tempered by no measured exposure data during pregnancy.

Santa Clara County, California: Investigators from the California Department of Health Services examined any relationship of pregnancy outcomes and cardiac abnormalities to contamination of a municipal water supply (CDHS, 1985; Deane et al., 1989). The contamination resulted from a November 1981 leak in an underground storage tank. The leak contaminated Well 13, which supplied drinking water to nearby residents. The well was removed from service in December; 1,1,1-trichloroethane (methyl chloroform) was measured at 1,700 ppb at the time of removal. Pregnancy outcomes were assessed in two census tracts among women who conceived between January 1, 1980, and December 31, 1981. One census tract included the community that had received contaminated water from Well 13. The other census tract, which lay outside the water company's service area, served as the control tract. During the period of interest, 250 women in the study area and 316 in the comparison area conceived. Structured interviews were conducted to ascertain pregnancy outcomes, and 781 birth records were examined. The study area was found to have a spontaneous abortion rate twice that of the control area (OR = 2.4, 95% CI 1.3 to 4.2), a congenital anomaly rate triple that of the control area (RR = 3.1, 95% CI 1.1 to 10.4), and an absence of low-birth-weight babies. The investigators concluded, "...the cause of the observed excess in adverse pregnancy outcomes has not yet been determined" (Deane et al., 1989).

A followup investigation focused on cardiac defects in infants (Swan et al., 1989). Using data from the California Birth Defects Monitoring Program, a county-wide study during the period 1981–1983 found an increased prevalence of cardiac defects in infants in the area serviced by the water company that operated the contaminated well. During the period of potential exposure (January 1981 through August 1982), 106 children with major cardiac anomalies were born in Santa Clara County; 12 of them were born to residents of the service area; six cases would have been expected in the county, yielding a relative risk of 2.2 (95% CI 1.2 to 4.0). However, when Shaw et al. examined the temporal and geographic patterns of the infants born with cardiac defects, they concluded the solvent leak was an unlikely cause of the excess number of cardiac anomalies.

San Francisco Bay Area: Researchers from the California Department of Health Services examined birth defects and birth weight in the five-county San Francisco Bay area (Shaw et al., 1992). The investigators reviewed case records of congenital malformations and birth weights for all live births and fetal deaths that occurred from 1983 through 1985. Congenital malformations in 5,046 children were identified from a total of 215,820 live births and compared with 28,085 infants without malformations drawn from the same geographic area. Birth weights of 190,400 children also were analyzed. Each mother's census tract of residence at time of delivery was used to categorize exposure potential. Each census tract was categorized into one of three groups, according to the presence or absence of sites of environmental contamination (e.g., landfills, chemical dumps, industrial sites, hazardous materials treatment and storage facilities) and whether information was available about potential human exposure to contaminants released from waste sites. An elevated risk (OR = 1.5; 95% CI 1.1 to 2.0) was found for infants with malformations of the heart and circulatory system in census tracts with sites having potential for exposure; no similar risk was found in census tracts with no sites. No increased risk was found for other malformations or for low birth weight.

California State: In a follow-up to the Shaw et al. study (Shaw et al., 1992), Croen et al. used data from two population-based, case-control studies conducted in California (Croen et al., 1997). They investigated the risk of residential proximity to hazardous waste sites on giving birth to infants with birth defects. The congenital malformations investigated consisted of neural tube defects (NTDs) (n = 507), conotruncal heart defects (n = 201), and oral cleft defects (n = 439). Cases were identified in the California Birth Defects Monitoring Program, a population-based registry of fetuses and infants with major structural malformations diagnosed between conception and the infant's first birthday. Controls (n = 972) were live-born infants without congenital malformation. Residential histories and extensive data on potential confounders were obtained from detailed interviews of mothers of cases and controls. All addresses where a woman lived during the periconceptional period were geographically matched to census tracts and latitude/longitude coordinates. Reviewers of birth defects data were blind to the residential proximity of cases and controls.

Hazardous waste sites (n = 764) were identified from California state and federal agency sites, and related to census tract and latitude/longitude coordinates. Detailed information on site-related contamination was obtained for all sites on the EPA's National Priorities List (NPL) (n = 105), and the exposure potential of each NPL site was assigned according to site-specific characteristics. Each site's exposure status for each birth defect case and control was assigned according to residential proximity to waste sites during the periconceptional period. Proximity was defined by (1) census tract, and (2) distance (one mile) from a waste site.

Croen et al. (1997) found several associations between waste site characteristics and the occurrence of specific kinds of birth defects. They found 1.5- to 5-fold risks for NTDs and conotruncal heart defects in association with both proximity measures, but no increased risks were observed for oral defects in association with either proximity measure.

The study of Croen et al. makes a substantial contribution to the scientific database on hazardous waste sites and associated birth defects. They enumerated several strengths and weaknesses in their study. Strengths involved cases of birth defects identified from a population-based, congenital malformation registry that uses active ascertainment from multiple sources. Investigators were able to define exposure according to a variety of geographic parameters, and detailed exposure information was available on environmental contamination. Weaknesses involved lack of direct measurements of individuals' exposure to contaminants from waste sites; wide variation in the quality of site-specific information across sites; and an average of four years lapse before interviews were completed with case and control mothers of children whose data were used in the study. On balance, this study gives important information about the association of hazardous waste sites and their impact on reproductive health.

Site-Specific Reproductive Effects Studies

Love Canal, New York: Love Canal, near Niagara Falls, New York, was the first large residential area known to be affected by buried hazardous waste. The community had been built over an abandoned chemical waste dump, and the resulting presence of hazardous substances in households raised concern for the impact on residents' health. The Love Canal episode helped shape the American public's perception of hazardous waste sites as a major environmental hazard. An environmental sampling survey of Love Canal in 1977 found that hazardous substances had migrated from the waste dump into the basements of nearby homes and storm sewers. More than 200 hazardous substances were found in the Love Canal dump site, including benzene and lindane.

Two studies have been conducted of Love Canal residents' reproductive outcomes and infants' development. One study compared growth of Love Canal children with a referent group (Paigen et al., 1987). Anthropometric measures of children, stature of parents, demographics, and health information were obtained for 493 children living near Love Canal and 428 children from census tracts matched to the Love Canal community. Children born and spending at least 75% of their lives in the Love Canal area (n = 172) had significantly (p = 0.004) shorter stature for age percentile (46.6 ± SE 2.2) than did 404 comparison children (53.3 ± SE 1.4). This difference could not be accounted for by differences in parents' stature, socioeconomic status, nutrition, birth weight, or chronic illness.

In another study, the incidence of low birth weight among white, live-born infants from 1940 through 1978 was studied in various sections of the Love Canal area (Vianna and Polan, 1984). Infants who weighed 2,500 grams or less at birth were considered low birth weight. Investigators gave particular attention to the swale areas of Love Canal because they received drainage and therefore may have concentrated hazardous substances in soil. A significant excess of low-birth-weight births in the historic swale area was found from 1940 through 1953, the period when various chemicals were dumped in the waste site. The investigators state, "For the period of active dumping (i.e., before 1954), the swale area's percentage of low weight births was higher than in upstate New York (z test, $p < 0.0001$) and the rest of the canal (z test, $p < 0.012$)." These studies of the stature and birth weight of infants born to Love Canal residents were important for validating the health concerns of the area's residents and for providing early evidence of the need to interdict releases from hazardous waste sites.

Woburn, Massachusetts: The health of children living in Woburn, Massachusetts, has been the subject of investigation for several years. In June 1979, two municipal wells were found to be contaminated with hazardous substances, principally trichloroethylene (TCE), tetrachloroethylene, chloroform, and other organic compounds. This discovery was associated in the public's mind with an elevation of leukemia in the children of area residents. The wells were closed and investigations of health impacts commenced (MDPH, 1996). In addition to investigating leukemia mortality in children, investigators examined adverse pregnancy outcomes and childhood disorders.

In 1982, investigators conducted a telephone survey of Woburn households about adverse reproductive outcomes (Lagakos et al., 1986). The survey gathered information on adverse pregnancy outcomes occurring to former and current family members between 1960 and 1982. The adverse outcomes included spontaneous abortions, and congenital malformations. Investigators estimated the percentage of contaminated water delivered annually to each household in the sample of households surveyed. The surveyed households provided information on 4,396 pregnancies during the specified period. No evidence was found to associate exposure to contaminated water and spontaneous abortion and low birth weight. For congenital malformations, no association was found between exposure to contaminated water and musculoskeletal and cardiovascular defects. A positive association was found between exposure to water contamination and eye/ear defects and CNS/chromosomal/oral cleft anomalies. Residents also reported an excess of kidney and urinary tract disorders.

Wayne Township, New Jersey: In New Jersey, a study was conducted of the effects on local residents' health of a thorium waste disposal site in Wayne Township, Passaic County, (Najem and Voyce, 1990). Soil samples taken on the waste site and nearby showed thorium, radium, and uranium contamination. Ambient air radiation levels were within background levels. No biologic sampling of residents occurred. A structured health status survey was conducted by the same interviewer who was blind to the exposure status of each person interviewed. The interview included 171 persons living near the waste site and 191 persons living nine to ten blocks from the site. Findings showed a higher prevalence of birth defects (RR = 2.1, 95% CI 0.62 to 7.25) in the presumed exposed group than in the comparison group. No consistent pattern of specific kinds of birth defects was found; defects included cerebral palsy, heart defects, inguinal hernia, foot deformation, and other malformations. These findings must be tempered by the relatively small group sizes and lack of documented radiation exposure.

Tucson, Arizona: In another site-specific study, an investigation was conducted of births in Tucson, Arizona, to evaluate the association between congenital cardiac malformations and parental contact with drinking water contaminated for several years, primarily with TCE released into groundwater from an NPL site (Goldberg et al., 1990). Using the investigators' own case registry, 707 children were identified with congenital heart disease who were born between 1969 and 1987, conceived in the Tucson Valley, and whose parents spent both the month before the first trimester of pregnancy and the first trimester of pregnancy in the Tucson Valley. Of the 707 cases, 246 lived in the area with contaminated water. The remaining 461 cases served as a case-comparison group. According to investigators, "The odds ratio for congenital heart disease for children of parents with contaminated water area contact during the period of active contamination was three times that for those without contact ($p < 0.005$) and decreased to near unity for new arrivals in the contaminated water area after well closure."

A unique corollary of the Goldberg et al. (1990) study is the collaborative toxicologic study of cardiac teratogenesis conducted by Dawson et al. (1990), who used a rat model to assess the teratogenic effects of TCE and dichloroethylene (DCE). On day seven of pregnancy rats were divided into five groups and each group was administered one of the following: TCE (1500 ppm in saline or 15 ppm in saline), DCE (150 ppm in saline or 1.5 ppm in saline), or physiologic saline (i.e., the control treatment). All treatments were administered by syringe through direct intrauterine exposure of fetuses. On day 22 of pregnancy, animals were killed to determine congenital heart anomalies in the fetuses. The prevalence of heart anomalies was 3% in the control group, 9% in the low TCE, 12.5% in high TCE, 14% in low DCE, and 21% in high DCE groups. These findings are quite interesting, even though the experimental technique (direct intrauterine exposure) was intentionally provocative, because the toxicologic findings support the human epidemiologic findings from Goldberg et al. (1990).

Gloucester County, New Jersey: The Lipari Landfill in New Jersey is number one on the EPA's NPL on the basis of its Hazard Ranking Score. It had received industrial waste, municipal waste, household wastes, and semisolid chemical wastes. The EPA estimated that the period of greatest waste dumping into the landfill occurred from 1967 through 1969. The landfill operated until 1971. The New Jersey State Department of Health examined birth weights of infants born to mothers living near the Lipari Landfill (Berry, 1994; Berry and Bove, 1997). Birth weights were grouped into five-year periods from 1961 to 1985. For this time span, birth weights for 2,767 births with gestational age greater than 27 weeks were compared with 7,389 comparison births. Birth weights were also compared for 2,563 term births (gestational ages between 37 and 44 weeks) and 6,840 comparison births.

Births were assigned an exposure area based on distance from the landfill to the mother's residence at time of birth. Residents living within 1 kilometer of the site were defined as "exposed." Births in towns near the Lipari Landfill but greater than 1 km away constituted the comparison group. The area nearest the landfill was divided into two sectors based on proximity to the landfill: Area 1A was the neighborhood immediately downwind and adjacent to the landfill; Area 1B was the remainder of Area 1, less Area 1A, and was generally more distant from the site (Figure 5.2). Significantly greater reductions (80–152 grams) in average birth weights occurred in Area 1 for the years 1971–1975 than in the comparison area for births with gestational age greater than 27

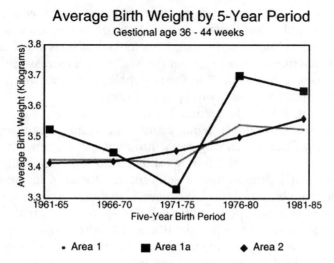

Figure 5.2. Reproductive outcomes for residents near the Lipari NPL site (Berry, 1994, with permission of author).

weeks. Term births in Area 1A showed a significant average reduction of 188 grams for years 1966–1975 than in the comparison area for the same period. Birth weight reductions occurred only during the period 1971–1975 when peak releases of volatile organic compounds (VOCs) and metals from the landfill were occurring. The investigator noted, "The excess risk appeared stronger in term births suggesting a mechanism of growth retardation rather than prematurity." No biologic exposure levels of contaminants were available for the study populations.

Montreal, Québec, Canada: The risk of giving birth to infants with low birth weight was examined among persons living near the Miron Quarry municipal solid waste landfill in Montreal, Québec (Goldberg et al., 1995b). The landfill, third largest in North America and located in a densely populated metropolitan area, was known to produce biogas that contained methane, carbon dioxide, VOCs, and sulfur compounds. Potential exposure to ambient air pollutants from the landfill was defined by proximity to the site. From 126,655 births available for analysis, case-control analyses were conducted to evaluate the risk of low birth weight (less than 2500 grams), very low birth weight (less than 1500 grams), preterm birth (fewer than 37 weeks), and small-for-gestational-age (less than third percentile). Findings indicated that low birth weight was significantly greater in the exposure zone proximal to the landfill (OR = 1.20; 1107 exposed cases; 95% CI 1.04 to 1.39). Excess risk was observed for small-for-gestational-age, but the association was not as strong as for low birth weight (adjusted OR = 1.09; 951 exposed cases; 95% CI 0.96 to 1.24). No significant associations were observed for very low birth weight or for preterm birth.

Silver Valley, Idaho: The ATSDR has conducted two health studies of persons in the Silver Valley of Idaho who in years past experienced known high exposure to lead and other metals from operation of the Bunker Hill primary smelter. Mining and smelting operations continued at the facility from 1886 until 1981. The Bunker Hill site is on the NPL. In one study, middle-aged women who had worked at the smelter were examined (Lee, 1997). Women first began working on the production and maintenance lines in

1972; lead screening in 1975 showed the average blood lead levels were 54 µg/dL in women working in production operations and 36 µg/dL in women who performed clerical work at the smelter complex. Investigators looked at body burdens of lead and responses to structured questionnaires about the rate of osteoporosis, reproductive health indices, and neurologic illnesses. The study consisted of 140 women workers and a comparison group of 121 women from Spokane, Washington,

From these groups, 108 workers and 99 comparison women were selected for measurement of lead concentration in bone, using the K-band X-ray fluorescence method (K-XRF). Measures of bone turnover, renal function biomarkers, liver function tests, and blood counts were included. Findings showed no significant differences between the exposed and comparison groups in self-reports of gastrointestinal disorders, liver disease, gallbladder problems, neurologic disease, diabetes, and affect. The mean bone lead in workers was 13.3 µg/g bone mineral; 2.6 µg/g for the comparison group. Corresponding blood lead means were 3.8 µg/dL and 1.6 µg/dL, respectively. The exposed group was more likely to report hypertension, chest pain, and heart palpitations. Regarding reproductive health, the exposed group's bone lead value was significantly correlated with the total number of pregnancies ($r = -0.21$, $p = 0.004$), and exposed women without hysterectomies stopped their menses at a significantly earlier age than did the comparison group (exposed = 46.6 years, comparison = 51.1 years, $p = 0.01$). Although the differences in reproductive health between former workers and the comparison group cannot be causally attributed to prior lead exposure, this study's measurement of bone and blood lead levels makes an important case that past exposure to lead had a significant latent effect on women's reproductive health.

Silver Valley, Idaho: The ATSDR conducted the second study of Silver Valley residents among young adults who had exposure to lead more than 20 years ago during operation of the smelter (ATSDR, 1996g). A health survey in 1974 of the area's children showed widespread lead poisoning. The exposed group consisted of 916 young adults who were current or past residents of the area near the smelter; a comparison group of 754 persons was recruited from the Spokane, Washington area. Structured interviews were conducted of all the study's participants, and 281 exposed persons and 287 comparison persons underwent biomedical testing. Exposure to lead was estimated by lifetime residential histories and measured by bone lead concentration using K-XRF measurements.

Findings showed a statistically significant increase in the prevalence of having difficulty conceiving children among residents of Silver Valley (males [OR = 2.3; 95% CI 1.12 to 4.72]; females [OR = 1.7; 95% CI 1.10 to 3.15]); the prevalence increased with increasing duration of residence in the Silver Valley. Age, sex, and education did not account for the increased prevalence of infertility. The leaded group compared with the reference group showed a significantly greater prevalence of symptoms of central and peripheral nervous system disorders; neurobehavioral testing indicated poorer performance by the exposed group on tests of motor function, cognitive function, balance, and vibration sensitivity. The average bone lead concentration was 4.6 µg/g bone mineral for 262 exposed persons (range: nondetectable–37 µg/g) and 0.60 µg/g bone mineral for 268 comparison individuals (range: ND–17.4 µg/g). The findings of impairments on neurobehavioral tests and reduced fertility are remarkable, given that lead exposure had occurred primarily more than 20 years before the study.

Camp LeJeune, North Carolina: A retrospective cohort study of exposure to VOCs in drinking water and a variety of adverse pregnancy outcomes was conducted at a military base in North Carolina (ATSDR, 1997d). Part of the military base was an NPL site, potentially contributing to contamination of the base's drinking water supply. The study examined birth records among residents of the base's family housing. One part of the study focused on mean birth weight and small-for-gestational-age as indicators of potential adverse reproductive effects. Birth certificates were evaluated for the period 1968–1985. Study cohorts consisted of infants born to 6,131 long-term, tetrachloroethylene (PCE)-exposed women, 141 short-term TCE-exposed women, 31 long-term TCE-exposed women, and 5,681 unexposed women. Findings for the PCE-exposed group as a whole showed no association between drinking water contamination and mean birth weight and small-for-gestational-age. However, for vulnerable subgroups within the PCE-exposed group, associations between PCE exposure and the study outcomes were noted for infants of mothers 35 years of age or older and infants whose mothers had histories of fetal deaths. For older mothers, the adjusted difference in mean birth weight for PCE-exposed births was 207 grams less than for unexposed births (90% CI –334 to –79), and the adjusted odds ratio for PCE exposure and small-for-gestational-age was 3.9 (90% CI 1.6 to 9.8). TCE-exposed groups were generally too small to permit a causal inference between exposure and reproductive outcomes.

Skarsborg, Sweden: In addition to risk factors for birth defects and other adverse reproductive effects associated with uncontrolled hazardous waste sites, reproductive health risks from municipal incinerator emissions have also been of concern. A study was conducted in Sweden to ascertain whether the incidence of cleft lip and palate had increased after the installation of incinerators to burn municipal waste (Jansson and Voog, 1989). The Swedish investigators examined the possible relation of dioxin emissions from municipal incinerators to the induction of elevated rates of cleft lip and/or palate. They conducted two kinds of investigations. In one, a case study of six infants born with cleft lip/palate found no association with maternal proximity to a municipal incinerator. The investigators reviewed Sweden's birth defects registry for all boroughs in Sweden with operating incinerators for the period 1975–1986. No increase in the rate of cleft lip and/or palate defects was found in boroughs after incinerators were installed. The birth defects part of the study is compelling because of the large number of infants with cleft lip and palate, but limited because measured dioxin levels were lacking in boroughs whose registry data were examined.

Summary of Birth Defects and Reproductive Disorders

Table 5.10 contains a summary of the 18 studies of birth defects and reproductive disorders described in this section and in the Love Canal section at the beginning of the chapter. A review of the table suggests these observations:

- Increased risk of congenital malformations in infants whose parents lived near hazardous waste sites was reported in studies from New York and California that used reproductive outcome surveillance systems (Geschwind et al., 1992; Shaw et al., 1992; Croen et al., 1997). Defects of the heart, neural tube, and oral cleft palate were the malformations most frequently reported in these studies.
- Congenital malformations of the heart were reported for infants in Arizona born to parents exposed to TCE and other VOCs in drinking water contaminated by an NPL site (Goldberg et al., 1990).
- Reduced birth weight was reported for infants whose parents lived near hazardous waste sites in New York, New Jersey, North Carolina, and Québec (Vianna and Polon,

Table 5.10. Key Studies of Reproductive Effects Associated with Hazardous Waste Sites or Incinerators.

Study Location (Reference)	Study Design (Period of Observation)	Number and Type of Subjects	Contaminants Exposure Measures	Major Health Endpoints	Outcome Reported
Love Canal, NY (Vianna and Polan, 1984)	Retrospective follow-up; 1940–1978	174 live births in uncontrolled hazardous waste site (HWS) area. Referent: (1) 443 live births in rest of HWS area (2) all live births in upstate NY	Benzene, lindane, 200 substances. Surrogate: proximity to HWS and at least 5 months residence	Low birth weight	Elevated incidence of low birth weight among exposed
Santa Clara County, CA (CDHS, 1985; Deane et al., 1989)	Retrospective follow-up; 1980–1981; 1981–1982	250 pregnancies in the study area, 316 in the control census tract; 781 birth records reviewed	Methyl chloroform Surrogate: residence in household receiving contaminated water	1980–1981: pregnancy outcome 1981–1982: congenital heart defects	1980–1981: excess of spontaneous abortion and congenital malformations 1981–1982: excess incidence of heart defects
Woburn, MA (Lagakos et al., 1986)	Retrospective follow-up; 1960–1982	4396 pregnancies among Woburn residents, 5018 residents 18 yrs or older; referent: internal	TCE & other VOCs Surrogate: households served by contaminated water supply	Adverse pregnancy outcomes and childhood disorders	Association with perinatal, deaths, eye/ear anomalies, CNS anomalies; kidney and urinary tract disorders
Love Canal, NY (Paigen et al., 1987)	Cross-sectional; 1980	493 children living near the Love Canal HWS Referent: 428 children matched in adjacent census tracts	Benzene, lindane, 200 substances Surrogate: proximity to the HWS	Anthropometric measurements	Increased prevalence of children with shorter height
Santa Clara, CA (Swan et al., 1989)	Retrospective follow-up (1981–1983)	106 children with major cardiac anomalies Referent: children in remainder of county	Methyl chloroform Surrogate: residence in area served by contaminated water supply	Major cardiac anomalies	12 children with major cardiac anomalies in service area; 6 expected in the county; excess not associated with water contamination
Skarsborg County, Sweden (Jansson and Voog, 1989)	Case-control and registry	Case-control study of 6 children born with cleft lip and/or palate; 57 cases in Skarsborg County registry	Incinerator emissions/Case control: maternal proximity to municipal incinerator and air dispersion modeling. Registry: No exposure estimate or surrogate	Cleft lip and/or palate in children born during 1975–1986	Small increased number of cleft lip and/or palate cases; no geographic relationship to incinerator locations

Location (Reference)	Study design	Population	Contaminant / Surrogate	Outcome	Results
Wayne Township, NJ (Najem and Voyce, 1990)	Cross-sectional	171 persons living near a waste site; referent: 191 persons living distant from the site	Thorium, radium, and uranium soil contamination Surrogate: residential proximity	A structured health status interview	Higher prevalence of birth defects, but no pattern in defects; low statistical power
Tucson, AZ (Goldberg et al., 1990)	Case-control; 1969–1987	707 children (246 in area of contaminated water) Referent: 461 cases distant from the contaminated area	TCE Surrogate: parental residence in area contaminated by TCE	Congenital heart defects	Children from contaminated area had 3 times risk of heart defects
New York State, less New York City (Geschwind et al., 1992)	Ecologic	9,313 newborns with congenital malformations Referent: 17,802 healthy children	Hazardous waste Surrogate: Maternal proximity to HWS	Congenital malformations	Increase in overall malformation rates; higher rates for malformations of integument system, nervous system, oral clefts, musculoskeletal system
Five-county San Francisco Bay area, CA (Shaw et al., 1992)	Ecologic	5,046 children with congenital malformations; 190,400 liveborn children	Hazardous waste Surrogate: Mother's census track of residence at time of birth	Congenital malformations and live birth weight	Elevated risk for malformations of heart and circulatory system; no risk for lowered birth weight
United States (Sosniak et al., 1994)	Quasi-case-control	1,733 "exposed" mothers and 15,674 "unexposed" survey respondents	Hazardous waste Surrogate: Maternal proximity to NPL sites	Very low birth weight, infant/fetal death, prematurity of birth, congenital malformations	No association of any outcome with proximity to HWS
Gloucester County, NJ (Berry, 1994; Berry and Bove, 1997)	Retrospective follow-up	2,767 births with gestational age greater than 27 weeks Referent: 7,389 comparison births	VOCs and metals Surrogate: maternal proximity to the HWS	Birth weight	Birth weight reductions during period of peak releases from the HWS
New York State, less New York City (Marshall et al., 1995, 1997)	Case-control	473 CNS cases, 3,305 MUS cases Referent: 12,436 normal births	Hazardous waste Surrogate: Maternal proximity to HWS	Congenital malformations (CNS and MUS)	No association with maternal proximity to HWS

Table 5.10. Key Studies of Reproductive Effects Associated with Hazardous Waste Sites or Incinerators (continued).

Study Location (Reference)	Study Design (Period of Observation)	Number and Type of Subjects	Contaminants Exposure Measures	Major Health Endpoints	Outcome Reported
Montreal, Québec, Canada (Goldberg et al., 1995a)	Case-control	126,655 births	Biogas (methane, CO_2, VOCs, sulfur compounds) Surrogate: 3 areas near the landfill; reference area	Low birth weight (LBW) (<2500 g), very low birth weight (<1500 g), preterm birth (<37 completed weeks), small-for-gestational-age (SGA) (<third percentile)	LBW and SGA rates were significantly elevated in the exposure zone near the landfill
California State (Croen et al., 1997)	Case-control	Live births: 507 neural tube defects, 201 heart defects, 439 oral cleft palate Referent: 972 matched cases	Hazardous waste Surrogate: Maternal proximity defined by (1) census tract, and (2) distance from waste site	Three congenital malformations: neural tube defect (NTD), heart, oral cleft palate	1.5- to 5-fold risks for NTDs and conotruncal heart defects for both proximity measures; no increased risk for oral defects
Silver Valley, ID (Lee, 1997)	Cross-sectional study of women who formerly worked at a smelter	140 women former workers and 121 comparison group women; 108 former workers and 99 comparison women were measured for bone lead levels	Lead Concentration of lead in bone using X-ray fluorescence (K-XRF)	The rate of osteoporosis, reproductive health indices, and neurologic illnesses	Bone lead correlated with fewer pregnancies; leaded group had earlier cessation of menses
Silver Valley, ID (ATSDR, 1996g)	Cross-sectional study of young adults for whom blood lead was measured more than 20 years earlier	Structured interviews conducted of all study participants (916 lead-exposed; 754 referents), and 281 exposed persons and 287 comparison persons underwent biomedical testing	Lead Estimated by lifetime residential histories and measured by bone lead concentration using K-XRF measurements	Structured interviews focused on reproductive and neurologic health (central nervous system [CNS] and peripheral nervous system [PNS])	Lessened fertility; increased prevalence of CNS and PNS disorders; impaired neurobehavioral performance
Camp Lejeune, NC (ATSDR, 1997d)	Retrospective cohort study of live births during 1968 through 1985	6,131 PCE-exposed women; 141 short-term TCE exposed; 31 long-term TCE exposed; 5,681 unexposed referents	Tetrachloroethylene (PCE) and TCE Measured levels of contaminants in water distribution system	Birth certificates were examined for mean birth weight and small-for-gestational-age	Reduced birth weight in infants whose mothers were 35 years or older or who had history of fetal deaths

1984; Paigen et al., 1987; Berry, 1994; Berry and Bove, 1997; Goldberg et al., 1995b; ATSDR, 1997d).

- Of particular note, reduction in birth weight was associated in Love Canal, New York, and Gloucester County (Lipari NPL site), New Jersey, with the period of known peak releases of contaminants from hazardous waste landfills (Vianna and Polan, 1984; Berry, 1994; Berry and Bove, 1997).

- Decreased fertility was reported in two studies of persons exposed earlier in life to large amounts of lead in soil and ambient air from a primary smelter that later became an NPL site (ATSDR, 1996g; Lee, 1997).

- The epidemiologic studies of adverse reproductive outcomes have all relied on parental residential proximity to hazardous waste sites as a surrogate for measured levels of exposure to specific hazardous substances. This, of course, limits the studies' strengths.

- Given the paucity of information about emissions of operating incinerators and potential exposure of human populations, additional research is warranted to characterize any adverse effects on reproductive health.

Cancer

More than one million persons in the United States learn each year that they have cancer. With the exception of cardiovascular disease, more deaths in the United States result from cancer than from any other cause. Not only does cancer cause great emotional and physical suffering, it also exacts a high fiscal cost in lost productivity and medical expenses. Current public health activities are focused on reducing the burden through prevention, early detection, and improved treatment of cancer (Williams et al., 1993).

The mechanism of normal cells becoming cancerous cells is not completely understood. Many agents directly cause or assist in the development of cancerous cells. Most cancer cases in the United States are believed to be associated with personal lifestyle factors such as diet and tobacco use; environmental factors such as radiation, sunlight, drugs, and other substances have also been associated with cancer induction (Williams et al., 1993). Epidemiologic evidence from occupational settings indicates many hazardous substances found at hazardous waste sites can cause cancer; however, whether they contribute to an increased cancer rate in communities depends on exposure and other factors. A partial list of chemical carcinogens was given and described in Chapter 4.

Most human environmental health studies that provide information on the carcinogenicity of environmental toxicants have been conducted on workers' exposure to substances in occupational settings. Before effective control measures were put into place, occupational workers were exposed to various toxicants at relatively high levels. Data from laboratory animal studies have been used to infer health effects in humans. However such data have not been as conclusive as the epidemiologic studies conducted on workers.

Cancer has been one of the major health concerns of individuals living in the vicinity of hazardous waste sites. It is difficult to study a population living near a hazardous waste site or similar point of chemical contamination and determine whether cancer rates are associated with exposure to substances released from the site. A major difficulty is not knowing the level and nature of individual exposure to a carcinogenic agent. Waste sites contain more than one chemical, making it difficult to associate health outcomes to a single exposure. Surrogate measures of exposure are usually based on proximity to the waste site, but they can result in serious misclassification bias. Often extensive confounding variables have to be accounted for before making any associations of the disease outcome to a given exposure from the site. Moreover, because of the long latency period of cancer development and the kind of behavioral risk factors associated with many cancers (e.g., tobacco use, alcohol consumption, and unhealthy diet), clearly associating an exposure to the presence of cancer is difficult.

New Jersey Counties: Two teams of investigators have reported increased frequencies of cancers in counties containing hazardous waste sites. One study focused on gastrointestinal (GI) cancer mortality in New Jersey counties and the relationship with environmental variables (Najem et al., 1983). GI cancer mortality was known to occur at a very high rate in New Jersey. The investigators compared the death rates in all 21 New Jersey counties with those for the nation for the period 1968–1977. They compared the number of observed deaths with the number expected. The expected number of deaths was obtained by applying the age-race-sex-specific death rates for each anatomical GI cancer site for the total United States to the age-race-sex-specific subpopulations for each of 21 New Jersey counties. The county race-sex-site-specific cancer rates were correlated with: the distribution of Chemical Toxic Waste Disposal Sites (CTWDS), annual per capita income, density of population, urbanization, and percentage of the population employed in chemical industries.

Findings showed that age-adjusted GI cancer mortality rates (all anatomical sites combined) were higher than national rates in 20 of New Jersey's 21 counties. Within specific sex and race groups, cancer mortality rates in the state during the period 1968–1977 significantly exceeded national rates for cancers of the esophagus (males), stomach (males and females), colon (white males and females), and rectum (white male, white female, nonwhite female). Environmental variables most frequently associated with GI cancer mortality rates were population density, degree of urbanization, and presence of toxic waste disposal sites (Najem et al., 1983). The investigators note that their findings are limited because of the dilution effects of studying a large group of people, many of whom have not been exposed to the risk factors (e.g., CTWDS) used in the analysis.

Clinton County, Pennsylvania: The Drake Chemical NPL site in Lock Haven, Pennsylvania, was known to be contaminated with beta-naphthylamine, a known human bladder carcinogen. Other contaminants included benzidine and benzene. To evaluate the prevalence of bladder cancer, investigators obtained county-wide, age-adjusted, sex-, race-, and site-specific cancer mortality rates for the years 1950 through 1979 (Budnick et al., 1984). No environmental or personal exposure data were available to the investigators, nor was there information on bladder cancer risk indicators (e.g., occupational history) for persons who died of bladder cancer. During the 1970s, bladder cancer deaths were significantly increased among white males in Clinton County, and a significantly greater number of other cancer deaths occurred in the general population of Clinton and three adjacent counties. The trend of increased bladder cancer mortality among Clinton County white males was not evident in the county's white female population, reducing the likelihood that a general environmental exposure to carcinogens had occurred in the county. The investigators concluded that county-wide cancer mortality rates were useful as a screening method for more in-depth investigation of specific hazardous waste sites. No association between the Drake Chemical site and county-wide cancer mortality rates was possible with this study.

New Jersey Municipalities: Clusters of cancer mortality in New Jersey municipalities were investigated by Najem et al. (1985). Their previous work had shown elevated cancer mortality rates in 20 of New Jersey's counties in comparison with U.S. national rates for several cancers (Najem et al., 1983). In the 1985 study, Najem et al. obtained death and birth certificates and other data from the vital statistics records of the state of New Jersey. The period 1968–1977 was selected for study because of the availability of vital statistics data. Thirteen anatomical major cancer sites were studied; annual age-adjusted mortality raters (per 100,000 population) were calculated for 194 municipalities (10,000 or more population) listed on death certificates as the community of usual residence. The investigators arbitrarily defined a cancer cluster if a community had: (1) two or more age-adjusted cancer rates that were at least 50% greater than corresponding national rates at a significance level of at least 0.01 and (2) one or more cancer rates that was significantly (set at $p < 0.0005$) higher than national rates. In 10 New Jersey counties, 23 municipalities met the criteria for cancer clusters; 16 of the 23 were located in the heavily industrial northeast area of the state. Of the cancers in the municipalities with clusters of excessive mortality, 72% were gastrointestinal cancers, especially the stomach, rectum, and colon. Correlation analyses indicated that most of the cancer rates were negatively associated with annual per capita income and positively with the density of chemical toxic waste disposal sites.

Woburn, Massachusetts: Leukemia in children living in Woburn, Massachusetts, has been the subject of investigation for several years. In June 1979, two municipal wells were found to be contaminated with hazardous substances, principally trichloroethylene (TCE), tetrachloroethylene, chloroform, and other organic compounds. This discovery was associated in the public's mind with an elevation of leukemia in the children of area residents. The wells were closed immediately, and investigations of health impacts began (MDPH, 1996). In the period 1969 through 1979, 12 cases of childhood leukemia were diagnosed in Woburn. Six of the cases were from an area served by the contaminated wells. By 1986, nine additional cases had been diagnosed.

Woburn, Massachusetts: A study of a leukemia cluster in children living in Woburn, Massachusetts, was conducted in response to the community's concerns that hazardous waste sites and contaminated municipal wells were responsible (Cutler et al., 1986). Two city wells found to contain chloroform, TCE, and tetrachloroethylene were closed in 1979. Industrial pollution was the source. The occurrence of childhood leukemia cases covered the period 1969–1979. The investigation confirmed an increase in incidence, which was distributed uniformly over the 11-year period. During that time, 1,748 children were age 14 years or younger; 12 cases of childhood leukemia occurred and nine had acute lymphocytic leukemia. Childhood leukemia incidence in Woburn in this period was significantly higher than expected: 12 cases were observed, 5.2 would be expected ($p = 0.007$). When analyzed by residence at the time of diagnosis, the leukemia children were clustered in eastern Woburn, near a pond. Investigators observed that the two contaminated municipal wells were possibly relevant to the leukemia cluster in the eastern part of Woburn, although investigators were unable to reconstruct any distribution of exposure to contaminants in the water supplied to homes.

Woburn, Massachusetts: Researchers from the Harvard School of Public Health investigated the cluster of childhood leukemia cases in Woburn (Lagakos et al., 1986). They obtained information on 20 cases of childhood leukemia (ages 19 and under) diagnosed in Woburn between 1964, the year the two contaminated wells (G and H) began pumping, and 1983. All 20 cases had been identified in tumor registries maintained by state or hospital tumor registries. For exposure assessment, the percentage of each household's annual water supply was estimated using water distribution and other data.

Two exposure metrics were used: (1) cumulative exposure from birth to contaminated water from the two contaminated wells and (2) a binary indicator of exposure (no exposure vs some exposure). Using either exposure metric, findings showed a positive association between exposure to water from wells G and H and the incidence of childhood leukemia. Lagakos et al. conclude that the cumulative and binary metrics statistically explain about four and six cases, respectively. Thus, five or six cases of childhood leukemia were still unexplained. Publication of the Lagakos et al. paper was accompanied by critiques of the paper in the same issue of the journal. These critiques shed additional light on the strengths and limitations of the Lagakos et al. study. Uncertain exposure characterization was mentioned by some commentators.

U.S. Counties: Like Najem et al. (1983, 1985), who evaluated New Jersey's cancer mortality rates, the EPA investigators examined national cancer rates in relation to hazardous waste sites (Griffith et al., 1989). Using the NPL, they identified 593 hazardous waste sites (HWS) in 339 U.S. counties in 49 states for which analytic evidence showed contaminated ground drinking water as the sole source of water supply. For each of the 339 counties, age-adjusted, site-specific cancer mortality rates for 13 major anatomical cancer sites were extracted from *U.S. Cancer Mortality and Trends 1950–1979* for white males and females in the decade 1970–1979. Each of the 339 counties was coded for each cancer as to whether the county had an excessively high number of deaths, and comparisons were made with the 2,726 counties without uncontrolled hazardous waste sites. The total number of non-HWS counties in the United States that had excess numbers of deaths was enumerated for each of the cancers selected for analysis. For example, 136 counties had excess numbers of deaths due to cancer of the pancreas; 20 of these counties contained uncontrolled hazardous waste sites. A chi-square analysis was used to determine the statistical significance of these kinds of comparisons.

In comparison with counties not having uncontrolled hazardous waste sites, significant associations ($p < 0.002$) for white males were shown between all counties having uncontrolled hazardous waste sites and excess deaths from cancers of the lung, bladder, esophagus, stomach, large intestine, and rectum. A similar comparison for white females showed significant associations for deaths from cancers of the lung, breast, bladder, stomach, large intestine, and rectum. The investigators observed, "It is important to note that individuals in the uncontrolled hazardous waste sites counties may have been exposed to chemical pollutants while working for companies that created the waste, through contamination of local food supplies, emissions into the ambient environment, and contaminated water." The actual population at risk of contact with substances released from uncontrolled hazardous waste sites into groundwater was unknown to the investigators.

The similarity in findings between the EPA national study and Najem et al.'s study of New Jersey cancer rates is interesting. Both investigations point to elevated cancer mortality rates in cancers of the gastrointestinal tract, an outcome that might represent human exposure to hazardous substances in drinking water supplies.

Northwestern Illinois: An investigation of bladder cancer rates in northwestern Illinois was conducted because the region was known to have areas of high mortality from this cancer (Mallin, 1990). Eligible cases were those first diagnosed with bladder cancer between 1978 and 1985. Age-adjusted, standardized incidence ratios were calculated for each county and 97 ZIP code areas within these counties. County results indicated no excess in bladder cancer. One ZIP code area had a significant excess in males (standardized incidence ratio = 1.5) and females (standardized incidence ratio = 1.9). This excess was found to be associated with one town in which one of four public drinking water wells had been closed because of contamination from VOCs. TCE concentrations in both wells exceeded safe drinking water standards; lesser concentrations of benzene, methylene chloride, tetrachloroethylene and other VOCs were also measured. The wells were within a half mile of a landfill site that had ceased operations in 1972. Investigators noted that none of the hazardous substances found in municipal well water was a known human bladder carcinogen.

Park County, Montana: The Montana Department of Health and Environmental Sciences and the ATSDR investigated a putative cluster of pancreatic cancer of men in Park County, Montana (Hutchinson, 1992). They did so because area residents were concerned about health effects from contamination emanating from a local railroad refueling and repair station, particularly an apparent increase in cancer mortality in the area. The observed number of pancreatic deaths from 1980 through 1989 among white men living in the area under investigation was compared with the age-adjusted number of deaths expected on the basis of U.S. age- and sex-specific pancreatic cancer morbidity rates (1980–1988). Eleven pancreatic cancer deaths were observed compared with 4.7 deaths expected. Smoking did not appear to account for this increase. The cluster of pancreatic cancer deaths appeared to be among persons who worked in the railroad industry, indicating that occupation may have been a risk factor. In addition, the addresses of family members seemed to cluster along a railroad right-of-way, suggesting the need to evaluate the role of environmental exposure as a contributing factor in the community's pattern of pancreatic cancer. As follow-up to this investigation, the ATSDR conducted a case-control study of pancreatic cancer in the 12 populated areas where railroad refueling facilities are located in Montana (Prausnitz and McGeehin, 1995). The study population included 479 case decedents and 881 controls. A review of death records and a spatial analysis of residential distance from the refueling stations showed no statistical relationship between residential proximity and pancreatic cancer mortality.

Upper Cape, Massachusetts: A case-control study of cancer rates was conducted in communities in the Upper Cape area of Massachusetts near the Massachusetts Military Reservation, a Department of Defense CERCLA site (Ozonoff et al., 1994). The military base was the site of gun and mortar discharges and had contaminated groundwater due to improper disposal of solvents, fuels, and organic chemicals. Incident cases of several cancers were identified from the state's cancer registry: breast (344 cases), colorectal (420), lung (226), bladder (79), kidney (42), pancreatic (434), brain (42), liver (6), and leukemia (44). The cases were matched to controls from among other living or deceased Upper Cape residents and Medicare beneficiaries.

Findings showed that breast, lung, and pancreatic cancer risks were increased for persons who had ever lived within 3 km of the base runways and were exposed to aircraft-related hazardous substances. The risks increased with exposure duration and residential proximity to the base. For persons who had lived for 20 or more years near the artillery ranges, the OR for lung cancer was 2.5 and for breast cancer 2.2. Investigators found only modest evidence to associate cancer risk with groundwater contaminants. This study's strength is its case-control design, but it is limited by a lack of individual exposure data.

Montreal, Québec, Canada: The incidence of cancer was studied among persons living near the Miron Quarry municipal solid waste landfill in Montreal, Québec (Goldberg et al., 1995b). The landfill, the third largest in North America, is located in a densely populated metropolitan area. It is known to produce biogas containing methane, carbon dioxide, VOCs, and sulfur compounds. Potential exposure to ambient air pollutants from the landfill was defined by geographic proximity to the site. Findings indicated elevated risks for men of cancers of the stomach (RR = 1.3; 95% CI 1.0 to 1.2), liver and intrahepatic bile ducts (RR = 1.3; 95% CI 0.9 to 1.8); and trachea, bronchus, and lung (RR = 1.1; 95% CI 1.0 to 1.2). For women, rates of stomach cancer (RR = 1.2; 95% CI 1.0 to 1.5) and cervix uteri cancer were elevated (RR = 0.9; 95% CI 0.9 to 1.0). The investigators concluded, "Because of the unavailability of detailed environmental data to define exposure areas and the absence of some important potentially confounding variables, we cannot conclude whether the excess risks for cancer observed here represent true associations with exposure to biogas."

Woburn, Massachusetts: The Massachusetts Department of Public Health conducted an expanded study of 21 cases of childhood leukemia (MDPH, 1996). Each leukemia case was matched with two cases of leukemia drawn from the Woburn area. Controls were matched for age, sex, and race. Using municipal water quality records, exposure values representing the relative exposure to water from two contaminated wells were generated for cases and controls. Investigators reported, "A relationship between exposure and leukemia was identified for exposure which occurred during the time the mother was pregnant with the study cases (OR = 8.33; 95% CI 0.73 to 94.67). A weaker association was identified for maternal exposure beginning two years before conception and ending at conception of the leukemia case. [...] a significant trend across exposure categories was identified for the period during pregnancy (p < 0.05) suggesting a dose response relationship for mothers who drank contaminated water during pregnancy." This study's findings support those of Lagakos et al. (1986) that less strongly associated consumption of contaminated water with excess leukemia cases in Woburn children. However, the number of leukemia cases is small and although the exposure indices are reasonable, the actual extent of exposure to TCE and other solvents will never be known.

United Kingdom: Some communities have expressed concern that not only uncontrolled hazardous waste sites but also incineration of waste materials can cause releases of hazardous substances into the environment. These concerns are most often voiced when dioxins/furans are thought to have been released from incinerators. Investigators in the United Kingdom analyzed the incidence of cancers of the larynx and lung in persons living near 10 incinerators used to burn waste solvents (Elliott et al., 1992). The precipitating event for the study was reports of a cluster of cases of cancer of the larynx in persons living near one particular incinerator. Cancer registration data for the years 1974–1987 and lag periods of 5 and 10 years were used between start-up of the incinerators and cancer incidence. Ratios of number of cancer cases observed to number of cases expected were calculated for persons living within 3 km and 3–10 km of each incinerator site and then aggregated over all sites. Expected numbers of cases of larynx and lung cancer cases were calculated using national rates, regionally adjusted. The investigators found none of the observed/expected ratios differed significantly from unity for either cancer or either lag period (i.e., 5 or 10 years). Moreover, no evidence was found of decreasing cancer risk with distance from the incinerators. The major limitation of this study is its inability to lag cancer beyond 10 years; any cancer induction associated with the solvent waste incinerators would more likely occur after 20 or more years.

In a follow-up study, Elliott et al. (1996) examined cancer incidence of over 14 million people living near the 72 municipal solid waste incinerators in the United Kingdom. The study area was selected to be within 7.5 km of the incinerators. Numbers of observed cancer cases drawn from the national cancer registry, covering the period 1974–1987, were compared with expected numbers calculated from regionally-adjusted national rates after stratification by a deprivation score based on 1981 census data. The deprivation score consists of statistics on unemployment, overcrowding, and social class of head of household. The study was conducted in two stages. In the first, a sample of 20 incinerators was selected, stratified by population size. The second stage included analyzing cancer incidence associated with the remaining 52 incinerators.

According to Elliott et al. (1996), risk of all cancers combined decreased ($p < 0.05$) as distance from incinerators increased. Findings from the second phase of the study showed cancer excess ranged from 37% for liver cancer to 5% for colorectal cancer in areas 0 to 1 km from incinerators. However, the investigators attributed these excess rates in part to residual confounding near the incinerators. The confounders included misdiagnosis of tumors and deprivation associated with residential proximity to incinerators. The investigators acknowledge the lack of exposure data in their analysis, and they note the need to further review a potential association between liver cancer risk and residential proximity to incinerators.

U.S. Hazardous Waste Sites: Laboratory data from ten cross-sectional studies that ATSDR conducted since 1991 were combined into one database and evaluated (Sarasua et al., 1997; Vogt et al., 1997). More than 50,000 immune test results from approximately 6,000 individuals constituted the database. Lymphocyte phenotypes were determined on a total of 5,868 individuals in these 10 studies, 3,812 from target areas (i.e., near specific CERCLA sites) and 2,056 from comparison areas at least 5 miles from the target areas. Samples from 11 individuals showed phenotypic characteristics similar to those seen in early B–cell chronic lymphocytic leukemia (B–CLL). Of this number, eight individuals came from the target populations and three from the comparison population, yielding a prevalence ratio (PR = 1.8, 95% CI 0.5 to 6.9) associating the B–CLL-like phenotype with residence in a target area. Of the 10 sites, four were contaminated predominantly with lead and other metals not known to be associated with induction of leukemia. The remaining six sites all had VOC contamination and other hazardous substances. The prevalence ratio for these six sites was 2.7 (95% CI 0.6 to 12.9). The investigators conclude, "These findings suggest that the presence of B–CLL-like LPT [lymphocyte phenotypes] reflects increased risk for B–cell lymphoproliferative disorders" (Vogt et al., 1997). Lack of more adequate information on individuals' exposure to VOCs limits this interpretation, but the findings raise compelling questions about the association between VOCs and the induction of leukemia.

Summary of Cancer Studies

Table 5.11 gives a summary of the 13 cancer studies described in this section and in the Love Canal section at the beginning of the chapter. A review of the table suggests these observations:

- Increased risks of bladder cancer and gastrointestinal tract cancers were reported in U.S. counties (Griffith et al., 1989), New Jersey counties (Najem et al., 1983), and New Jersey municipalities (Najem et al., 1985) having hazardous waste sites.
- Increased risk of childhood leukemias was reported in a population exposed to TCE and other VOCs in municipal water supplied in Woburn, Massachusetts (Lagakos et al., 1986; MDPH, 1996).
- Increased risk of bladder cancer was found in populations living near specific hazardous waste sites in a town in Illinois known to have TCE and other VOCs in municipal wells (Mallin, 1990).

Table 5.11. Key Cancer Studies Associated with Hazardous Waste Sites or Incinerators.

Study Location (Reference)	Study Design (Period of Observation)	Number and Type of Subjects	Contaminants Exposure Measures	Major Health End Points	Outcome Reported
Love Canal, NY (Janerich et al., 1981)	Retrospective follow-up (census tract); 1955–1977	700 census tract residents Referent: NY state population	Benzene, lindane, 200 substances Surrogate: proximity to dump site	Liver cancer, lymphomas, and leukemias	No increase in cancer incidence
New Jersey Counties (Najem et al., 1983)	Mortality rate analysis; 1968–1977	All 21 counties in New Jersey Referent: U.S. cancer mortality rates	Hazardous waste Presence of toxic disposal facilities as one of several risk factors in regression analysis	Age-adjusted gastrointestinal (GI) cancer mortality rates	Elevated cancer mortality rates for cancers of esophagus, stomach, colon, and rectum in 20 of 21 counties
Clinton County, PA (Budnick et al., 1984)	Mortality rate analysis; 1950–1979	Clinton County and three adjacent counties Referent: 1. State of Penn. 2. U.S. rates	β-naphthylamine, benzidine, benzene Surrogate: residence in the area	Bladder cancer mortality	Increased county-wide bladder cancer mortality rate in male population after 1970; not thought related to environmental contamination
Municipalities in New Jersey (Najem et al., 1985)	Cancer mortality rate analysis; 1968–1977	194 municipalities with populations of 10,000 or more	Chemical wastes Surrogate: density of chemical toxic waste disposal sites (CTWDS)	Age-adjusted cancer mortality rates for 13 anatomical cancer sites	Excess gastrointestinal cancer mortality rates in cluster areas; correlates with density of CTWDS
Woburn, MA (Cutler et al., 1986)	Case-control; 1969–1979	All leukemia deaths of Woburn citizens from 1969 through 1978	TCE and other VOCs Surrogate: proximity to two contaminated wells and a pond	Leukemia deaths in children 15 years and younger	Increased incidence distributed uniformly over the 11-year period; 12 cases were observed, 5.2 would be expected
Woburn, MA (Lagakos et al., 1986)	Case-control 1964–1983	20 childhood leukemia cases of children who lived in Woburn; referent: 164 Woburn children	TCE and other VOCs Exposure metrics: cumulative exposure from birth and binary exposure status	Childhood leukemias	Association with estimated exposure to contaminated municipal water

Location (Reference)	Study design; period	Population / cases	Exposure	Outcome	Results
U.S. Counties (Griffith et al., 1989)	Mortality rate analysis	593 NPL sites in 339 counties Referent: 2,726 non-HWS counties in the United States	Hazardous waste Counties with hazardous waste sites (HWS) and with analytic evidence of contaminated ground drinking water	Cancer mortality rates for 13 cancers	Excess deaths in HWS counties for cancers of bladder, stomach, large intestine, and rectum
Winnebago County, IL (Mallin, 1990)	Incidence cases of bladder cancer in northwestern Illinois; 1978–1985.	725 incident cases in 6 northwestern counties; 21 cases in the cluster community (12.1 expected)	TCE, other VOCs Surrogate: 1 town known to have contaminated municipal wells	Age-adjusted, standardized incidence ratios for bladder cancer	Significant bladder cancer excess in a town with VOCs in a municipal well
United Kingdom (Elliott et al., 1992)	Ecologic	390 cases of larynx cancer with a 5-year lag; 70 cases with a 10-year lag; cases of lung cancer are not given	Incinerator emissions from 10 waste solvent incinerators Surrogate: distance of cancer cases from individual incinerators	Cancers of larynx and lung	No relationship between cancer rates and distance from incinerators
Upper Cape, MA (Ozonoff et al., 1994)	Case-control; 1988–1991	Incident cases of cancer: breast (344), colorectal (420), lung (226), bladder (79), kidney (42), pancreas (43), brain (42), liver (6), leukemia (44); Matched controls from Upper Cape residents and Medicare beneficiaries	Organic chemicals, solvents, and fuels in groundwater; airborne emissions from guns and mortars, including 2,4-dinitrotoluene	Incident cancers	Breast, lung, and pancreatic cancer risks were increased for persons who had ever lived within 3 km of gun and mortar positions.
Montreal, Québec, Canada (Goldberg et al., 1995b)	Incident cases of cancer in the Québec Tumour Registry; 1981–1988	Males: 9,334 cancer cases; 18,405 reference cases Females: 9,345 cancer cases; 19,134 reference cases	Biogas (methane, CO_2, VOCs, sulfur compounds) Surrogate: 3 areas near the landfill; reference area	17 anatomical sites of cancer in men; 20 cases of cancer in women	Elevated risks in men for cancers of the stomach, liver, and lung. For women, rates of stomach and cervical cancers were elevated
Woburn, MA (MDPH, 1996)	Matched case-control; 1969–1986	21 diagnosed cases of childhood leukemia Referent: 42 controls from Woburn	TCE and other VOCs Exposure values for each leukemia case and controls were generated from water quality data and water distribution records	Childhood leukemia	Strong relationship between putative exposure and risk of childhood leukemia

Table 5.11. Key Cancer Studies Associated with Hazardous Waste Sites or Incinerators (continued).

Study Location (Reference)	Study Design (Period of Observation)	Number and Type of Subjects	Contaminants Exposure Measures	Major Health End Points	Outcome Reported
Park County, MT (Hutchinson, 1992; ATSDR 1995c)	Mortality rate analysis (1992); case-control (1995c); 1980–1989	Mortality rate analysis: 11 cases; case-control: 479 cases, 881 controls	VOCs, diesel fuel, TCE, DCE, PCE Surrogate: residential proximity to railroad refueling stations	Pancreatic cancer	Mortality rate analysis: elevated rate of cancer; case-control: no relationship between cases and proximity to refueling stations
United Kingdom (Elliott et al., 1996)	Ecologic	Cancer incidence data from the national cancer registration system, covering the years 1974–1987	Incinerator emissions from 72 municipal waste incinerators Surrogate: distance of cancer cases from individual incinerators	Cancers thought possibly associated with incinerator emissions	Statistically significant ($p<0.05$) decrease in cancer risk with increased distance from incinerators; increased risk of liver cancer, but was confounded by case misclassification
U.S. Hazardous Waste Sites (Sarasua et al., 1997; Vogt et al., 1997)	Laboratory analysis of data from immune tests	5,868 individuals in 10 cross-sectional studies; 3,812 from target areas, 2,056 from comparison areas	VOCs at 6 sites; metals at 4 sites	B-cell chronic lymphocytic leukemia	Increased risk for B-cell lymphoproliferative disorders

- Increased risk of stomach and other cancers was reported in a population exposed to ambient air contaminants that included VOCs from a solid waste landfill in Montreal, Québec (Goldberg et al., 1995b).
- An approximately three-fold increased prevalence was found of lymphocyte phenotypes with characteristics similar to those seen in early B–cell chronic lymphocytic leukemia. Individuals at risk lived near a hazardous waste site with VOC contamination.
- None of the cancer studies summarized in this chapter have had data on individuals' levels of exposure to specific hazardous substances, thus limiting the ability to generalize findings.

Immune Function Disorders

The immune system identifies and provides a host defense against foreign substances. Following exposure to environmental toxicants, the immune system can show important changes. It reacts differently to toxic substances than do other organ systems in the body; the immune response can be either augmented or depressed as a result of exposure. Hypersensitivity, autoimmune conditions, or immunodeficiency can be the consequence of such exposure (Chou and Metcalf, 1993).

Although studies have shown that hazardous substances can cause changes in the relative ratios of the various cells of the immune system, these changes have only rarely been associated with clinical manifestations or diseases. Effects on immune functions following exposure to chemical toxins have been reported, but few reliable studies on humans exist, according to Chou and Metcalf (1993).

Most human studies of the effects of substances on immune function involve occupational exposures or reactions to medication. However, very little is known about how the immune system is affected by exposure to environmental toxicants at the low levels often released from hazardous waste sites. Changes resulting from low-level exposures to environmental toxicants might give investigators important clues in understanding the body's response to these exposures and the early recognition of adverse health outcomes.

Times Beach, Missouri: In 1971, dioxin-contaminated sludge wastes were obtained from a hexachlorophene production facility in Verona, Missouri, and were mixed with waste oil (Hoffman et al., 1986). Several areas in eastern Missouri were contaminated with dioxins when the oil mixture was applied to roads and other surfaces to control dust. Times Beach, Missouri, was a small community in which the contaminated oil was applied for dust control. Two studies of Times Beach residents were conducted to investigate the potential effects of dioxin exposure on the immune system. In 1984 medical examinations were conducted on 30 persons thought likely to have low exposure to dioxin and on 64 "high risk" individuals (Knutsen, 1984). The examinations included individuals' reactions to various substances (e.g., tetanus, diphtheria) known to provoke a delayed-type hypersensitivity reaction (DHR). No statistically significant alterations of cellular immunity were found. A follow-up study was conducted 12 years later on Times Beach residents; 82 were presumed exposed to dioxin and 40 were controls (Webb et al., 1987). No significant alterations in DHR or numbers of lymphocytes were found, although total numbers of platelets were significantly increased.

Gray Summit, Missouri: One of the areas most heavily contaminated by dioxin in eastern Missouri was the Quail Run mobile home park in Gray Summit, Missouri, which was found to have soil contamination levels of dioxins up to 2,200 parts per billion (ppb). Investigators performed comprehensive examinations of 154 exposed and 155 unexposed persons in the mobile home park area (Hoffman et al., 1986). Results of liver function tests suggested possible subclinical effects. The exposed group had an increased frequency of anergy, leading the investigators to comment, "These findings suggest that long-term exposure to 1,2,7,8-tetrachlorodibenzo-p-dioxin is associated with depressed cell-mediated immunity, although the effects have not resulted in an excess of clinical illness in the exposed group." No body burden levels of dioxin were measured. However, a follow-up study conducted three years later of 28 persons from the Quail Run cohort and 15 controls did not confirm that the cohort had suppressed DHR responses (Evans et al., 1988). The investigators speculated that the difference in outcomes between the two studies of Quail Run residents was due to procedural differences or recovery of immune function in the three years between the studies.

Woburn, Massachusetts: Health studies of an elevated leukemia rate and TCE contamination of two municipal wells in Woburn, Massachusetts were described in the previous section. Other studies of the Woburn leukemia cluster included an assessment of immune function (Byers et al., 1988). Medical and laboratory tests were conducted on 25 surviving family members of patients having leukemia, with particular emphasis on the immune system. White blood cell counts and lymphocyte levels were determined using conventional clinical laboratory histochemical staining techniques. Total lymphocytes and lymphocyte subpopulations in the exposed population were analyzed on three occasions. The samples were obtained from 23 nonleukemic adult subjects five years after closure of the contaminated wells; tests were repeated one month and again 17 months later. Comparison subjects were 30 healthy laboratory workers living in Boston. Findings showed damage to the immunologic system as manifested by altered ratios of T lymphocyte subpopulations, increased incidence of auto-antibodies, increased infections, and recurrent rashes. The small sample size limits the study's ability to show whether the symptoms of immunologic damage are statistically significant, but the use of well-established lymphocyte tests is a positive feature of the study's design.

Tucson, Arizona: In Tucson, Arizona, the prevalence of symptoms of systemic lupus erythematosus (SLE) and of fluorescent antinuclear antibodies (FANA) was investigated in a community population exposed to TCE and other contaminants in well water contaminated with leachate from an NPL hazardous waste site (Kilburn and Warshaw, 1992). SLE is a multisystem inflammatory disorder characterized by the production of a wide spectrum of specific autoantibodies and by activation of B–lymphocytes. According to the investigators, the resident population was exposed to TCE in concentrations from six to more than 500 ppb for up to 25 years. In 1981 the city closed several contaminated wells, and additional wells were closed in 1984 (Rao and McGeehin, 1996). Kilburn and Warshaw state, "The client group included several patients with systemic lupus erythematosus (SLE) who had been exposed to TCE and other chemicals that contaminated the culinary water of Southwest Tucson." To explore the prevalence of SLE in the TCE exposed group, the investigators administered the American Rheumatism Association (ARA) SLE questionnaire, and measured fluorescent antinuclear antibodies. The exposed group consisted of 146 subjects and another 196 subjects tested one year later. The reference group was 158 residents of southwest Phoenix, matched with the Tucson group, and without TCE exposure.

Findings showed the frequencies of all 10 symptoms in exposed subjects were elevated above the average of the comparison group; differences for five of the symptoms were statistically significant: arthritis for less than three months, Raynauds' Phenomenon, skin lesions, pleurisy signs or symptoms, and seizure or convulsion. The frequency of FANA titers greater than 1:80 was approximately 2.3 times greater in women but equally as frequent in men as in laboratory controls. Kilburn and Warshaw conclude, "Long-term, low-dose exposure to TCE and other chemicals in contaminated well water significantly increased symptoms of lupus erythematosus as perceived by the ARA scores and the increased FANA titers." Given the gravity of systemic lupus erythematosus, this is an important study, only diminished somewhat by reference to the exposed group as a "client group." This suggests possible bias in subject selection and on their self-reporting of symptoms. Moreover, the investigators do not indicate that they were blind to study participants' exposure histories.

Gainesville, Georgia: The prevalence and incidence of SLE in an African-American community in Gainesville, Georgia, was investigated by Kardestuncer and Frumkin (1997). The community was located near various industrial plants and, moreover, had been built over a municipal landfill. No environmental contamination or industrial emissions data were available to the investigators. Community residents expressed concern that exposure to hazardous substances was occurring, resulting in excess rates of cancer, renal failure, and lupus. A retrospective cohort study was conducted to obtain incidence and a cross-sectional study to obtain prevalence of SLE. Investigators focused initially on an alleged excess rate of lupus. A community survey was administered to community residents. The questionnaire included questions about lupus and other health problems. A case of lupus was defined by the investigators as a person who had received a physician's diagnosis of lupus. When available, medical records of lupus patients were reviewed to confirm the patient's self-report of lupus.

Findings were based on 81 households and information from 246 current and past individuals. No comparison community was included in the investigation. The prevalence of lupus was three cases/300 persons, which is a six-fold increase ($p = 0.015$) greater than the highest prevalence reported for an African-American population. The incidence of lupus was three cases/4,709 person-years (which is 63.7 cases/100,000 person-years). This is nine-fold increase ($p = 0.0001$) in incidence. Although the study suffers from lack of environmental contamination and personal exposure data, the findings raise troubling questions about whether SLE is associated with exposure to hazardous substances like those commonly found in hazardous waste.

Kidney Dysfunction

The function of the kidney is to eliminate wastes and to maintain the composition and volume of extracellular fluid for the body (Taylor and Faroon, 1993). To accomplish these functions, the kidney receives 25% of the circulatory blood flow via the arterial circulation system. As a consequence, the kidney is sensitive to a wide variety of bloodborne toxic agents. The pathologic changes to the kidney resulting from exposure to renal toxicants can be reversible, permanent, or lethal. Damage to the kidney and to renal functions depends on the severity of the toxic insult and the susceptibility of the individual exposed.

The kidney can be damaged by environmental toxicants through several mechanisms: direct cellular injury, obstruction of urine flow, and alteration of renal blood flow (Taylor and Faroon, 1993). In addition, exposure to environmental toxicants can result in reversible or permanent changes in renal function depending on the agent and the magnitude of the toxic insult.

Midwestern United States: The ATSDR's National Exposure Registry (NER) (Gist et al., 1994, 1995) contains data on noncancer chronic disease morbidity of relevance to the impact of hazardous waste sites on human health. The NER consists of subregistries of persons having documented exposure to substances to which the ATSDR has assigned priority and for which gaps exist in information about the substance's toxicity. The subregistry of persons exposed to TCE as a contaminant in private well water consists of data from 4,280 persons located in 13 sites in Indiana, Illinois, and Michigan (Burg et al., 1995). Levels of TCE varied from site to site as did levels and numbers of co-contaminants. However, the ATSDR considered TCE as the primary contaminant across all 13 sites. The duration of registrants' exposure to TCE varied from months to several years. It is important to note that exposure to TCE ceased for all registrants when they were provided an alternative water supply, in most cases, a municipal water system.

Using a structured questionnaire, TCE registrants are surveyed periodically to obtain their self-reported health status. Respondents' health status was compared with responses in the National Health Interview Survey (NHIS), which is a population-based survey of the American population conducted annually. Four demographic characteristics—sex, age, race, and education level—are considered in the NHIS and TCE subregistry comparisons. The samples were comparable, except for the race. The TCE subregistry has fewer persons from minority groups, and therefore, statistical analyses were based only on data from 3,914 white respondents. When compared with the NHIS population, the TCE subregistry population had a higher rate of reported adverse health outcomes. For adults this included stroke, liver disease, diabetes, anemia and other blood disorders, kidney disease, urinary tract disorders, and skin rashes. Children in the TCE subregistry population reported elevated rates of hearing impairment and speech impairment (Burg et al., 1995).

These findings for TCE do not prove a causal relationship between TCE exposure and adverse health effects; more definitive epidemiologic investigations will be required to strengthen causal inference. However, because many of the chronic health problems reported by the TCE registrants seem to have toxicologic plausibility (Burg et al., 1995), these morbidity data have serious implications concerning any association between long-term, low-level exposure to TCE and impacts on human health.

Three Lakes Municipal Utilities District, Texas: Using procedures similar to those in the TCE exposure subregistry, Burg and Gist (1998) describe health findings from the ATSDR benzene subregistry, comprised of 1,143 persons served by a municipal water supply contaminated with benzene at concentrations up to 66 ppb. Statistically significant higher rates of anemia, blood disorders, ulcers, gall bladder trouble, stroke, urinary tract disorders, skin rashes, diabetes, kidney disease, and respiratory allergies were found during one or more data collection periods. Notable was the increased reporting of blood disorders because of benzene's known hematopoietic toxicity.

New York State: In another study, the ATSDR researchers evaluated the relationship between residential proximity to hazardous waste sites and the risk of end-stage renal disease (ESRD) (Hall et al., 1996a). ESRD is characterized by total renal failure, making hemodialysis or organ transplantation necessary. Although ESRD is known to be caused by diabetes mellitus and polycystic kidney disease, many cases have no known origin and may be related to environmental factors. Because several metals (e.g., lead and cadmium) and hydrocarbon solvents are known to cause kidney disease, a case-control study was conducted of ESRD rates for persons living near hazardous waste sites. The study was conducted of residents in 20 counties of New York State, excluding New York

City, and covered ESRD cases reported to the federal government's Health Care Financing Administration (HCFA) in 1992 and 1993. Diagnosed cases of ESRD are reported to the HCFA because almost all ESRD cases are covered by Medicare, which is administered by the HCFA.

After excluding ESRD cases caused by diabetes and other known causes, 216 case-control pairs were analyzed by administering a standard questionnaire and determining each case's proximity to known hazardous waste sites. An elevated odds ratio (OR = 1.40; 95% CI 0.92 to 2.11, p = 0.13) was found for persons with ESRD who ever lived within one-mile radii of hazardous waste sites. This elevated odds ratio was not statistically significant. The small number of ESRD cases reduced the statistical power of the study.

Liver Dysfunction

The liver is a unique structure whose functions are essential for life. Because it receives venous blood from the intestine and has diverse enzymatic content, the liver has a wider variety of physiologic functions than does any other organ in the human body. The liver's main functions are production and excretion of bile, storage of nutrients, metabolism, and detoxification and clearance of unneeded compounds and toxicants. Because of its many functions, a single liver cell (hepatocyte) can be likened to a factory (it makes many chemical compounds), a warehouse (it stores glycogen, iron, and some vitamins), a waste disposal facility (it excretes bile pigments, urea, and various detoxication products), and a power plant (its catabolism produces considerable heat) (Wilson and Straight, 1993).

The liver has many enzymes capable of metabolizing toxicants, but sometimes this metabolism can produce metabolites of greater toxicity than the parent toxicant. Because the liver is the first organ to be affected by foreign chemicals that are absorbed into the bloodstream from the gut, it is often exposed to relatively high concentrations of toxic substances before they reach other organs. When liver damage or dysfunction does occur, all organ systems of the body can be affected. Generally, liver damage from toxicants is dose related and usually manifests after a relatively short latency period. Liver injury depends on the dose received by the liver, the section of the liver affected, and how long the individual is exposed to the toxicant. Hepatotoxins can cause liver cell necrosis (cell death) or biochemical changes; these changes can be manifested by impaired formation of bile, disturbances of lipid metabolism, altered enzyme activity, or by interfering with hepatic blood flow.

Hardeman County, Tennessee: The first investigation of the effects on liver function of hazardous substances released from a waste site was probably conducted by Meyer (1983). A hazardous waste dump in Hardeman County, Tennessee, was estimated to contain between 16 and 25 million gallons of solid and liquid waste. In 1978, more than a dozen chlorinated organic compounds were found in wells in the area of the dump. High levels of carbon tetrachloride (median level = 1500 μg/L) and chloroform (median = 140 μg/L) were measured. Blood samples assayed for liver enzymes were obtained in November 1978 from 36 persons who had consumed contaminated well water; 49 persons were similarly tested in January 1979; 31 persons were common to both tests. The January 1979 testing included physical examinations and medical histories. Comparison groups for both test periods were selected from 57 persons with measured uncontaminated water supplies. The November 1978 testing showed an increased elevation of exposed persons' alkaline phosphatase and SGOT liver enzymes and significantly lower levels of albumin and total bilirubin. No differences in liver function were found between exposed persons and controls for the January 1979 test. Meyer does not account for the different findings between November and January, although one can presume that contact with contaminated water in wells had ceased. Meyer concluded, "During this study comprehensive evaluation of that population (i.e., the exposed group) revealed multiple symptoms, evidence of hepatomegaly and elevated liver function tests apparently caused by ingestion of water contaminated by numerous organic chemicals, many of which are known to be hepatotoxins." It was noted that the symptoms and abnormal liver function tests in all but one case returned to normal during the period of November 1978 through March 1979.[6] This reversal of liver function effects after cessation of exposure to well water contaminants adds support to the findings of a subclinical transitory liver insult.

Wayne Township, New Jersey: In New Jersey, a study was conducted of the effects of a thorium waste disposal site in Wayne Township, Passaic County, on local residents' health (Najem and Voyce, 1990). Soil samples taken on the waste site and nearby showed thorium, radium, and uranium contamination. Ambient air radiation levels were within background values. No biologic sampling of residents occurred. A structured health status survey was conducted face-to-face by one interviewer. The interviewer was blind to the exposure status of each person interviewed. Interviewed were 171 persons living near the waste site and 191 persons living nine to ten blocks from the site. Findings showed a higher prevalence of liver diseases (RR = 2.3, 95% CI 0.35 to 12.0) among the presumed exposed group than in the comparison group. These findings must be tempered by the lack of individual radiation exposure measurements and by relatively small group sizes.

Midwestern United States: From the ATSDR National Exposure Registry, the TCE subregistry population compared with the reference database described previously, had a higher rate of reported adverse health outcomes. This included liver disease in adults (Burg et al., 1995).

Lung and Respiratory Disease

The major function of the respiratory system is to transfer oxygen from the atmosphere to the cells and, in turn, transfer carbon dioxide from the cells back to the atmosphere. The anatomy of the respiratory system can be divided into three functional regions: the nose and nasopharynx, the trachea and bronchi, and the respiratory bronchioles and alveoli. Taken together, these structures are called the respiratory tract. As the lung exchanges gases between the atmosphere and the blood, it may be exposed to toxicants in the environment.

Direct action of inhaled toxicants (gases, solids, or liquid aerosols) on the lung can be acutely and chronically important to health (Metcalf and Williams, 1993). Environmental toxicants can cause damage to the lung and respiratory system, including irritation, necrosis and edema, fibrosis, and allergic responses. The most common effect of chemical substances on the respiratory tract is irritation. Toxicity can result from the irritant and/or corrosive properties of a chemical, which can be reversible, or can, in more extreme cases, cause permanent disability or death. Various materials can cause cellular

---

[6] The March 1979 date is not explained by Meyer (1983).

damage (i.e., necrosis) (Metcalf and Williams, 1993). Breakdown of the epithelium leads to edema by allowing fluid to leak into the lungs. Fibrosis is a nonspecific inflammatory response of the alveolar wall. Physiologic changes include a decrease in lung volume, compliance, and diffusing capacity. Allergic diseases depend on the body's immunologic response to environmental stimuli. In the case of asthma, the reaction is mediated by an immunoglobulin, immunoglobulin E (IgE). Asthma is characterized by reversible airway obstruction. Persons with asthma have hyper-responsive airways, which means that nonallergic stimuli (e.g., viral infection, irritants, cold air, exercise) also can produce asthma attacks.

The levels of hazardous substances in air are not often sampled in areas near hazardous waste sites; groundwater, soil, and surface water being the pathways most often sampled. Lack of air toxics data has therefore limited the conduct of epidemiologic studies on lung and respiratory disease in waste site locales.

Montchanin, France: French investigators searched for health effects potentially attributable to an industrial waste landfill in Montchanin, France. Subjects were 694 local residents living in three parts of town. Measurement of the consumption of prescription drugs was used as a surrogate for morbidity data (Zmirou et al., 1994). In France, drug prescription and purchase data are available for more than 99% of the population. A slight trend was noted in the consumption of drugs prescribed for ear, nose, and throat, and pulmonary ailments. The trend was not statistically significant.

In a companion work, following closure of the waste site, the same set of investigators conducted a case-control study two years later to assess the short-term health impact of residual emissions of VOCs from the landfill (Deloraine et al., 1995). Physicians in the community administered the same questionnaire to their patients that sought health information on patients' health status. The respondents were separated into two groups: 432 patients with conditions putatively associated with emissions from the waste site; and 384 control cases. Individuals' exposure to VOCs was assessed using a combination of modeling of airborne toxicants together with personal activity patterns. Findings showed nonspecific irritative respiratory symptoms occurred more frequently among the exposed subjects after adjusting for such confounders as age, smoking, and alcohol use. Greater exposure was associated with a higher frequency of respiratory symptoms. The investigators do not relate their health findings to specific VOCs. Moreover, selection bias inherent in local physicians' selection of patients limits this study.

Southern New Jersey: A lack of elevated illness rates was noted in a southern New Jersey rural community adjacent to an NPL site containing VOCs in industrial waste (Najem et al., 1994). Structured questionnaires were administered to 676 persons living within one mile of the waste site and presumed to be exposed to documented levels of contaminants released into groundwater and soil. A comparison group consisted of 778 persons from the same community, but at a distance at least three miles from the NPL site. No significant differences were found between the two groups in the respondents' rates of cancer, respiratory disease, liver disease, and skin diseases. A significantly higher prevalence was found of respiratory diseases (RR = 1.9; 95% CI 1.1 to 3.3), and seizures (RR = 4.3; 95% CI 1.1 to 13.9). After adjusting for various confounding factors like cigarette use and water consumption patterns, the differences in rates of respiratory illness and seizures disappeared. However, the investigators conclude, "...the contaminated private water supply and soil may have had an association with the prevalence of these diseases."

Neurotoxic Disorders

The human nervous system performs three general functions. It receives and stores information, it responds to external stimuli, and it communicates and coordinates the activities of all other organ systems (Chou and Hutchinson, 1993). Because the nervous system plays an essential role in regulating metabolism and other vital body functions, damage to the nervous system can have profound consequences. Moreover, abnormal functioning in other organ systems can modify neurologic functioning, with similar consequences.

Neurotoxicity is the capacity of chemical, biologic, or physical agents to cause adverse functional or structural changes in the nervous system. Massive neurologic injury can result in incoordination, dementia, convulsions, paralysis, and coma. Even slight damage to the nervous system can produce memory loss, impaired ability to reason, motor dysfunction, and communication problems. Children and developing fetuses are of special concern because damage to the developing nervous system can cause great functional consequences and be long-lasting. Fetal Alcohol Syndrome is an example of neurotoxic damage caused to a fetus by a toxicant (ethanol) and which has profound consequences in lowered birth weight or in more extreme cases, of mental retardation.

Woburn, Massachusetts: Feldman et al. (1988) investigated the effect on cranial nerves of consuming drinking water in Woburn, Massachusetts, contaminated with TCE. Electrophysiologic measurement of the blink reflex can quantify the nerve conduction velocity in the reflex arc involving the trigeminal and facial nerves. Using electrophysiologic methods, the investigators measured the blink reflex in a population (n = 21) that had alleged chronic exposure to TCE at levels 30–80 times the EPA's Maximum Contaminant Level of 5 ppb. The study population was self-referred to the investigators for neurologic evaluation. Feldman et al. note that chronic exposure to TCE had ended six years before electrophysiologic testing; therefore, their clinical protocol was designed to assess possible residual neurotoxic effects of chronic environmental exposure to TCE. A group of 27 unexposed individuals was selected for comparison with the study group. A highly significant lengthening (p < 0.0001) in the means of conduction latency components of the blink reflex was found compared with the unexposed group. Although sample sizes are small and the subjects were self-selected, the blink reflex is a highly stable measurement of a known anatomical pathway. The lack of actual exposure data diminishes the study's findings, but the methodology is commendable.

Tucson, Arizona: Tests of neurophysiologic (NPH) and neuropsychologic (NPS) functions were administered in Tucson, Arizona, to 170 persons exposed to well water contaminated with TCE released into groundwater from a hazardous waste site (Kilburn and Warshaw, 1993a). The 170 subjects were selected from a larger cohort of 1,600 "clients." The TCE levels ranged from 6 to 500 ppb over a period up to 25 years. The exposed subjects were compared with 68 subjects from the exposed subjects' geographic area, presumably not exposed to TCE. Exposed subjects were statistically significantly impaired on NPH tests. The impairments included performance on tests of body sway and choice reaction time. A significant difference in the blink reflex of exposed subjects was found compared with the referent group. This study was evidently conducted as part of litigation, given that "clients" constituted the pool of exposed subjects. The potential bias of testing persons with vested interests in the study's outcome diminished the study's findings.

Muscle Shoals, Colbert County, Alabama: Neurobehavioral testing and postural sway were used to compare persons living near an industrial plant in Muscle Shoals, Colbert County, Alabama, with a reference group (Kilburn and Warshaw, 1993b; Kilburn et al., 1994). The exposed group consisted of 117 persons randomly selected from 500 residents whose well water was contaminated with PCBs and VOCs from an aluminum die-casting plant. The study group had lived within 2.4 km of the plant for at least four years between 1956 and 1981. The exposed group were all plaintiffs in litigation against the die-casting company. A comparison group numbered 46 persons. Analyses conducted in the 1980s of the well water found PCBs up to 3,500 ppm and TCE levels that exceeded 5 ppb.

The neurobehavioral battery of tests included visual reaction time, blink reflex latency, color discrimination, intelligence, memory, and motor function. A force platform and a device that measured head movements were used to measure postural sway. Investigators were blind to the exposure group being tested. Findings showed neurophysiologic impairment, and cognitive and psychomotor dysfunction and affective disorders, particularly depression and excessive frequency of symptoms, in the exposed group when compared with the reference group. The exposed group's balance was significantly inferior to that of the reference group, both by measuring head tracking ($p < 0.03$) and force exerted on the force platform ($p < 0.01$). Both measurement methods showed impaired balance associated with years of exposure to PCBs and TCE.

This study conveys important information about the possible effects on the nervous system of PCBs and VOCs in well water. The neurobehavioral tests, including postural sway measurement, used by the investigators are commendable because of the wide range of tests of neuropsychologic and neurophysiologic functions. The study's findings are tempered by the nonrandom sample of exposed subjects who were all litigants in a lawsuit, and by the lack of individual exposure data.

Baton Rouge, Louisiana: Signs of neurotoxicity and neurobehavioral impairment were examined by Kilburn and Warshaw (1995) in a sample of residents near a former plant in Baton Rouge that reprocessed motor oil from 1966 to 1983 and later became an NPL site. Neurophysiologic and psychologic tests and a symptom questionnaire were administered to 131 subjects presumed exposed to releases from the plant and accompanying uncontrolled waste site. The subjects were matched to 66 presumed unexposed referents who lived 35 km from the plant and waste site. The referents were randomly selected and matched to the exposed group for age, sex, and education. The investigation was conducted as part of litigation brought by residents living near the oil reprocessing plant. Tests were administered to all study participants by persons blind to the participants' exposure status.

The exposed group was significantly different from the referent group on tests of body balance, reaction time, cognitive function, eye-hand coordination, and affect. The differences were all in the direction of impaired or reduced performance by exposed subjects. The exposed group's symptom frequencies and affect scores were indicative of psychologic depression.

The investigators were unable to demonstrate by regression analysis any association of exposed subjects' neurobehavioral tests with various surrogate indices of exposure; for example, duration of presumed exposure and residents' distance from the oil processing site. This study demonstrated a difference between two groups (presumed exposed and referent) on tests of neurobehavioral function, but the lack of actual exposure data and environmental contamination data limit the study's findings.

Midwestern United States: From the ATSDR National Exposure Registry, the TCE subregistry population had a higher rate of reported adverse health outcomes compared with the reference database as described previously. For adults this included stroke, diabetes, anemia and other blood disorders, and skin rashes. Children in the TCE subregistry population reported elevated rates of hearing impairment and speech impairment (Burg et al., 1995).

Woburn, Massachusetts; Alpha, Ohio; Twin Cities Army Depot, Minnesota: Neuropsychologic evaluations and neurologic examinations were administered to persons exposed to TCE in drinking water supplies in three cities (White et al., 1997). The persons evaluated were referred to the investigators as part of litigation. Examined were 28 persons from Woburn, 12 persons from Alpha, and 14 from an area near the Twin Cities Army Depot. TCE concentrations in drinking water supplies ranged from 63 to 2440 ppb across the three groups. Exposure durations ranged from 0.25 to 25 years. Results from neuropsychologic evaluations and neurologic examinations were compared with the investigators' laboratory reference data.

Findings showed a high rate of cognitive deficits associated with solvent exposure. Rates of mild–moderate encephalopathy were high in persons tested within the three groups: 86% in the Woburn group, 75% in the Alpha group, and 100% in the Minnesota group. Of special note, individuals who were exposed during childhood (under age 18 years) showed greater ranges of cognitive deficits than did individuals who were exposed to TCE as adults.

Summary of Noncancer, Nonreproductive Health Studies

A summary of 14 key health studies that investigated health outcomes other than reproductive outcomes (Table 5.10) and cancer (Table 5.11), is given in Table 5.12. The number and depth of the studies for each of the five priority health conditions is less than that for reproductive outcomes and cancer. This disparity results mostly from investigators' responsiveness to communities' health concerns; cancer and birth defects are feared by individuals and therefore represent concerns expressed to government agencies.

Symptom-and-Illness Prevalence Studies

Some investigators have examined the kind and frequency of symptoms of illnesses reported by persons living near hazardous waste sites. Symptom-and-illness prevalence studies are typically conducted by administering a structured questionnaire to a group of persons exposed to an environmental hazard and comparing their responses with those of a referent group or to a national illness prevalence database.

Londonderry Township, Pennsylvania: In Pennsylvania, community concerns led the state health department to conduct a health investigation of residents exposed to contaminants in well water (Logue et al., 1985). TCE, other VOCs, PCBs, and pesticides were found in wells serving a residential area near a former military base in Londonderry Township. TCE was measured in 14 homes in concentrations ranging from 2.6 to 140 ppb. The affected households had a potential exposure duration of 23 years. A health interview survey was administered by questionnaire, which asked about 14 diseases or medical conditions and 15 health symptoms. Residents were asked to refrain from using the contaminated water for the three weeks preceding the health survey. The exposed

Table 5.12. Key Studies of Selected Health Outcomes Other than Reproductive Dysfunction, Cancer, and Symptom-and-Illness Prevalence.

Study Location (Reference)	Study Design (Period of Observation)	Number and Type of Subjects	Contaminants Exposure Measures	Major Health Endpoints	Outcome Reported
Love Canal, NY (Picciano, 1980)	Cross-sectional; 1980	36 self-selected Love Canal residents; referent: historical laboratory data	Benzene, lindane, 200 substances Surrogate: residence in contaminated area	Cytogenetic assays	Abnormal "acentric" chromosome fragments
Hardeman County, TN (Meyer, 1983)	Cross-sectional; 1978, 1979	1978: 36 persons with access to contaminated water, 1979: 49 persons with such access; 57 persons with no known access to contaminated water	Carbon tetrachloride and chloroform Measured levels in well water	Laboratory measures of liver function	Transient abnormalities in liver function
Love Canal, NY (Bender and Preston, 1983; Heath et al., 1984)	Cross-sectional; Dec. 1981–Feb. 1982	46 current or past residents of Love Canal; 25 referents from Niagara Falls, NY	Benzene, lindane, 200 substances Households with prior measured environmental contamination	SCE and chromosome aberrations	No differences of note
Times Beach, MO (Knutsen, 1984; Webb et al., 1987)	Cross-sectional; unstated (Knutsen)	Knutsen: 64 at-risk persons, 30 at lower risk Webb et al.: 82 exposed, 40 controls	Dioxins Environmental contamination and residence used as surrogates	Measures of delayed-type hypersensitivity reactions; cell counts	No significant alterations of cellular immunity (Knutsen, 1984; Webb et al., 1987); total numbers of platelets were significantly increased (Webb et al., 1987)
Gray Summit, MO (Hoffman et al., 1986; Evans et al., 1988)	Cross-sectional; 1971–1983	Hoffman et al.: 154 exposed residents of a trailer park; 155 unexposed; Evans et al.: 28 persons from the Quail Run cohort and 15 controls	2,3,7,8-TCDD Surrogate: residence in area with soil contamination	Tests of liver and immune function; neurologic tests	Association found with cell-mediated immunity; follow-up study did not confirm

Table 5.12. Key Studies of Selected Health Outcomes Other than Reproductive Dysfunction, Cancer, and Symptom-and-Illness Prevalence (continued).

Study Location (Reference)	Study Design (Period of Observation)	Number and Type of Subjects	Contaminants Exposure Measures	Major Health Endpoints	Outcome Reported
Woburn, MA (Byers et al., 1988)	Cross-sectional; 1969–1979	23 nonleukemic adult subjects 5 years after closure of the contaminated wells; referents: 30 healthy laboratory workers in Boston.	TCE and other VOCs Surrogate: residence in Woburn area served by contaminated water supply	Medical and laboratory tests with particular emphasis on the immune system	Altered ratios of T lymphocyte subpopulations, increased incidence of auto-antibodies, increased infections, and recurrent rashes
Woburn, MA (Feldman et al., 1988)	Clinical case-control; 1981?	21 persons exposed 6 years earlier to TCE in drinking water; referents: 27 individuals with no known exposure to TCE	TCE and other VOCs Surrogate: residence in area served by contaminated water	Measure of blink reflex	Significant difference in blink reflex function
Passaic County, NJ (Najem and Voyce, 1990)	Cross-sectional; unstated	171 persons living near the waste site; referent: 191 persons living distant from the site.	Thorium, radium, and uranium soil contamination Surrogate: residential proximity	A structured health status interview	Higher prevalence of liver diseases and birth defects
Tucson, AZ (Kilburn and Warshaw, 1992)	Cross-sectional; 1987, 1988	146 subjects and another 196 subjects tested 1 year later; referents: 158 residents of southwest Phoenix	TCE Surrogate: residence serviced by contaminated water supply	SLE questionnaire, and fluorescent antinuclear antibodies	Increased symptoms of lupus erythematosus and increased FANA titers
Tucson, AZ (Kilburn and Warshaw, 1993a)	Cross-sectional; 1986, 1987, 1989	170 persons exposed to contaminated water; 68 comparison subjects not exposed to TCE	TCE Surrogate: residence serviced by contaminated water supply	Neurobehavioral tests and questionnaire	Impaired performance on tests of body sway and choice reaction time; significant difference in the blink reflex

Muscle Shoals, Colbert County, AL (Kilburn and Warshaw, 1993b; Kilburn et al., 1994)	Cross-sectional; 1956–1981	117 persons living near a metal casting plant; 46 referents living distant from the plant	PCBs and VOCs in well water	Neurobehavioral tests and postural sway using a force platform and a head tracking device	Neurophysiologic impairment and cognitive and psychomotor dysfunction, affective disorders; exposed group's balance was significantly inferior to the reference group
Montchanin, France (Zmirou et al., 1994; Deloraine et al., 1995)	Retrospective cohort; Jan. 1987–June 1988	694 local residents living in three parts of town; referent: national database on consumption of prescribed drugs.	Unspecified Surrogate: residence in area near landfill	Consumption of prescriptive drugs was used as a surrogate for morbidity data.	A nonsignificant trend in the consumption of drugs prescribed for ear, nose, and throat and pulmonary ailments
Montchanin, France (Deloraine et al., 1995)	Case-control; 1990	432 patients; 384 controls	Fugitive emissions of VOCs from closed landfill; modeling of airborne emissions	Health symptoms questionnaire	Nonspecific irritative respiratory symptoms occurred more frequently among exposed subjects; greater exposure was associated with a higher frequency of respiratory symptoms
Baton Rouge, LA (Kilburn and Warshaw, 1995)	Matched cohort; 1966–1983	131 persons presumed exposed to site releases; 66 unexposed living distant from the site	Toluene, MEK, lead Surrogate: residential proximity to waste site compared with reference community	Neurophysiologic and psychologic tests and a symptom questionnaire	Poorer performance on tests of body balance, reaction time, cognitive function, eye-hand coordination, and affect
Midwestern, U.S. (Burg et al., 1995)	Exposure subregistry; 1989–1990	3,914 persons living near 13 NPL sites; referent: national database	TCE and other VOCs Documented presence of contaminants in water supplies	Registrants' self-reported health status via face-to-face interview and telephone questionnaire	Higher rates in adults of stroke, diabetes, anemia, kidney disease, urinary tract disorders, skin rashes; in children, hearing and speech impairments

Table 5.12. Key Studies of Selected Health Outcomes Other than Reproductive Dysfunction, Cancer, and Symptom-and-Illness Prevalence (continued).

Study Location (Reference)	Study Design (Period of Observation)	Number and Type of Subjects	Contaminants Exposure Measures	Major Health Endpoints	Outcome Reported
New York State Counties (Hall et al., 1996a)	Case-control; 1992–1993	216 case-control pairs	Hazardous waste Surrogate: residential proximity to waste site	End-stage renal disease	Elevated OR, not statistically significant
Woburn, MA; Alpha, OH; Twin Cities Depot, MN (White et al., 1997)	Clinical evaluation; unstated	Woburn, MA: 28 persons examined; Alpha, OH: 12; MN: 14; laboratory reference data	TCE in drinking water Levels measured in water supplies	Neuropsychologic evaluation and neurologic examination	High rate of cognitive deficits and mild–moderate encephalopathy
Gainesville, GA (Kardestuncer and Frumkin, 1997)	Retrospective cohort and cross-sectional; 1980–1995	81 households; 246 African-American individuals; national reference data	Residential proximity to industrial operations; community built on municipal landfill	Systemic lupus erythematosus	Incidence of lupus was 9-fold national rate for African-Americans; prevalence was 3-fold greater than national rate
Three Lakes Municipal Utilities District, TX (Burg and Gist, 1998)	Exposure subregistry; 1991–1994	1,143 persons served by a municipal water supply; referent: national database	Benzene Documented presence of contaminants in water supplies	Registrants' self-reported health status via face-to-face interview and telephone questionnaire	Higher rates of anemia, blood disorders, ulcers, gall bladder trouble, stroke, urinary tract disorders, skin rashes, diabetes, kidney disease, and respiratory allergies

group consisted of members of the 14 households (61 persons) with documented use of contaminated water. The comparison group consisted of 23 households (66 persons) from a part of the Township without well water contamination. Findings indicated the exposed group reported significantly higher prevalence rates for three symptoms (eye irritation, diarrhea, sleepiness); no significant differences were noted for other symptoms and diseases. The investigators identify exposure to TCE at levels greater than 28 ppb as possibly associated with the elevated rate of the three symptoms. This study is notable because of the documented exposure of the study group.

Lock Haven, Pennsylvania: The Drake Chemical NPL site was the subject of several health studies and a worker surveillance program (Chapter 6). The company operated from 1962 until 1981. Chemicals used, manufactured, and/or stored on the Drake site include the human carcinogens beta-naphthylamine, benzidine, and benzene (Logue and Fox, 1986). An environmental sampling program found a wide variety of organic and inorganic compounds both on-site and off-site. The Pennsylvania Department of Health conducted a cross-sectional health survey of 179 residents living within two to four blocks of the Drake Chemical NPL site (Logue and Fox, 1986). A comparison group consisted of 151 individuals from outlying areas of the city or township. A 10-year length-of-residency requirement was imposed on both groups. The prevalence of 14 self-reported diseases or medical conditions and of 15 self-reported health symptoms was compared between the study and referent groups. Findings showed the exposed group reported significantly higher rates of sleepiness and skin problems. The groups were comparable on all other symptoms and diseases, including cancer, liver disease, birth defects, reproductive problems. Because no exposure data were available, and the extent of reporting bias was unknown, the higher reporting rates for sleepiness and skin problems could have been spurious findings.

Hamilton, Ontario, Canada: One of the earliest and still most impressive studies of residents' health problems associated with a hazardous waste site was conducted by Hertzman et al. (1987). This study is noteworthy for its attention to sound epidemiologic design, efforts to reduce confounding factors in the conduct and interpretation of the study, and use of methods to assess the strength of their findings.

Hertzman et al. conducted morbidity surveys of workers and residents from the Upper Ottawa Street Landfill site in Hamilton, Ontario. The health of workers at the landfill was investigated first as a means of designing an investigation of residents living near the landfill (Chapter 6 for workers' survey findings). The landfill had operated from the early 1950s until 1980 when it was closed. Health survey interviews were conducted of groups of landfill area households based on proximity to the landfill. A comparison community, "less than" five miles from the landfill, was selected based on matching for socioeconomic characteristics. No data were available to document community exposure patterns during the period of peak disposal activity at the landfill. Interviews were conducted in-home, and therefore no blinding of interviewers was possible. The landfill target group consisted of 614 households, and 636 households were in the comparison group. The number of persons interviewed in each household is not stated. Findings showed positive associations between residence adjacent to the landfill and five of six organ-system groups: respiratory (bronchitis, cough, shortness of breath), skin (dry, itchy skin), narcotic symptoms (fatigue, lethargy, dizziness, drowsiness), mood symptoms (anxiety or depression, irritability), and eye irritation. Each association was strongly positive, with relative risk greater than 1.50 and $p < 0.001$.

This study is unique within the literature on hazardous waste site studies, because the investigators used five criteria for examining causation (strength of association, gradient, temporality, specificity, and consistency). These criteria led Hertzman et al. to conclude, "Evidence is presented that supports the hypothesis that vapors, fumes, or particulate matter emanating from the landfill site, as well as direct skin exposure, may have led to the health problems found in excess."

Lowell, Massachusetts: A symptom prevalence study was conducted of a neighborhood exposed to airborne hazardous substances released from a defunct hazardous waste facility in Lowell, Massachusetts (Ozonoff et al., 1987). Only levels of benzene, toluene, TCE, and tetrachloroethylene were detected in ambient air in the area near the waste site. The study populations were defined by their proximity to the waste site. The exposed group included households in blocks within 400 meters of the site. The comparison area was a ring with innermost boundary at least 800 meters from the site and outermost boundary at most 1,200 meters from the site. Structured interviews were conducted of 1,049 individuals living in the exposed area and 948 persons in the comparison area. Findings showed residents living near the waste site reported more respiratory complaints (wheezing, shortness of breath, chest discomfort, persistent colds, coughs), irregular heartbeat, and more constitutional complaints (fatigue, bowel dysfunction) than did the comparison group. The health complaints were most frequent in persons who lived closest to the site. Recall bias may have confounded the findings to some extent, according to the investigators. No biologic measurements of exposure were conducted.

Riverside County, California: A health survey of 2,039 persons in 606 households near the Stringfellow NPL site in Riverside County, California, was conducted to assess whether rates of adverse health outcomes were elevated (Baker et al., 1988). Investigators administered a structured questionnaire, examined medical records of reported cancers and pregnancy outcomes, and reviewed birth and death certificates. Three study groups were selected based on proximity to the NPL site: 302 households (987 persons) in Pyrite, an area nearest the Stringfellow site; 101 households (328 persons) in West Glen Avon, an area south of the site where exposure potential was less likely; and 203 households (724 persons) in Rubidoux, a city three miles from the waste site, which served as the comparison community. All three groups were found to be similar with respect to mortality, cancer incidence, and pregnancy outcomes. The mean number of symptoms reported per person was 4.1 in Pyrite, 4.1 in West Glen Avon, and 2.9 in Rubidoux (Pyrite vs. Rubidoux, $p < 0.0001$; West Glen Avon vs. Rubidoux, $p = 0.0003$). Every odds ratio was greater than 1.0 for both Pyrite and West Glen Avon compared with Rubidoux; eight symptoms had odds ratios greater than 1.5: blurred vision, pain in ears, daily cough for more than one month, nausea, frequent diarrhea, unsteady gait, and frequent urination. Baker et al. concluded, "The apparent broad-based elevation in reported diseases and symptoms may reflect increased perception or recall of conditions by respondents living near the site." In other words, the investigators attribute the elevated reporting rates in the Stringfellow group to recall bias.

Fullerton Hills, California: As part of research in fulfillment of a graduate school degree, a study was conducted of the McColl NPL site's community in California (Lipscomb, 1989). The study was designed to assess the presence of symptoms and odor complaints among the current neighborhood population. Results from a study conducted seven years earlier by the California Department of Health Services were used to select three zones in the neighborhood on the basis of odor complaints. The three zones were high rate of odor complaints (57 persons), lower odor complaint zone (66 persons), and a comparison zone (70 persons) with few odor complaints. The questionnaire survey contained questions on environmental worry and perception of odors. One-year symptom period prevalence rates were calculated by exposure area. Findings showed elevated symptom reporting was strongly associated with several measures of environmental worry. When the association between symptom and "exposure" was stratified by level of worry, the association was strongest and remained significant only in the high-worry stratum. The investigator concluded from her study, "...these findings suggest that the concern and worry associated with living in a community with a waste disposal site, rather than the exposure to chemicals from the site, is responsible for the excess symptoms reporting observed in this study."

Galena, Kansas: Neuberger et al. (1990) used a structured questionnaire to survey health problems in persons living in Galena, Kansas, the site of a mining waste NPL site. Lead and cadmium contamination in soil were of particular concern to the investigators. Age- and sex-specific illness rates in Caucasians (n = 206) were compared with similar rates in two referent towns (n = 298 total), and rates in all three groups were compared with illness rates established for the U.S. population. A series of univariate and multivariate analyses led the investigators to conclude that an excess prevalence of chronic kidney disease (females age ≥ 65), heart disease (females age ≥ 45), skin cancer (males age 45–64), and anemia (females age 45–64) existed in Galena. Multivariate analyses revealed statistically significant associations of stroke, chronic kidney disease, hypertension, heart disease, skin cancer, and anemia with variables related to Galena exposure (length of residence in Galena, presence of mine tailings, private well water use). No biologic measurements of Galena residents' exposure to metals was conducted, and the group sizes were relatively small.

California: Investigators examined illness symptom rates in communities around five hazardous waste sites in California (Neutra et al., 1991). No evidence was found that rates of cancer and birth defects were elevated. However, the prevalence of subjective health symptoms of illness was elevated in reference to comparison communities. The investigators concluded that toxicologic response and mass psychogenic illness were not valid explanations of the elevated symptom rates. Excess reporting of symptoms was observed associated with those persons who complained of odors from the sites and who expressed worry about the effects of environmental chemicals. Using this observation, the investigators hypothesized that higher symptom rates in persons living near hazardous waste sites were because of stress-mediated mechanisms or behavioral sensitization.

Caldwell County, North Carolina: The ATSDR has conducted several symptom- and disease-prevalence studies of populations living near hazardous waste sites. In July 1991, the ATSDR conducted a retrospective cross-sectional symptom-and-disease prevalence study of 713 residents living within 1.5 miles of a hazardous waste incinerator in Caldwell County, North Carolina, and 588 residents of a comparison community (Straight et al., 1993). Residents had complained of emissions from the incinerator and alleged health problems they associated with uncontrolled releases from the incinerator. A structured questionnaire was administered to ascertain the prevalence of self-reported symptoms and diseases. No personal or area exposure data were available. Findings showed residents of the target area were about nine times more likely to report symptoms of recurrent wheezing or cough following a respiratory insult, adjusted for cigarette smoking, asthma, and environmental concern (OR = 9.0; 95% CI 2.8 to 29.1). Significantly increased ORs for chest pain, poor coordination, dizziness, and irritation were found among persons living less than 0.9 mile from the incinerator when compared with those living more distant.

Houston, Texas: The respiratory health effects of substances putatively released from two NPL sites were a primary concern of an ATSDR study. A cross-sectional health study of the Southbend subdivision, near Houston, Texas, and adjacent to the Brio Refining Co., Inc. and Dixie Oil Processors, Inc. NPL sites (Miller et al., 1995; Miller and McGeehin, 1997) was conducted in response to environment data that indicated low level emissions of VOCs were associated with the NPL sites. From 1989 through 1993, environmental sampling found low levels of benzene, toluene, xylene, 1,1,1-trichloroethane, trichlorethylene, naphthalene, and mesylene in ambient air. A standardized, in-person questionnaire was used to collect information from 414 Southbend residents and 360 persons in the comparison group, which consisted of individuals living in another part of Houston. The questionnaire inquired about individual health characteristics and characteristics of the household environment. Interviewers who administered the questionnaire were not blind to the exposure status of study and comparison subjects. Blood and urine specimens were collected for a battery of medical tests designed to assess the hepatobiliary (liver), renal (kidney), immune, and hematopoietic (blood) systems of the study participants.

Southbend participants had higher odds ratios that were statistically significant for 11 of 14 self-reported symptoms with onset after participants moved into their current homes than did comparison area participants. Prevalence of all 4 self-reported respiratory symptoms present at the time of the study and adjusted for smoking status were significantly higher for the Southbend participants. The symptoms were higher occurrence of cough, presence of phlegm or mucus, chest wheezing, and shortness of breath. Other health outcomes significantly elevated in Southbend participants included eczema or other skin problems, periods of anxiety, nervousness, or depression, and ulcer or other stomach disease. Some biomarker levels were significantly different between study and comparison groups, but their clinical significance was unclear (Miller et al., 1995).

This cross-sectional study has several important limitations, as acknowledged by its investigators. No actual exposure data to VOCs were available, and participation rates were low enough to limit any generalization to the Southbend community at large. Moreover, no validation of the self-reported symptoms was incorporated into the study. Nonetheless, the study included a large number of potential at-risk members of a community with putative exposure to hazardous substances, and the use of a standardized questionnaire adds credibility to the findings.

Tucson, Arizona: The ATSDR conducted a study of persons living near the Tucson International Airport NPL site in Tucson, Arizona (Rao and McGeehin, 1996). It was discovered in 1981 that TCE released from an industrial operation had contaminated several Tucson municipal wells at TCE levels above 5.0 μg/L, the state's action level. The contaminated wells were closed in 1981; additional wells were closed in 1984 and residents were placed on uncontaminated municipal water supply, although state and federal authorities believed private, possibly contaminated, wells were still in use. In 1994, at the request of a local community group, a cross-sectional symptom-and-disease prevalence health survey was conducted. The target group was located within the area of the NPL site; a comparison group was selected from a community five miles distant and was without known exposure to TCE in water supplies. Structured health survey interviews were conducted in randomly selected households; 350 persons were interviewed in the target area, and an identical number in the comparison area. Participation rates were 82 and 80%, respectively, which was considered adequate. After adjusting for several confounding variables, statistically significant odds ratios (ORs) for two symptoms: poor coordination (OR = 3.6) and skin rashes (OR = 3.3) and three diseases: skin problems (OR = 5.0), anxiety or nervousness (OR = 3.5) and cancer (OR = 3.4) were found in excess in target area residents. The lack of documented individual exposure data and medically unconfirmed disease reported by study participants were major limitations of this survey.

Sacramento, California: Operations at McClellan Air Force Base, Sacramento, California, resulted in extensive groundwater contamination from volatile organic compounds, metals, and other contaminants (Jones and McGeehin, 1996). Local residents expressed concern that this contamination had impacted their health. In the mid-1980s, 550 homes west of the base were switched to a municipal water supply. In 1994, the ATSDR conducted a cross-sectional health survey of 453 residents living west and within 1.5 miles of the base and 202 residents from a comparison area. A questionnaire was administered to determine the prevalence of self-reported health problems in the study groups. For the target area, each of the 14 symptoms assessed had odds ratios greater than unity (ORs > 1.0), indicating higher symptom occurrence in the target area. Of the 14 symptoms, four (numbness, loss of consciousness, severe headaches, tingling in the hands or toes) had ORs ≥ 2.0, indicating that they were at least twice as likely to be reported in the target area as in the comparison area. Recall or reporting bias did not appear to account for the elevated reporting of health symptoms. No environmental exposure measurement data are cited in the report.

Grand Island, Nebraska: A symptom-and-disease prevalence study of area residents was conducted for another military base, the Cornhusker Army Ammunition Plant, Grand Island, Nebraska (Hamar et al., 1996). Operations of the base had led to groundwater contamination with chemical compounds used in developing explosives, hexahydro-1,3,5-trinitro-1,3,5-triazine (RDX) and 2,4,6-trinitrotoluene (TNT). Levels of contaminants in groundwater were of public health concern. A study was conducted to examine whether the contamination was associated with residents' health problems. In all, 600 randomly selected Grand Island residents (300 target, 300 comparison) were administered a standardized questionnaire. A subset of both groups was administered a neurobehavioral battery of tests to ascertain neurologic effects. Findings showed seven self-reported health outcomes were about twice as likely (ORs of 1.7 to 2.2) to be reported by target area participants as by comparison area respondents: neurologic problems, weakness or paralysis of limbs not due to stroke, urinary-tract disease, numbness or sensation of pins in fingers or toes, trouble sleeping, trouble remembering, and irritated eyes. No significant differences in reproductive history were observed. Neurobehavioral test results did not differ significantly between the target and comparison groups.

Possible Reporting Bias

Several investigators have examined factors they postulate might account for elevated reporting rates of symptoms and illnesses expressed by persons living near hazardous waste sites. These factors include hypochondriasis of persons living near hazardous waste sites, stress and autonomic mechanisms, environmental worry, and odor perception.

Kingston, Queensland, Australia: The findings from a community health survey of persons living near a hazardous waste site in Kingston, Queensland, Australia, showed no significant differences between two groups of persons living near the waste site and a comparison community (Dunne et al., 1990). One group (n = 147 persons) lived within 300 meters of the site; a second group (n = 110 persons) lived within 300–1,000 meters of the site. A comparison community (n = 105 persons) was located 16 km from the waste site. No significant associations were found between residents' reports of disease rates and proximity to the site. Although persons in both groups in the waste site area reported higher rates of symptoms of general poor health, high levels of stress and anxiety, and a higher incidence of miscarriages, these were independent of proximity to the waste site and duration of residence when referenced to the comparison group. Moreover, symptom prevalence and perceived decline in recent health status correlated with stress and anxiety measures.

Calcasieu Parish, Louisiana: Some investigators have postulated that persons who reside near hazardous waste sites may report increased numbers of health complaints because of reporting bias. Roht et al. (1985) evaluated two potential indices of reporting bias, hypochondriasis and respondents' opinions about the environmental effects of waste sites. A household health survey was conducted of residents living near two operating hazardous waste disposal sites in Calcasieu Parish, Louisiana, in 1981–1982. Because of news media coverage of the waste sites, and because environmental contamination caused by releases from the waste sites was judged to be without health effect, investigators speculated that reporting bias would be more prevalent in the hazardous waste communities. An unexposed community was included for comparison. Sample sizes for the two waste site communities were 279 and 110 and for the comparison community, 115. The health questionnaire included questions from a hypochondriasis index.

Findings showed both waste site communities reported more symptoms than did the comparison community; one community reported significantly more eye, respiratory, and upper and lower GI symptoms than did the comparison community. No differences in overall self-reported chronic illnesses were found among the three communities. Elevated hypochondriasis scores were associated with symptom reports from both waste site communities. The investigators caution against use of health surveys that do not control for reporting bias.

California Communities: The possible interaction between symptom prevalence and odor–worry interaction (Shusterman et al., 1991) and stress (Neutra et al., 1991) was examined by investigators from the California Department of Health Services who examined health survey data for 2,000 persons who had participated in studies of populations near three hazardous waste sites in California. The three sites contained acid petroleum waste (McColl), municipal and sewage waste, paint and petroleum sludge (Operating Industries), and synthetic rubber residues and DDT (Del Amo/Montrose). All three health surveys solicited information about the frequency or severity of several of common physical symptoms. Additional questions were asked about the frequency of perception of environmental odors and respondents' degrees of environmental worry. Shusterman et al. determined by retrospective analysis that 15 symptoms were ascertained in a comparable manner across all three studies. The investigators chose two symptoms on an *a priori* basis that they believed were potentially related to autonomic or stress-induced mechanisms (headache and nausea) and two as potential irritative processes (eye soreness or irritation, throat soreness or irritation). The investigators stratified prevalence data for the above symptoms by self-reported frequency of odor perception and by self-reported degree of environmental worry. Findings showed, "...significant positive relationships between the prevalence of each of the index symptoms (headache, nausea, eye and throat irritation) and both frequency of odor perception and degree of worry."

Shusterman et al. opine that odors serve as a sensory cue for the manifestations of stress-related illness (or heightened awareness of underlying symptoms) among persons concerned about the quality of their environment. Neutra et al. (1991) noted that excess symptoms in persons living near waste sites was found primarily in those persons who complain of odors or who are worried about environmental chemicals. Neutra et al. suggest, "...the possibility that autonomic, stress-mediated mechanisms or behavioral sensitization are active in the genesis of these symptoms."

Summary of Symptom-and-Illness Prevalence Studies

A summary of 18 symptom-and-illness prevalence studies is given in Table 5.13. Symptom-and-illness prevalence studies have several limitations and shortcomings. Recall bias can be a problem; that is, persons' memories of health problems and related events can be faulty. Self-reporting of adverse health events has been thought by some to be subject to some degree of bias for reason of economic gain or attention. On the positive side, this kind of study can be useful in designing hypothesis-generating investigations; for example, investigating a pattern of respiratory health complaints through pulmonary-function testing.

Stress and Hazardous Substances

Does residence near a hazardous waste site produce psychologic effects in the residents? In particular, is stress manifested in residents who are impacted by existing uncontrolled hazardous waste sites, proposed sites, or emergency releases of hazardous substances? These questions are important because of the well-known link between chronic stress and adverse health impacts. A search of the scientific literature provides some clues, but no coherent science, to answering these questions. Research on the stressful effects of environmental hazards is emerging. This research is often described in terms of **environmental stress**. In this context, Baum et al. (1985) state, "Stress is a process by which environmental events threaten, harm, or challenge an organism's

Measurement of Stress (Baum et al., 1985)
- Self-report measures
  - Interviews
  - Questionnaires
- Behavioral measures
  - Observations
  - Performance tests
- Biochemical measures
  - Psychoendocrine assays
- Psychophysiologic measures
  - Dermal, cardiovascular, respiratory

Table 5.13. Key Symptom-and-Illness Prevalence Studies of Communities Near Hazardous Waste Sites.

Study Location (Reference)	Study Populations	Contaminants Exposure Measures	Health Findings
Love Canal, NY (Paigen et al., 1985)	Cross-sectional: 523 Love Canal and 440 comparison children from the Niagara Falls area.	Benzene, lindane, other substances Surrogate: (1) distance of residence from the canal and (2) the proximity of homes to "wet" areas in the Love Canal area	Increased prevalence of seizures, learning problems, hyperactivity, eye irritation, skin rashes, abdominal pain, and incontinence. Each health problem showed evidence of a dose-response for both surrogate measures of exposure
Londonderry Twp, PA (Logue et al., 1985)	Cross-sectional: 14 households (61 persons) exposed to contaminated well water; 23 comparison households (65 persons)	TCE and others Contaminated well water; measured levels of TCE	Significantly higher prevalence rates for 3 symptoms (eye irritation, diarrhea, sleepiness); possibly related to TCE > 28 ppb or other factors
Calcasieu Parish, LA (Roht et al., 1985)	279 persons and 110 persons in two waste site communities; 115 persons in a comparison community	Unstated No exposure data were collected; prior environmental testing showed no contaminants of health concern in ambient air or water	Both waste site communities reported more symptoms than the comparison community; one community reported significantly more eye, respiratory, and upper- and lower-GI symptoms than did the comparison community
Drake Chemical NPL site, Lock Haven, PA (Logue and Fox, 1986)	Cross-sectional: 179 persons living 2–4 blocks from the NPL site; 151 comparison persons living distant from the site	Off-site: Arsenic, pentachlorophenol, dichlorobenzene, trichlorophenylacetic acid No exposure data; proximity to the site was the exposure surrogate	Increased prevalence of sleepiness and skin problems
Hamilton, Ontario, Canada (Hertzman et al., 1987)	Retrospective follow-up: 614 households near the landfill; 636 comparison households; number of persons interviewed not stated; 1976–1980	Industrial waste No data available on community exposure patterns; residential proximity to site was surrogate exposure measure	A positive association between residence adjacent to the landfill and five of six organ-system groups: respiratory, skin, narcotic and mood symptoms and eye irritation
Lowell, MA (Ozonoff et al., 1987)	Cross-sectional: 1,049 persons in exposed area; 948 persons in comparison area; 1983	Airborne VOCs Surrogate-exposed group: persons living within 400 meters of site; comparison area: residence 800 to 1,200 meters from the site	Residents living near the waste site reported more respiratory complaints, irregular heartbeat and heart trouble, constitutional complaints, and anemia and other blood disorders

Location (citation)	Study design/population	Exposure measures	Results
Riverside County, CA (Baker et al., 1988)	Cross-sectional: 987 persons living near the site; 328 persons living somewhat farther away from the site; 724 persons in a comparison city; 1983	"Toxic waste" Surrogate measures: residential proximity to the site; current household water use; length of residence near the waste site	8 symptoms had odds ratios greater than 1.5: blurred vision, pain in ears, daily cough for more than one month, nausea, diarrhea, frequent diarrhea, unsteady gait, and frequent urination
Galena, KS (Neuberger et al., 1990)	Retrospective follow-up: 206 whites near a heavy metals mining waste site; 298 whites from two comparison towns; 1980–1985	Lead, cadmium, chromium Proximity to waste site; Pb and Cd in soil; Cr in drinking water	Excess prevalence of morbidity for chronic kidney disease (females age 65), heart disease (females age 45), skin cancer (males age 45–64), and anemia (females age 45–64)
Fullerton Hills, CA (Lipscomb, 1989)	Three zones adjacent to the site formerly assessed to have high rate of odor complaints (57 persons), lower odor-complaint zone (66 persons), and a comparison zone (70 persons) with few odor complaints.	Proximity to waste site; no biologic measurements of exposure	Elevated symptom reporting was strongly associated with several measures of environmental worry. When the association between symptom and "exposure" was stratified by level of worry, the association was strongest and remained significant only in the high-worry stratum
Kingston, Queensland, Australia (Dunne et al., 1990)	147 persons living within 300 m of the site; 110 persons living within 300–1,000 m; 105 persons living 16 km from the site	Chemical wastes Surrogate: residential proximity to the waste site	No significant associations were found between residents' reports of disease rates and proximity to the site
Southern California (Shusterman et al., 1991; Neutra et al., 1991)	More than 2,000 adult respondents to health surveys of persons living near 3 hazardous waste sites	Sludges, synthetic rubber residues, DDT (Shusterman et al., 1991); Airborne aromatic and chlorinated solvents (Neutra et al., 1991) Residential proximity to waste sites	Significant positive relationships between the prevalence of each index symptom (headache, nausea, eye and throat irritation) and both the frequency of odor perception and degree of worry (Shusterman et al., 1991); autonomic, stress-mediated mechanisms or behavioral sensitization postulated to be active in the genesis of the symptoms (Neutra et al., 1991)

Table 5.13. Key Symptom-and-Illness Prevalence Studies of Communities Near Hazardous Waste Sites (continued).

Study Location (Reference)	Study Populations	Contaminants Exposure Measures	Health Findings
Caldwell County, NC (Straight et al., 1993)	713 residents living within 1.5 miles of a hazardous waste incinerator; 588 residents of a comparison community	Unstated incinerator emissions Surrogate: residential proximity to the incinerator	Residents of the target area were about nine times more likely to report symptoms of recurrent wheezing or cough following a respiratory insult; significantly increased ORs for chest pain, poor coordination, dizziness, and irritative symptoms in persons living less than 0.9 mile from the incinerator
Southern NJ (Najem et al., 1994)	676 persons living within 1 mile of the site; 778 persons living 3 or more miles from the waste site	VOCs Surrogate: residential proximity to the waste site	No significant differences between the two groups were found in the respondents' rates of cancer, respiratory disease, liver disease, and skin diseases
Houston, TX (Miller et al., 1995)	414 Southbend residents and 360 persons in the comparison group located in another part of Houston	Benzene, toluene, xylene, 1,1,1-trichloroethane, trichloroethylene, naphthalene, and mesiylene in ambient air Surrogate: residence adjacent to waste site	Higher occurrence of cough; presence of phlegm or mucus; chest wheezing and shortness of breath; eczema or other skin problems; periods of anxiety, nervousness, or depression; and ulcer or other stomach diseases
Tucson, AZ (Rao and McGeehin, 1996)	350 persons in target area; 350 in comparison area	TCE Proximity to waste site; TCE contamination in two municipal wells in prior use	Self-reported increased prevalence of poor coordination; skin rashes; and three diseases: skin problems, anxiety or nervousness and cancer, in target-area residents

McClellan Air Force Base, Sacramento, CA (Jones and McGeehin, 1996)	453 residents living west and within 1.5 miles of the base and 202 residents from a comparison area	VOCs, metals Surrogate: residential proximity to the military base	4 symptoms (numbness, loss of consciousness, severe headaches, tingling in the hands or toes) had ORs ≥ 2.0. Two illnesses (bowel disease or intestinal problems and ulcers or other stomach diseases) were significantly higher in the target area
Grand Island, NE (Hamar et al., 1996)	600 randomly selected Grand Island residents (300 target, 300 comparison)	Compounds used in developing explosives: hexahydro-1,3,5-trinitro-1,3,5-triazine (RDX) and 2,4,6-trinitrotoluene (TNT) Residence in contaminated area	7 health outcomes were about twice as likely in target area: neurologic problems, weakness or paralysis of limbs not due to stroke, urinary-tract disease, numbness or sensation of pins in fingers or toes, trouble sleeping, trouble remembering, and irritated eyes

existence or well-being, and by which the organism responds to this threat." The environmental events to be described in this section will be specific to uncontrolled releases of hazardous substances.

Measurement of stress in individuals and community populations consists of four basic approaches, as noted in the inset box. According to Baum et al. (1985), self-report measures involve direct questioning of persons about their feelings, beliefs, opinions, and reactions to environmental conditions. Behavioral measures can consist of observing persons who live in presumed stressful environmental conditions (i.e., coping strategies), or by administering performance measures to ascertain the effects of stress on some ability or skill during or after exposure to the stressor of concern. Biochemical measures of stress provide estimates of psychoendocrine response to stressors. Psychophysiologic measurements assess the response of peripheral nervous system response by select organ systems; for example, dermal, cardiovascular, and respiratory responses. These same psychophysiologic measures are used in the polygraph machine, better known as the "lie detector." A contemporary study of environmentally induced stress will use one or more of the basic approaches listed in the inset box.

Material that follows contains summaries of stress in communities impacted by technology disasters that released radiation into the environment, a chemical spill, a leaking waste site, and a proposed landfill that faced community opposition.

Three Mile Island, Pennsylvania: In March 1979, human error occurred during operation of the Three Mile Island (TMI) nuclear power plant in Pennsylvania. Conditions inside the plant's nuclear reactor were out of control for many hours. This led to doomsday forecasts of impending massive releases of radiation, which, however, did not occur. The reactor's condition was stabilized, and the plant's operations were shut down for several years before reopening.

The TMI event was, and remains, a signal story in the area of environmental policy. President Carter appointed a commission to investigate the episode and to advise on nuclear reactor policies. The community living near the plant was temporarily relocated, many engineering and human health investigations were performed, and the public's belief that nuclear power plants were unsafe was reinforced through news media reports.

The incident at TMI has been the most intensively researched technologic mishap in history insofar as psychologic effects are concerned (Hartsough and Savitsky, 1984). These studies began soon after the nuclear plant's reactor was stabilized. Some investigators interviewed a sample of community residents, using standardized psychologic scales of distress. One investigator sampled catecholamine levels in urine as an indicator of stress levels.

A review of psychologic-effects studies of TMI residents led reviewers to form several conclusions (Hartsough and Savitsky, 1984). Stress levels in the area neighboring to TMI increased sharply because of the nuclear incident. Differential stress levels existed; mothers of young children suffered the greatest levels of psychologic stress. Evidence for TMI-induced chronic stress was equivocal.

Livingston, Louisiana: In 1982 a major derailment of a freight train occurred in Livingston, Louisiana. Approximately 20 tank cars containing hazardous substances ruptured, which led to fires, explosions, and evacuation of the town. A major part of the town's residents were relocated for 14–17 days (Gill and Picou, 1991). No deaths, serious injuries, or substantial property damage occurred, even though the derailment was a major hazardous substances event. After the train derailment, a class action legal action was initiated. A district court ordered an assessment of the impact of the derailment on the Livingston community. The assessment included an evaluation, conducted 20 months after the derailment, of psychologic and sociologic impacts.

Interviews were conducted of 233 randomly selected persons who lived in 133 households in the Livingston area (Gill and Picou, 1991). The interview format included questions to indicate social psychologic stress: individuals' perceptions of future risks, relocation desires, and quality of life. A comparison community was selected that had similar demographics and was located by a railroad. Findings showed elevated stress and strongest concerns about health risks in the Livingston residents who lived closest to the derailment site and members of families who were separated at the time of the derailment. The health concerns most often expressed were worry about future cancer risks and concern that drinking water was contaminated. This study supports other investigations of stress in community populations that were impacted by technology disasters. However, the investigators do not indicate how they controlled for bias in a population involved with litigation that could potentially benefit them. Moreover, no environmental exposure data were used to classify persons at potential health risk.

Mid-Atlantic States: A study compared symptoms of chronic stress in persons impacted by a natural disaster with symptoms of persons living near a leaking hazardous waste site (Baum et al., 1992). Investigators compared 23 flood victims nine months after the flood occurred with 27 persons living near a hazardous waste site in the mid-Atlantic region of the United States after the EPA announced to the public that the waste site was among the most hazardous in the country.

The study population consisted of 77 persons: 23 living in homes that had been flooded, 27 living near the leaking landfill; and 27 persons in a comparison group whose homes had never been flooded and who lived distant from any known hazardous waste site. Commendably, the investigators used concurrent assessment of self-report, behavioral, and biochemical indices of stress. Findings showed that persons living near the hazardous waste site were more anxious, depressed, and alienated, less able to perform challenging tasks, and more aroused than were flood victims or the comparison group.

Fernald, Ohio: As in the mid-Atlantic region episode of the leaking uncontrolled hazardous waste site, residents of Fernald, Ohio, were evaluated for symptoms of stress after hearing about radiation released from an adjacent nuclear weapons plant (Green et al., 1994). The study group was 50 individuals who were litigants in a class action suit brought against the plant's operator. Part of the Fernald residents' antipathy was linked to a revelation that the federal government had for decades intentionally withheld information about environmental contamination caused by the weapons plant.

To the 50 litigants, interviewers administered survey instruments consisting of a psychiatric evaluation form, impact-of-event scale, symptom checklist, coping strategies inventory, and inventory of stressors associated with the Fernald facility. No comparison group was evaluated. Findings indicated that the Fernald group showed elevations in symptoms of anxiety, depression, belligerence, and somatic concerns. The outcome of this study fits with findings of psychologic distress reported in other studies of persons impacted by technology disasters and uncontrolled releases of hazardous substances. Unfortunately, the study design did not control for bias in the litigants who constituted the studied group.

Rainbow Valley, Arizona: An interesting complement to the Baum et al. (1992) study of stress levels in persons living near an existing uncontrolled hazardous waste site was conducted by Bachrach and Zautra (1985), who evaluated stress in an Arizona community targeted to receive a hazardous waste facility. The Arizona State Legislature's decision to select Rainbow Valley as the facility's location was actively opposed by many local residents.

A sample of 99 Rainbow Valley residents was surveyed to ascertain the magnitude of psychologic distress in the community. Respondents included 70 adults from randomly selected households and 29 residents who had participated in meetings or activities related to siting the facility in Rainbow Valley. No comparison group was used in the study. Trained interviewers administered questionnaires to assess respondents' coping mechanisms, demoralization, and perception about the proposed hazardous waste facility. Findings indicated strong feelings of demoralization and distrust of government. Highly demoralized residents were less likely to be involved in community activities. This study contributes to an appreciation of the impact that siting unwelcome waste facilities can have on community stress. Unfortunately, lack of a comparison group severely limits the study's findings.

The five studies summarized here of psychologic stress in community populations impacted by hazardous substance releases are illustrative of other similar investigations. In the aggregate, these five studies demonstrate that technology disasters and uncontrolled hazardous waste releases can produce short-term psychologic distress. Whether chronic stress is also a manifestation is equivocal.

CORRELATIVE ENVIRONMENTAL CONTAMINATION STUDIES

Some substances known to be released from hazardous waste sites have also been associated with adverse human health effects when released into environmental media as hazardous waste other than from sites. When specific substances and levels of environmental contamination are similar, human health implications from these other releases may predict possible impacts from hazardous waste sites.

For example, Cantor et al. (1987) found increased risk of bladder cancer with increasing intake levels of beverages made with tap water. The risk gradient with intake was restricted to persons with at least 40 years' exposure to chlorinated surface water and was not found among long-term users of nonchlorinated groundwater. Findings from another study of contaminants in public drinking water supplies showed an association between 1,2-dichloroethane content and cancers of the colon and rectum, and between nickel and cancers of the bladder and lung (Isacson et al., 1985). The investigators suggest that low levels of chemical contaminants were not causal factors but rather indicators of possible anthropogenic (i.e., related to human activities) contamination of other types.

In other studies, findings from a case-control study showed the frequency of spontaneous abortion, compared with controls, was associated with detectable public-drinking-water levels of mercury, high levels of arsenic, potassium, and silica, among other factors (Aschengrau et al., 1989). In another study, rates of leukemia and non-Hodgkin's lymphoma (NHL) in 75 towns in New Jersey showed incidence of all leukemias and NHL in women was associated with levels of TCE in drinking water (Cohn et al., 1994). The most commonly occurring contaminants were TCE, tetrachloroethylene, and 1,1,1-trichloroethane. This investigation was possible because New Jersey requires all public community water systems to monitor for 14 VOCs. Regression analysis showed an age-adjusted rate ratio (RR) for total leukemia among females of 1.43 with a 95% CI = 1.07 to 1.90 when incidence in towns with the highest strata of TCE exposure was compared with that in towns with no detectable TCE exposure. Non-Hodgkin lymphomas among women were associated with the highest TCE stratum (RR = 1.36; 95% CI = 1.08 to 1.70). The study was of ecologic design, so findings are subject to some degree of exposure misclassification because individuals' information on degree of personal exposure was imprecise.

A study of the possible relation between incidence of total leukemia and VOC contamination of New Jersey drinking water found an increased risk among females with increasing levels of nontrihalomethane VOC contamination (Fagliano et al., 1990). A population-based, case-control study was conducted of bladder cancer and drinking water disinfection methods in Colorado municipal water supplies (McGeehin et al., 1993). Chlorination and chloramination of drinking water supplies were compared for bladder cancer association. Chloramination leads to lower concentrations of trihalomethanes (THM) than does chlorination. Years of exposure to chlorinated surface water were significantly associated with increased risk for bladder cancer.

A population-based, case-control study evaluated the relationship between cases of cancer (bladder, kidney, leukemia) and exposure to tetrachloroethylene from public drinking water (Aschengrau et al., 1993). The contaminant had leached from the plastic lining of drinking water distribution pipes. Exposure to tetrachloroethylene was associated with an increased risk of leukemia, and evidence showed that the increased risk was dose related. No increased risks of kidney or bladder cancers were noted, although the small numbers of these cancer cases limited the study.

The association between community drinking water quality and the occurrence of late, adverse pregnancy outcomes was investigated in a case-control study in Massachusetts (Aschengrau et al., 1993). The frequency of stillbirths was increased for women exposed to chlorinated surface water and for women exposed to detectable lead levels; the frequency of cardiovascular defects was increased relative to detectable lead levels (OR 2.2; 95% CI 0.9 to 5.7).

SURVEILLANCE SYSTEMS

Public health surveillance is defined as, "...the ongoing, systematic collection, analysis, and interpretation of health data essential to the planning, implementation, and evaluation of public health practice, closely integrated with the timely dissemination of these data to those who need to know..." (Thacker and Berkelman, 1992). A public health surveillance system collects data to monitor the occurrence of specific diseases (called disease surveillance) or the distribution of exposure to potential hazards (called hazard surveillance). Surveillance systems provide early warning of situations in which epidemiologic investigations or other public health actions should be taken. Disease surveillance locates problems, determines how frequently problems occur, plots whether the problems are increasing or decreasing, and evaluates whether prevention efforts have been effective. An example of disease surveillance is the reporting of cancer cases to state health departments.

One source defines hazard surveillance as, "...the assessment of the occurrence of, distribution of, and the secular trends in levels of hazards (toxic chemical agents, physical agents, biomechanical stressors, as well as biologic agents) responsible for disease and injury" (Wegman, 1992). An example of a hazard surveillance system is the Toxics Release Inventory, a data-reporting system that CERCLA established requiring those who emit hazardous substances into the environment to report to the EPA any quantities in excess of regulatory-specified levels.

The ATSDR's surveillance program focuses on exposure to substances at hazardous waste sites and includes disease surveillance systems that follow populations exposed to hazardous substances where they live (i.e., site-specific surveillance) or where they work (site remediation workers, Chapter 6). It includes hazard surveillance of emergency events in which hazardous substances are released into the environment. The ATSDR National Exposure Registry, which contains chemical-specific subregistries, maintains a listing of persons exposed to hazardous substances. Chemicals are selected from the ATSDR/EPA priority list of hazardous substances (Table 4.2).

Site-Specific Disease Surveillance

A site-specific disease surveillance program related to hazardous waste issues was implemented for workers of the former Kilsdonk and Drake Chemical Companies, Lock Haven, Pennsylvania. The companies made specialty chemicals from 1948 to 1981, and workers were exposed to beta-napthylamine (BNA), a potent human bladder carcinogen. The state banned the use of BNA in 1962, and the chemical facility and site became an NPL site in 1982. In 1983, the Pennsylvania Health Department established a disease surveillance program that contained a health effects survey and annual bladder cancer screening for former chemical workers and their families (Leviton et al., 1991). A primary benefit of regular screening for bladder cancer is the early identification of abnormal bladder pathology, an outcome that comports with efficacious medical treatment. Although the number of workers in the surveillance program is unstated, Leviton et al. comment that by 1991, 26 individuals had positive screening findings and are being monitored closely for abnormal cytologies.

Emergency Event Hazard Surveillance

Applying a broad definition of "waste," substances released into the environment when controlled industrial processes fail or from transportation mishaps also represent hazardous waste problems and lead to emergency responses. The consequences to human health of emergency chemical releases can be substantial, depending on the released substance's toxicity, the susceptibility of exposed populations, and the duration of exposure. Generally, the human health implications of chemical emergencies are different from those posed by releases from hazardous waste sites. In emergencies, acute exposure and short-term health effects are the primary concern; for hazardous waste sites, chronic, generally low-level exposures predominate, and chronic adverse health effects are the focus of concern.

Knowledge of the frequency and nature of uncontrolled chemical releases that lead to emergency response is important for planning and prevention. For example, surveillance-based data about which substances are most frequently released are very useful for training emergency response personnel. In 1990, the ATSDR initiated the Hazardous Substances Emergency Events Surveillance (HSEES) system with five state health departments (Hall et al., 1994, 1995, 1996b). Each department actively seeks emergency event information from other state agencies and local departments within its state. The HSEES system now includes 14 states.[7]

---

[7] AL, CO, IA, MN, MS, MO, NC, NH, NY, OR, RI, TX, WA, WI. Minnesota and Mississippi were not included in the 1994 database.

As of January 1, 1993, the ATSDR defined hazardous substance emergency events as uncontrolled or illegal releases or threatened releases of hazardous substances or the hazardous by-products of substances. Events involving petroleum products exclusively are not included because they are exempted from coverage under CERCLA (ATSDR, 1995b). Events are included in the HSEES when federal, state, or local law requires that the amount of substance released (or that might have been released) be removed, cleaned up, or neutralized. Events are also included when release of a substance was only threatened, but this threat led to an action (e.g., evaluation) that could have affected the health of employees, responders, or the general public. Victims are defined as individuals who suffer at least one injury, or death, as a consequence of the event (ATSDR, 1995b).

The 14 states participating in the HSEES system in 1996 reported a total of 5,502 events for the year, with 79% occurring at fixed facilities and 21% were related to transportation (ATSDR, 1996h). Only 18% of events occurred during the weekend, and in 96% only a single substance was released. The two categories of substances most often released from fixed facilities were Other Inorganic Substances (25% of events) and Volatile Organic Compounds (19%). Corresponding categories for transportation events were Pesticides (15%), Acids (12%), and Volatile Organic Compounds (12%). Of the 5,502 events, 390 involved victims, of which there were 1,620. In all events, 543 events led to evacuation of people, and 33 deaths occurred; 17 persons died in fixed-facility events and 16 died in transportation events. Injuries involved employees (51%), general public (30%), and responders (19%).

Risk factors for hazardous substance releases that result in injuries and evacuations were identified from emergency event surveillance data of nine states (Colorado, Iowa, North Carolina, New Hampshire, New York, Oregon, Rhode Island, Washington, Wisconsin) (Hall et al., 1996b), covering the years 1990 through 1992. A total of 3,125 releases of hazardous substances were reported to the HSEES system in this period. During 467 of these releases, 1,446 persons were injured. Of 2,545 releases for which information was available, 400 incidents resulted in evacuations. Of 2,391 releases, 77% occurred at fixed facilities, and 23% were transportation related. The substances most often released or involved with releases were acids, ammonia, pesticides, and volatile organic compounds. Hall et al. cite data from New York State that showed 33% of the releases were due to equipment failure, 14% to human error, and 10% to deliberate dumping of hazardous substances.

These hazard surveillance system data indicate that the health impact of emergency chemical releases extends to plant employees, responders, and the general public. Such data can be used in support of primary prevention interventions like improved equipment maintenance programs, employee and employer training and education, transportation improvements, and increased security in areas where dumping of substances has been shown to be a problem.

OTHER REVIEWS OF HEALTH EFFECTS LITERATURE

In addition to the preceding summary of human health findings, several comprehensive reviews of the literature on hazardous waste and human health have been published. These reviews vary in currency, content, and breadth of analysis. This section summarizes findings from key reviews conducted to date.

Andelman and Underhill

The Fourth Annual Symposium on Environmental Epidemiology held in 1983 led to *Health Effects from Hazardous Waste Sites*, edited by Andelman and Underhill (1987). In the preface the editors observed, "Assessing the adverse human health effects of chemical exposure from waste disposal sites and other point sources is at best difficult. But with the thousands of hazardous waste sites in the United States and abroad, it is necessary that we become more knowledgeable about this ubiquitous problem." The book describes the scope of the hazardous waste problem, exposure assessment, health effects determination, and case studies of communities assessed for the impact of substances released from hazardous waste sites. Any scholar interested in environmental health will profit from this book because its contributors attempt to present a unified epidemiologic and environmental approach to ascertaining an emergent public health problem, hazardous waste sites.

Regarding the public's health concerns, the key chapter in the book is "Evaluating Health Effects of Exposure at Hazardous Waste Sites: A Review of the State-of-the-Art, with Recommendations for Future Research," by Marsh and Caplan (1987). They discussed in depth the approaches to ascertaining human health impacts of sites and developed 15 recommendations they believe are necessary if the human health impacts of hazardous waste are to be adequately addressed. These recommendations were grouped according to the three fundamental phases of health effects evaluations: exposure assessment and definition of exposed and unexposed populations, measurements of health outcomes, and determination of the exposure-health outcome relationship. Examples of individual recommendations within each phase are: Determine in general populations the baseline exposure levels of chemicals likely to be found at waste sites; Develop tests to preclinically recognize the effects of chemicals in human populations; Develop exposure-specific and site-specific registries of exposed populations and suitable control populations with provisions for short-term monitoring and long-term follow-up. The ATSDR and the EPA have implemented many of the 15 recommendations in part or whole.

Marsh and Caplan concluded, "In summary, the overall paucity of published research coupled with the largely inconclusive findings that have been produced to date again underscore the need and importance of developing research programs that will lead to a more complete understanding of the health problems derived from hazardous waste sites."[8]

The Grisham Review

In 1986, a panel organized by the Board of Directors of Universities Associated for Research and Education in Pathology assessed the scientific evidence of association between demonstrated exposure to chemicals from disposal sites and occurrence of human diseases and disorders (Grisham, 1986). The assessment was conducted as a contract for the American Industrial Health Council, Chemical Manufacturers Association, National Agricultural Chemicals Association, National Association of Print-

---

[8] Excerpted with permission from *Health Effects from Hazardous Waste Sites*, Andelman, J.B. and D.W. Underhill, Eds., 1987, Lewis Publishers. Copyright CRC Press, Boca Raton, Florida.

ing Ink Manufacturers, National Paint and Coatings Association, and the U.S. Chamber of Commerce.

The panel critiqued 29 investigations of the health of populations near chemical disposal sites. They found few published scientific reports of health effects clearly attributable to chemicals from uncontrolled waste sites, noting only one report that in their view offered convincing evidence of statistically significant and biologically plausible cause-and-effect relationships between specific chemicals and human disease.

The panel developed ten recommendations as part of their review. From an historical perspective, it is noteworthy that some recommendations were particularly visionary. Quoting from the panel's report:

- Additional prospective and retrospective health studies using sensitive methods should be scheduled according to the precedence indicated for those chemical disposal sites on the EPA's list of priority sites. The populations that have been exposed should be monitored for at least 20 years to detect any excess incidence of cancer or other chronic disease.
- An alert and informed citizenry should continue to be involved in alerting scientists and government of issues relating to disposal sites. Principal investigators should keep local communities informed of the progress and conclusions of their work.
- Because of the numerous uncontrolled candidate sites identified for remedial action and the critical need for setting priorities, each site should be placed in perspective from the viewpoint of its potential effects on human health by the responsible public health agencies of the federal government.

The ATSDR has adopted these recommendations in part or whole and made them part of its public health programs.

National Research Council

In 1991 the National Research Council (NRC) published a comprehensive review of the scientific literature on human health implications of hazardous waste (NRC, 1991b). An expert committee conducted the review and consulted the salient literature on hazardous waste and its human health implications. The NRC concluded,[9] "In spite of the complex limitations of epidemiologic studies of hazardous-waste sites, several investigations at specific sites have documented a variety of symptoms of ill health in exposed persons, including low birth weight, cardiac anomalies, headache, fatigue, and a constellation of neurobehavioral problems. It is less clear whether outcomes with a long delay between exposure and disease have occurred, because of complex methodologic problems in assessing these outcomes. However, some studies have detected excesses of cancer in residents exposed to compounds, such as those that occur at hazardous waste sites."

The NRC observed that the overall impact of hazardous wastes in the U.S. environment is unknown because of limitations in identifying, assessing, or ranking hazardous-waste site exposures and their potential effects on the public's health. Among several recommendations, the NRC suggested additional effort to conduct epidemiologic investigations of communities around hazardous waste sites.

---

[9] Excerpted with permission from *Environmental Epidemiology, Vol. 1,* pp. 19-20. Copyright 1991 by the National Academy of Sciences. Courtesy of the National Academy Press, Washington, DC.

ATSDR Congressional Testimony

In preparing testimony to be given in congressional hearings on CERCLA reauthorization, the ATSDR reviewed the health effects literature pertaining to hazardous waste sites and reached the following conclusion, "When evaluated in aggregate (i.e., by combining health data from many populations around hazardous waste sites), proximity to sites is associated with a small to moderate increased risk of some kinds of birth defects and, less well documented, some specific kinds of cancers. Health investigations of communities around some individual hazardous waste sites have found increases in the risk of birth defects, neurotoxic disorders, leukemia, cardiovascular abnormalities, respiratory and sensory irritation, and dermatitis. However, no adverse health findings have been reported in many studies. These apparent discrepancies may be because of factors such as limitations inherent to epidemiologic methods, differences in environmental conditions in the communities investigated, and the small numbers of persons evaluated at some sites" (Johnson, 1993).

The weight of evidence in the preceding sections bearing on cancer and adverse reproductive outcomes led to the following finding reported to Congress (Johnson, 1995; Johnson, 1997):

> Although epidemiologic findings are still unfolding, *when evaluated in aggregate* (i.e., by combining health data from many Superfund sites), proximity to hazardous waste sites seems to be associated with a small to moderate increased risk of some kinds of birth defects and, less well documented, some specific cancers.

The Sever Review

Under the sponsorship of the Chemical Manufacturers Association, Sever reviewed the evidence on environmental contamination and human health effects (Sever, 1997). He reviewed several methodology problems that cause difficulties and shortcomings in environmental epidemiologic investigations, including investigations of health effects of substances released from hazardous waste sites. Sever argues that failure to assess exposure accurately enough to differentiate groups by level of exposure has reduced the effectiveness of epidemiologic investigators of populations putatively exposed to releases from waste sites. He discusses the methodologic problem of dealing inappropriately with health outcomes; that is, investigating health outcomes that may not have a toxicologic basis for the exposures that occurred.

Sever confines his review of health effects of environmental contamination to the literature on birth defects, birth weights, and self-reported symptoms. Included in his review of adverse reproductive outcomes are the studies of Love Canal births (Vianna and Polan, 1984), New York state hazardous waste sites (Geschwind et al., 1992; Marshall et al., 1995, 1997), New Jersey drinking water study (Bove et al., 1995), and California hazardous waste site study (Shaw et al., 1992). He observes, "...our review shows suggestive evidence for associations between either proximity or exposure to hazardous waste sites and congenital malformations and effects on birth weight. We consider the evidence to be suggestive, rather than convincing, of effects."

Included in his review are symptom-and-disease prevalence studies. Considered in his review are the reports by Straight et al. (1993), Shusterman et al. (1991), Ozonoff et al. (1987), and Hertzman et al. (1987). Sever observed that none of the studies took into account odor perception and recall bias in evaluating subjects' symptoms. He concluded, "...in the studies reviewed, there is little evidence to suggest associations between self-reported symptoms and disease." This statement seems tempered by later statements in his paper. He noted that Hertzman et al. (1987) made a strong case for the adverse [health] effects being caused by exposure to hazardous substances associated with the site investigated. Sever observed that the increased occurrence of respiratory symptoms reported in the Ozonoff et al. (1987) and Straight et al. (1993) studies was consistent with a potential airborne route of exposure to hazardous substances.

Overall, the Sever review is thorough and contains good advice on how to improve environmental epidemiologic investigations of populations exposed to hazardous substances in the environment.

The General Accounting Office (GAO) Health Information Study

In 1995, at the request of the chairman of the Senate's Committee on Small Business, the GAO conducted a study on current health risks posed by CERCLA sites (GAO, 1995b). More specifically, the GAO was asked to provide the best information on two questions: (1) the extent to which sites may pose health risks under current land uses, as opposed to the risks that may develop if land uses change in the future; the nature of current risks; and the kinds of environmental media that pose these risks, and (2) whether the EPA's short-term response actions to mitigate health risks from CERCLA sites have reduced risks under current land uses. The GAO study was apparently conducted under severe time pressure, because mention is made of "time constraints" that shaped their study methodology.

The GAO conducted its study using an EPA database called the Responsive Electronic Link and Access Interface. The GAO refers to this database as, "...the most comprehensive automated information available as of early 1995." The EPA database was created in 1993 and, at the time of the GAO's analysis, contained health risk data on 225 nonfederal CERCLA sites. The GAO does not describe the nature of the sites; that is, the mix of NPL sites and removal sites. Moreover, no information is provided on how the EPA selected the 225 sites for inclusion in their database.

According to the GAO, the EPA assesses the cancer risk, as well as the risk for other adverse health conditions (i.e., noncancer health risks), posed by the contaminants in each site's relevant environmental media (e.g., groundwater, air). The assessed risks are used to establish whether the site warrants cleanup action. The EPA considers the risk of cancer serious enough to warrant site cleanup if the risk indicates more than a 1 in 10,000 probability that exposure to the site's contaminants may cause an individual to develop cancer (GAO, 1995b). For noncancer health effects (e.g., birth defects), the EPA considers the risk serious enough to warrant cleanup if the risk assessment indicates that exposure to the site's contaminants might exceed human exposure levels of concern.

The GAO methodology used in conducting its study was apparently confined solely to an examination of the EPA risk assessment database and an acceptance of the EPA's site findings. The analysis consisted of categorizing each of the 225 sites according to the two questions asked by the Senate requestor.

Using the EPA's database, the GAO reported that 32% of the 225 sites posed serious health risks under the land use current at the time of the EPA's site risk assessment. Of the 225 sites, 53% did not pose risks warranting cleanup action under current land use, but posed such risks under the EPA's projections for future land use. The remaining 15% of sites did not pose health risks serious enough to warrant cleanup action under either current or future land use. Furthermore, for the 71 sites (i.e., 32% of 225 sites) posing health risks under current land use, (1) at 77% of sites, a single environmental medium, usually groundwater or contaminated soil, posed the health risk, and (2) at the remaining 23% of sites, multiple environmental media posed the health risk. For these 71 sites, the EPA's data indicated that 28% posed cancer risk, 30% posed noncancer risk, and 42% posed both cancer and noncancer risk.

Several problems exist with the aforementioned conclusions, stemming primarily from limitations in the GAO's method and the EPA database available to them. Concerning their method, the GAO did not incorporate any data from the ATSDR. This is remarkable, given that the ATSDR was created under CERCLA by Congress as the government entity to respond to public health impacts of hazardous waste sites and related issues. It is noteworthy that the ATSDR has a national database called HazDat (Chapter 2) that contains environmental and health data on more than 1,300 CERCLA sites, but the GAO did not use this database in their analysis. Moreover, the GAO apparently did not try to verify the adequacy of the one EPA database on which they based their study. The GAO addresses neither the currency and quality of the EPA site database, nor the criteria the EPA used to determine current and future land use of individual sites. The 225 sites in the EPA's database may not be representative of CERCLA sites generally. Given these limitations, the GAO study must be viewed as a limited effort that contributes to, but does not answer, to what extent hazardous waste sites impact human health.

The GAO is considered the primary auditing resource for the U.S. Congress, one of several organizations Congress created to provide analysis and advice. Other examples are the Congressional Research Service and the Congressional Budget Office. The GAO audits various federal government programs at the request of Congress, or more precisely, at the request of individual members or committees of Congress. The GAO has gained great credibility with both Congress and the general public because they are viewed as independent of the government agencies they audit. However, it is noteworthy that the GAO studies on technical subjects are not subjected to independent scientific peer review, as would be expected of other scientific publications.

Viscusi and Hamilton Review

Two Duke University economists conducted a comprehensive evaluation of site-specific risk assessments performed by the EPA for NPL sites (Viscusi and Hamilton, 1994). An EPA baseline assessment (Chapter 2) of the risks at each site is the first step in a process that leads to options for managing those risks. The investigators' database consisted of risk assessments for 78 NPL sites. They focused on cancer risk expressed in each risk assessment and arrived at several quite interesting findings. For instance, the magnitude of risk presented by CERCLA sites exceeded estimated risks for other activities falling under federal cancer regulations. The dominant risks arose from consideration in risk assessments about future land use scenarios. Viscusi and Hamilton note, "Consideration of the risk assessments for Superfund sites indicates, however, that it is not the existing risks that are most significant...Indeed, these future risks account for 90% of all the risk-weighted pathways for the Superfund sites in our sample."

ESTIMATES OF ADVERSE HEALTH EFFECTS COSTS

The preceding sections summarized key findings on associations between exposure to substances released from uncontrolled hazardous waste sites and adverse human health effects. Given these findings, what is the economic cost of the adverse health effects? Medical epidemiologists at the ATSDR collaborated with economists at the University of Tennessee's Joint Institute for Energy and the Environment to develop a methodology to estimate the costs (Lybarger et al., 1998).

The approach taken by the investigators consisted of identifying adverse health effects associated with specific hazardous substances (e.g., VOCs), examining environmental contamination data in the HazDat system to identify CERCLA sites where the specific contaminant had been measured in environmental media at levels of human health concern, calculating the excess number of adverse health effects (e.g., birth defects) in populations living near the identified CERCLA sites, and relating data on health care costs and lost productivity costs to the estimated excess cases of adverse health effects.

Adverse Health Effects Costs—To identify the adverse health effects for which cost impacts would be estimated, Lybarger et al. reviewed the scientific literature and determined that birth defects (neural tube defects, cardiac malformations, and cleft lip and palate) were associated with residential proximity to hazardous waste sites. Using data from the ATSDR National Exposure Registry's TCE subregistry, Lybarger et al. determined that stroke, diabetes, urinary tract disorders, eczema and skin disorders, and anemia were reported in excess in adults, and speech impairment and hearing impairment were reported in excess in young children. The work of Bove et al. (1995), who had studied birth defects and maternal exposure to VOC-contaminated public drinking water systems in New Jersey, was then used to extrapolate to VOCs released from hazardous waste sites. The adverse health effects analysis therefore led to focusing on VOCs in groundwater.

NPL Sites and Population Estimate—The size of the population exposed to VOCs was estimated by Lybarger et al. using data from the HazDat database. Hazardous waste sites included in the analysis were those with confirmed presence of TCE, trichloroethane, or benzene in a completed exposure pathway, and where the ATSDR considered the site to be a human health hazard through public health assessment or similar health review. This led to the selection of 258 NPL sites with completed exposure pathways (Chapter 2). The investigators used geographic information system techniques to estimate the size of the population within a one-half mile border of all boundaries for each site. The longitude and latitude information was inadequate for 33 sites, and a circle of one-half mile radius around those sites was used to estimate the population size.

Average Costs—The estimates of health conditions (other than birth defects and stroke) were measured by their excess prevalence. Published average annual inpatient and outpatient costs and lost productivity were estimated for each health condition. The ATSDR exposure registry data were used to estimate the average annual frequency of hospitalization and outpatient services among individuals having one of the health conditions shown in Table 5.14 other than birth defects and stroke. Cost estimates for the occurrence of birth defects (neural tube defects, cardiac malformations, cleft lip and palate) were derived from published studies on such costs. The cost estimates for birth defects included the direct medical costs over an individual's lifetime and the indirect cost of the health condition due to lost productivity from morbidity or premature mortal-

Table 5.14. Estimated Annual Cost of Select Adverse Health Conditions (Lybarger et al., 1998).

Health Condition	Number of Excess Cases	Cost per Case[a]	Annual Costs for Medical Complications, Long-Term Care, Lost Productivity[a]	Total Annual Cost[a]
Anemia	27,500	$350	[b]	$9.6M
Birth defects:				
Neural tube defects	26	$360,000	[b]	$9.4M
Cardiac malformations	135	$390,000	[b]	$52M
Cleft lip and palate	49	$140,000	[b]	$6.9M
Diabetes	8,600	$860	$43M	$50.4M
Eczema and skin diseases	20,600	$160	[b]	$3.3M
Hearing impairment	5,400	$400	[b]	$2.2M
Speech impairment	6,400	$400	[b]	$2.6M
Stroke	8,600	$7	$120M	$180M
Urinary tract disorder	37,800	$380	[b]	$14M
Total				$330M

[a] In 1995 dollars.
[b] Costs are included in the annual cost.

ity. Lybarger et al. note that deriving the cost estimate for stroke was complicated because even though an acute stroke event may have taken place several years ago, its effects are usually long-term, if not fatal. On a prevalence basis, the costs of treating the current effects of strokes that occurred in prior years are part of the current year's cost of stroke. Information in the ATSDR's National Exposure Registry for persons exposed to TCE was used to estimate excess cases of stroke, for which average annual inpatient costs, outpatient costs, and lost productivity cost estimates were derived using published health care cost data.

Findings—The estimated population living within one-half mile of the 258 NPL sites was approximately 1.72 million persons. Table 5.14 contains the cost estimates for the adverse health conditions included in the investigators' analysis. As shown in the table, the total annual cost estimate approximates $330 million. Data in Table 5.14 indicate that more than half of this estimate of annual costs is attributed to the cost of medical complications, long-term health care, and lost productivity from stroke.

It is important to put the figure of $330 million into context. Lybarger et al. comment, "The costs that were calculated give a general indication of the annual economic burden in the absence of any remediation or public health intervention programs. That is, estimates are an indication of the size of the problem that would have persisted had there been no Superfund legislation. However, because remediation has taken place at many sites, the risks of exposure at these sites have been reduced and the expected health burden associated with future exposure to VOCs at such sites has decreased to a corresponding degree."

Several limitations are present in the Lybarger et al. study. First, a known causal association is lacking between the ascribed adverse health effects (e.g., birth defects) and persons' exposure to substances released from the NPL sites analyzed by the investigators. Second, the health effects findings from the ATSDR National Exposure Registry are self-reported data, which are subject to reporting bias and other uncertainties. Third, personal exposure data were lacking; rather, investigators relied on an exposure surrogate—residence within one-half mile of each NPL site selected for evaluation. Fourth, the cost data pertain to a specific set of 258 sites; whether similar conditions of environmental contamination apply at other sites was not part of the investigators' analysis.

On the other hand, the Lybarger et al. analysis is very conservative in the context of overall health costs possibly associated with NPL sites. Only 258 NPL sites were selected for study because they had the strongest environmental contamination data and the sites represented completed exposure pathways. Sites with VOC contamination represent the largest group of NPL sites; hence any number beyond the 258 sites evaluated would obviously add health costs. Moreover, the health data were conservative because some known costs were not included; for example, costs for special education for children born with birth defects. Last, health costs associated with other kinds of site contaminants; for example, lead, and other social considerations, such as psychologic effects of living near a waste site and quality of life concerns, were not factored into the Lybarger et al. analysis.

It would be a mistake to compare this annual cost estimate with the costs of site remediation on an annual basis. Too many uncertainties exist in the health effects databases and in the number of sites for which the health effects data apply. Notwithstanding the limitations of the Lybarger et al. study, the costs associated with adverse human health effects appear to be considerable.

GUIDANCE ON WHEN TO CONDUCT HEALTH STUDIES

The ATSDR, in collaboration with its Board of Scientific Counselors, has adopted guidelines for considering health studies in communities that might be exposed to hazardous substances (ATSDR, 1996i).[10] The guidelines divide health studies into two basic types, those that are primarily exploratory in their approach (Type-1 studies), and those that require rigorous scientific methods to evaluate specific exposure-outcome relationships (Type-2 studies). Specific criteria are provided for determining when to do a health study, determining what type of study to do, and ensuring that a study is of high quality.

Type-1 health studies explore or generate hypotheses about exposure-outcome associations and address specific exposures, community health concerns, or specific information needs. Type-2 health studies are specifically designed to test scientific hypotheses about the association between adverse health outcomes and exposure to hazardous substances in the environment.

Many approaches might be considered when addressing health concerns or the needs of a community living near a hazardous waste site. As appropriate, these approaches might include different types of health studies or other public health activities. Site-specific circumstances (substance[s], exposure pathways, levels of exposure, health outcomes, and population at risk) and existing knowledge of the exposure and health outcome relationship will influence the need for and type of health study. In addition, whether adequate characterization of human exposure is available at a sufficient level to assess health effects should be determined before a health study is considered.

---

[10] This section summarizes an ATSDR document found on the agency's Internet website.

Major differences exist between various types of health studies and the level of scientific rigor needed to ensure quality. Type-1 studies can use various investigational approaches to explore health concerns or potential exposures. The approaches might include descriptive studies, surveillance activities, exploratory data analyses, and exposure investigations. These studies are often conducted to determine whether a need exists for a more definitive study. Type-2 health studies require a higher level of scientific rigor to evaluate specific exposure-outcome relationships; these studies use primarily the case-control or cohort approach. Case-control studies determine differences in exposures and risk factors for two groups of study subjects—persons with a specific illness (cases) and those without the illness (controls). Cohort studies compare the differences in illness occurrence in exposed and unexposed (reference) populations followed over a specified period of time.

SITE ASSESSMENTS

A public health assessment of a hazardous waste site provides information on the health hazard ranking of the site, community education needs, presence of hazardous substances, evidence of completed pathways of exposure, population demographics, and community health concerns. The assessment may conclude that health studies would not be appropriate or recommended for a specific site. In situations where health studies are determined to be appropriate, further consideration must be given to determining the type of study to be conducted and ensuring its quality.

Other reasons may also determine which sites can be considered for health studies. Health studies might be initiated before completing a public health assessment because an urgent health threat, or exposure situation, or both are present. The ATSDR research program on priority health conditions might identify specific health outcomes and contaminants or exposures that require additional health studies to assess the relationship between exposure and adverse health effects. Research needs might require multiple communities or regions of the United States to be included in studies of rare health outcomes. In addition, multisite studies might use the same study protocol to conduct studies at several sites that have similar contaminants and human exposure pathways.

When to Proceed with a Health Study

Before a health study can be recommended for a particular site, seven factors described in this section should be considered. The ATSDR uses the factors to set priorities for which health studies to conduct and when. Each factor should be considered when determining the relative importance and appropriateness of a proposed health study.

Public health significance—Public health significance is a key factor in considering the merits of a proposed health study. Issues to consider include the hazard ranking of the site, toxicity of the hazardous substance(s), pathways of human exposure, severity and biologic plausibility of the health outcome, need for new information (beyond what is already known or what has already been done), size and susceptibility of the population affected, ability to prevent or mitigate exposure or health outcomes, and relevance to other sites with similar contaminants and exposure pathways.

Community perspective and involvement—Community involvement is critical to the success of any proposed health study. Using an assessment of community needs and concerns, the ATSDR will usually initiate a formal community involvement activity. Various community involvement methods (Chapter 7) can be used for health studies. Issues for consideration include the ability to involve key community stakeholders, an understanding of community health concerns, an understanding of the approach and limitations of proposed activities, and community support for the study being conducted.

Scientific importance—Scientific importance is closely related to public health significance. Attention should be given to the ability to provide new knowledge or information about an exposure-outcome relationship, to address specific exposures or outcomes that have not been adequately studied, to allow new laboratory tests or study methods to be used or evaluated, to generalize to other situations or populations, and to provide confirmation or additional support to a preliminary hypothesis or theory.

Ability to provide definitive results—Because health studies can sometimes result in inconclusive findings, an important consideration is how definitive the study might be in providing scientifically useful results related to specific exposure-outcome relationships. Care must be taken to obtain appropriate exposure measures, document health outcomes and exposure, use adequate control or comparison populations, obtain community support to improve the participation rate, state clearly the study objectives and specific hypothesis to be tested, have sufficient statistical power to detect predicted effects, obtain data on important potential confounders, and evaluate a dose-response relationship or gradients of exposure.

Resources—Resources are critical to the support, conduct, and completion of any proposed health study. These include the availability of qualified personnel and technical support, the ability to obtain necessary data and health information, an appropriate project time line and budget, and the proper administration and project management.

Contribution to program goals—The contribution of a health study to program goals is important. Thought must be given to how the proposed health study addresses the sponsoring organization's program goals and complements other program activities and priorities.

Authority and support—Local, state, and federal health agencies must be involved early in discussions about potential health studies. Can support or technical assistance requested by the local or state health agency be provided? Can local and state health agencies address community health concerns? Are appropriate agencies with legislative and regulatory requirements, such as the EPA, involved?

When Health Studies Should Not Be Done

The seven factors in the previous section cover a wide range of important issues that directly affect the feasibility and value of any health study being considered. These considerations for health studies have to be applied on a case-by-case basis, because information and circumstances differ by site. Generally, Type-1 health studies would not be performed when insufficient information is available or other factors exist that severely limit the ATSDR's ability to provide new and useful information on the health or exposure status of the community. Type-2 health studies would not be conducted when information is insufficient or exposure documentation is limited, or when other factors exist that severely affect the ATSDR's ability to evaluate specific exposure-outcome relationships. The next section provides additional guidance on when studies are appropriate and what study attributes are considered necessary. When the additional guidance or attributes are not met, health studies would not be recommended.

Deciding When to Conduct Health Studies

In most situations, environmental contaminant and exposure information is limited for populations living near hazardous waste sites, and health outcome information is frequently incomplete or unknown. In other situations, sites have well-documented contaminants and identified potential exposure pathways. For yet other sites, environmental data do not support any human exposure pathways of concern. Each of these three scenarios is shown in Table 5.15 with a decision analysis approach and resultant actions or further considerations.

SYNTHESIS

Beginning in the 1970s, concern arose over the potential harm to human health of substances released from hazardous waste sites and chemical spills. The episode at Love Canal, New York, contributed greatly to the public concern that migratory hazardous waste was causing human illness and disease. The public's health concern in turn contributed to the enactment of laws and regulations to manage current hazardous waste and to remediate the legacy of hazardous waste left in the environment.

Almost 20 years have passed since CERCLA became law. Were the public's fears realistic that hazardous waste is a risk factor for disease? This chapter contains, in the author's opinion, a summary of health studies that provide an answer to that question. In the 1980s, reviews of the human health implications of hazardous waste sites generally described how to conduct environmental epidemiologic studies of populations living near sites, but found only limited evidence at that time for relating hazardous waste releases to effects on human health. Reviewers cautioned that individuals' exposure to hazardous substances was likely to be low and therefore unlikely to cause illness or disease.

The report *Environmental Epidemiology: Public Health and Hazardous Wastes, Vol. 1*, released in 1991 by the National Research Council (NRC) was the first review to accept that adverse health effects had occurred in populations living near some hazardous waste sites. The NRC cautiously noted that more epidemiologic studies were needed and strongly advocated the collection of exposure data on at-risk populations.

The approximately 60 health studies summarized in this chapter include and extend the body of literature available to the NRC in 1991. In particular, the ATSDR, as a consequence of additional CERCLA mandates and funding, has conducted or sponsored the development of much of the enhanced knowledge on human health effects associated with hazardous waste sites. What does this body of more than 60 human health studies say about the impact of hazardous waste sites on human health?

The most compelling health findings are those from studies of reproductive outcomes in populations living near certain kinds of hazardous waste sites. The weight of evidence associates select birth defects and reduced birth weight of infants born to parents who lived near sites. The release of VOCs into groundwater seems a common factor in studies of increased rates of birth defects and lower birth weight. The birth defects most often reported are malformations of the heart, neural tube, and oral cleft palate. There also is troubling evidence that human fertility in adults can be reduced from exposure as children to high levels of lead. This is significant because lead is the toxicant most commonly found in completed exposure pathways at hazardous waste sites (Chapter 2).

The most convincing evidence that associates residential proximity to hazardous waste sites and adverse reproductive outcomes comes from studies of residents at

Table 5.15. Decision Logic on When to Conduct a Health Study of a Hazardous Waste Release (ATSDR, 1996i).

I. Contaminants are sufficiently documented by type, media, and concentration. Potential human exposure pathways have been determined and an exposed at-risk population can be identified.			
	A. Evidence of human exposure is documented at a sufficient level of concern.	1. The association between exposure and health effects is already established.	Provide services that reduce or eliminate exposure, identify or prevent adverse health outcomes, and improve quality of life.
		2. The association between exposure and health effects is not already established.	Consider health studies that provide new knowledge about human health effects and exposures to specific hazardous substances. Studies help identify risk factors or recommend actions to prevent or mitigate adverse health outcomes.
	B. Evidence of human exposure or exposure at a sufficient level of concern is not documented.	1. Consider community health concerns for important or biologically plausible health outcomes.	[When appropriate] Provide support to the community that addresses its health concerns and site-specific issues. [Else] Site will remain under periodic review by the ATSDR.
		[When feasible] 2. Conduct an exposure investigation to determine whether human exposure has occurred at a sufficient level of concern.	If findings are positive and support human exposure...[go to IA] If findings are negative or do not support human exposure...[go to IBI]
		[When feasible] 3. Determine whether site information can provide enough source, production, or release data to suggest current or past human exposure.	If data are sufficient to support human exposure or reconstruct exposure or dose...[go to IA] If data are insufficient or do not further support exposure...[go to I.BI]

Table 5.15. Decision Logic on When to Conduct a Health Study of a Hazardous Waste Release (ATSDR, 1996i) (continued).

II. Documentation of contaminants is incomplete, a complex mixture exists requiring some surrogate measure, or the potential exposure pathways are unknown.	*[When appropriate]* A. Review additional environmental sampling data when they become available or conduct additional focused sampling when indicated (could require the EPA or state involvement). B. Consider community health concerns for important or biologically plausible health outcomes.	If sampling data better define the contaminants and potential exposure pathway...[go to I] If sampling data provide little new information or do not change level of uncertainty...[go to II.B] *[When appropriate]* Provide support to the community that addresses its health concerns and site-specific issues. *[Else]* Site will remain under periodic review by the ATSDR.
III. Documentation is sufficient with few contaminants identified and the environmental data do not support any exposure pathways of concern.	Consider community health concerns for important or biologically plausible health outcomes.	*[When appropriate]* Provide support or identify additional support from another agency to address the needs or concerns of the community. *[Else]* Site will remain under periodic review by the ATSDR.

Love Canal, New York, and Gloucester County, New Jersey (the Lipari NPL site). At both sites, state health department investigators found reduced birth weight in infants born to parents who lived near the waste sites compared with infants from comparison areas. This reduction in birth weight occurred during periods of documented peak releases of hazardous substances from the two sites.

The association between increased cancer rates and exposure to substances released from hazardous waste sites is less well documented than for reproductive outcomes. The evidence to date suggests areas with contaminated groundwater may increase the risk of cancers of the gastrointestinal tract and urinary bladder. If true, this would be consistent with studies of contaminated municipal drinking water supplies in New Jersey and Iowa that generally found a small to moderate increased risk of these kinds of cancer.

The reports of association between exposure to TCE and other VOCs and leukemia are troubling and potentially very significant. This assertion derives from the Woburn, Massachusetts, studies of leukemia rates in children and the laboratory study that found an approximately 3-fold increased prevalence of lymphocyte phenotypes with characteristics similar to those seen in early B–cell chronic lymphocytic leukemia.

There is another basis for concern that substances released from hazardous waste sites may increase the risk of cancer in exposed populations. This is the knowledge that half of the substances found in completed exposure pathways are known human carcinogens or reasonably anticipated to be (Table 4.3). Whether the degree and duration of human contact with the carcinogens in completed exposure pathways has been sufficient to elevate cancer rates is unknown, but should be pursued as a matter of investigation. Of special relevance is the continued examination of exposure registries of persons with documented exposure to carcinogens like benzene, dioxin, and other CERCLA priority hazardous substances to identify elevated cancer rates.

Health outcomes other than cancer and reproductive health have been investigated in populations living near some hazardous waste sites. The number and breadth of these studies per priority health condition is less than for studies of cancer and reproductive outcomes. Reports exist of lupus, liver and kidney diseases, abnormal immunologic characteristics, and alterations of nervous system responses associated with residential proximity to hazardous waste site. These studies are all site-specific, thereby limiting the ability to generalize to other hazardous waste sites. The contaminants most often associated with these noncancer, nonreproductive outcome studies are various VOCs, particularly TCE and similar hydrocarbon solvents. There is great need to extend these kinds of studies. Highest priority should be given to elaborating the reports of lupus and immune function disorders, given the significance of the immune system to protecting an individual's health.

Although several symptom-and-illness prevalence studies exist and most report an elevated rate of health problems in surveyed populations, the worth of such reports is problematic. This is because reporting bias (memory recall), problems in selection of some subjects surveyed, poor questionnaire design, and the possible contribution of "environmental worry" all add uncertainty to this kind of study.

Health studies of populations living near hazardous waste sites have typically involved relatively low level exposure to hazardous substances, but often over an exposure period of years. Moreover, at-risk populations in these populations include groups that are at elevated health risk because of factors that include young age, pregnancy, existing infirmities, low economic status, and lifestyle issues such as tobacco smoking. The populations at risk around hazardous waste sites therefore present special challenges to health investigators in terms of study design. One very useful resource available to public health authorities is an exposure registry, as described in this chapter. This is because exposure registries permit the tracking of health status in many groups within a population over a prolonged period of time, permitting the identification of emerging health problems that may be of a latent nature; that is, occurring long after individuals' exposure to hazardous substances has ceased.

The exact economic cost of adverse health effects in populations exposed to substances from hazardous waste sites is unknown. However, preliminary estimates from 258 sites that leaked VOCs into groundwater suggest an economic cost of approximately $330 million annually for those sites. It would be a mistake to compare this annual cost estimate with the costs of site remediation on an annual basis. There are simply too many uncertainties in the health effects databases and the number of sites for which the health effects data apply.

This chapter covers uncontrolled releases of hazardous substances. Although the emphasis has been on releases of substances from hazardous waste sites, other uncontrolled releases of substances also present a hazard to the public's health. Surveillance systems that accrue data on the number and consequences of emergency chemical events confirm the significant impact of these events on human health because of the large number of events and the severity of health impacts.

Chapter Six

Workers' Health and Safety

What health and safety hazards are faced by workers who remediate hazardous waste sites and those who respond to chemical emergencies? Do different kinds of waste site activities add to workers' safety risk? What CERCLA requirements apply to workers' health and safety? This chapter describes hazards of site remediation work, operations that increase workers' risk, and the OSHA standard on workers' health and safety.

Hazardous waste sites and other uncontrolled releases of hazardous substances into the environment have the potential to cause adverse health effects in community populations. The scientific data supporting this assertion were presented in Chapters 3, 4, and 5. In addition to the community residents already discussed, other persons are also at health and safety risk from contact with hazardous substances. In particular, workers whose occupation brings them into contact with hazardous waste sites, waste management facilities, and sites where chemical emergencies occur, face safety and health hazards.

This chapter highlights health and safety hazards faced by site remediation workers and others who come into contact with hazardous waste because of their occupation. Included are waste site remediation workers, incineration workers, and emergency responders. Estimates of the numbers of these workers are presented here. An overview is given of the applicable occupational safety and health regulations and the surveillance and training programs that help protect waste-site remediation workers from the hazards of their work.

OCCUPATIONAL HEALTH AND SAFETY RISKS

Care must be exercised that workers who remediate hazardous waste sites are not harmed by the conditions of their work. Primary protection measures include personal protective equipment, engineering controls, site monitoring equipment, and training and education of workers and managers. Secondary protection measures include periodic medical evaluation of remediation workers and workers' health surveillance. Adequate programs of workers' health protection requires both primary and secondary measures.

One comprehensive study has been conducted on the nature of site remediation work and emergency responding. The investigators used payroll records and ancillary data from site-remediation contractors and government agencies to estimate the number of remediation and emergency response workers and the kind of work performed at sites (Ruttenberg et al., 1996). They noted that construction laborers, industrial laborers, transportation workers, and emergency responders are the workers most often engaged in remedial actions and hazardous waste activities. Their study of 17 hazardous waste sites showed that three crafts accounted for more than 60% of the payroll: operators of equipment, laborers, and truck drivers. The major crafts represented at the sites studied are shown in Figure 6.1 as percentages of a site's gross payroll. The investigators note that the mix of labor crafts varies with the site being remediated and the stage of remediation. Each craft, as explained subsequently, brings its own level of health and safety risk.

The sections that follow describe some of the health and safety risks faced by hazardous waste workers.

Occupational Health Risk

The removal from hazardous waste sites of soil or water contaminated with toxic substances could certainly adversely affect the health of cleanup workers if the necessary precautions are not taken. Work conditions can include strenuous work, thermal stress, and exposure to toxic chemicals, biologic hazards, and radiation. The principles of exposure assessment and toxicology discussed in Chapters 3 and 4 apply to conditions faced by hazardous waste workers. In a public health context, prevention of workers' exposure to hazardous substances is the surest way to avoid adverse health effects.

Site Remediation Workers

Because large-scale, site-specific remediation work is a relatively recent enterprise, scientific literature on potential health implications for workers is sparse. According to one source, data from clinical centers that perform health surveillance examinations on hazardous waste workers have not shown health abnormalities related to the workers' activities (Favata et al., 1990). From a public health perspective, this is a very desirable outcome. However, the investigators caution that the apparent lack of health problems in hazardous waste workers may be because most of the workers examined had performed feasibility work (i.e., inspection and assessment), not actual remediation of sites.

In a study of Italian workers who performed cleanup operations at the highly dioxin-contaminated Seveso site after an industrial release, no meaningful differences in biochemical outcomes were found between remediation workers and a comparison group (Assennato et al., 1989). These gratifying results may be attributable to the safety measures taken by workers during the cleanup operations, but could also be a consequence of insufficient elapsed time to evidence any potential health problems.

Emergency Response Workers

Highway patrol officers are often members of emergency response teams that respond to hazardous materials spills. A study was conducted of illness rates for California Highway Patrol officers who responded to 223 hazardous materials (HazMat) events in 1984 (English et al., 1989). The records of 655 officers who responded to spills that involved acutely toxic

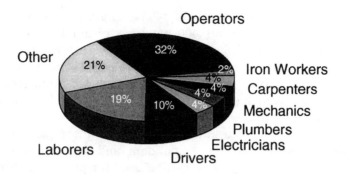

Figure 6.1. Crafts as a percentage of gross pay (Ruttenberg et al., 1996, with permission of author).

chemicals were identified from records kept on each spill. Officers who responded to spills not involving acutely toxic chemicals were the comparison group (n = 338). Work records were examined to tally sick leave taken and reasons for absence on the day of response to the hazardous materials spill and for each of the subsequent seven days. No significant differences in absenteeism or illness were found between the two groups, perhaps indicating that safety protocols and procedures for officers were effective.

Generally, the human health implications of chemical emergencies are different from those posed by releases from hazardous waste sites. The primary concern for chemical emergencies is acute exposure and short-term health effects, whereas the primary concern in hazardous waste release is for chronic adverse effects on health because chronic, generally low-level exposures predominate.

Knowing the frequency and nature of uncontrolled chemical releases that lead to emergency response is important for planning and prevention. For example, surveillance-based data like those maintained in the ATSDR's Hazardous Substances Emergency Events Surveillance system (HSEES) are very useful in training emergency response personnel. Data in HSEES helped identify risk factors for hazardous substance releases that resulted in injuries and evacuations in the 14 states participating in 1996 (Chapter 5).

These 14 states reported a total of 5,502 events that year, with 79% occurring at fixed facilities and 21% being transportation related (ATSDR, 1996h). Only 18% of events occurred during the weekend. In 96% of the events only a single substance was released. The two categories of substances most often released from fixed facilities were Other Inorganic Substances (25% of events) and Volatile Organic Compounds (19%). Most frequent categories for transportation events were Pesticides (15%), Acids (12%), and Volatile Organic Compounds (12%). Of the 5,502 events, 390 resulted in a total of 1,620 victims, 33 deaths resulted, and 543 events led to evacuation of people. Of the 33 deaths, 17 occurred in fixed-facility events, and 16 in transportation events. Most frequent victims by population group were employees (51%), general public (30%), and responders (19%).

Risk factors for hazardous substance releases that result in injuries and evacuations were identified from nine states' (Colorado, Iowa, North Carolina, New Hampshire, New York, Oregon, Rhode Island, Washington, Wisconsin) emergency event surveillance data (Hall et al., 1996b), covering the years 1990 through 1992. A total of 3,125 releases of hazardous substances were reported to the ATSDR's HSEES system during this period. During 467 of these releases, 1,446 persons were injured. Of 2,545 releases for which information was available, 400 incidents resulted in evacuations. Of 2,391 releases, 77% occurred at fixed facilities, and 23% were transportation related. The substances most often released or involved with releases were acids, ammonia, pesticides, and volatile organic compounds. Hall et al. cite data from New York State that showed 33% of releases were due to equipment failure, 14% to human error, and 10% to deliberate dumping of hazardous substances.

These surveillance system data indicate that the health impact of hazardous waste releases extends to plant employees, responders, and the general public. Data from emergency event surveillance systems can be used for primary prevention interventions like improved equipment maintenance programs, employee and employer training and education, transportation improvements, and increased security in areas where dumping of substances has been shown to be a problem.

Health Care Providers and Allied Personnel

Hospital staff provide medical care for victims of chemical emergencies. Hospital planners and administrators must take appropriate actions to protect hospital personnel and facilities against hazards posed by chemically-contaminated patients. Federal regulations apply. The Occupational Safety and Health Administration's (OSHA's) Process Safety Management standard requires hospital staff to wear splash suits and positive pressure respirators when decontaminating premises. Depending on the kinds of chemical contamination, different levels of personnel protection are needed. A recent survey of emergency room physicians led to the assertion that "Most U.S. hospitals put their emergency room staffs at risk because they are not prepared to treat victims of hazardous materials accidents" (Levitin and Siegelson, 1996). Two-thirds of the physicians surveyed thought their hospitals had proper protective equipment, but few respondents actually knew where the equipment was located, and only two of 45 hospitals surveyed actually had protective equipment assigned to the emergency departments.

The investigators advocate that hospitals should prepare to decontaminate victims in their emergency rooms and should not rely on ambulance crews or hazard materials (HazMat) responders (Levitin and Siegelson, 1996). They stress the need for improved training and education for emergency room health care personnel and hospital administrators. The investigators noted that the medical community still lacks knowledge about the extent of the HazMat threat to health care providers.

Waste Disposal Workers

Although waste disposal in general is not strictly a matter of hazardous waste disposal or emergency response, it is relevant to what happens at some hazardous waste sites being remediated. Waste generation will always be a reality. All human activities generate waste in some form or another. Indeed, the very state of being alive produces personal wastes that include carbon dioxide, fluids, and solids that must be removed from the body if life is to continue.

How waste is managed is critical to human well-being. Some examples illustrate the point. Improper disposal of raw sewage has historically caused outbreaks of cholera and other diseases. Medical waste became an environmental hazard when it washed ashore on Atlantic Ocean beaches in the early 1990s. Household waste can attract vermin if not disposed of properly. The disposal of household and industrial wastes is particularly important because of their sheer volume and potential hazard to human and ecologic health. The four primary ways to manage solid waste are: incineration, recycling-composting, landfilling, and source reduction/reuse of waste (Gochfeld, 1995). Incineration is used to reduce the volume of waste and destroy some of its harmful constituents. However, incineration can produce harmful emissions as air pollutants if precautions are not taken, and residual bottom ash itself requires waste disposal.

Moreover, as discussed in Chapter 1, household and some industrial wastes are found in municipal landfills. Therefore, in addition to waste site remediation work, the permitted disposal of hazardous waste by incineration, on-site destruction, or other permitted method of disposal can bring workers into contact with hazardous substances. Without adequate safeguards and work practices, workers' exposure to the wastes could lead to adverse health effects. Although no extensive occupational health literature on the subject exists, the following studies illustrate some hazards faced by waste disposal workers.

Landfill workers were included in a morbidity survey of the health impacts of the Upper Ottawa Street Landfill in Hamilton, Ontario (Hertzman et al., 1987). The health of workers at the landfill was investigated first as a means to design an investigation of residents living near the landfill (Chapter 5). The landfill had operated from the early 1950s until 1980 when it was closed. The exposed group consisted of 149 former workers at the landfill; 169 sanitation workers formed the comparison group. A structured health questionnaire, covering the years 1965–1980, was administered in 1983 to each person in the study. A review of medical records verified the subjects' reports of health problems. Investigators were not blind to subjects' group categories, thus introducing a possible source of investigator bias. No personal exposure data were available, nor was information provided about the landfill workers' work conditions. Findings showed that landfill workers more frequently reported chronic bronchitis, daily cough, combined respiratory problems, narcotic symptoms (headaches, dizziness, lethargy, balance), and mood disorders than did sanitation workers. Workers' exposure data and the addition of another comparison group not involved in sanitation work would have strengthened the study because some sanitation work involves exposure to hazardous substances.

Mexican workers at a newly opened landfill were evaluated for uptake of hazardous substances and for adverse health effects. Twenty-two workers whose duties required them to handle hazardous wastes were compared with manual laborers from an adjacent community. No protective measures or equipment were used by the landfill workers. The landfill workers had elevated levels of arsenic in their urine and hair; other metals were not elevated in biologic samples. No patterns of adverse health effects were found, which is not surprising given the newness of the landfill operations (Díaz-Barriga et al., 1993). In Nigeria, macrocytic anemia and leucopenia were noted in workers who had contact with radioactive waste when compared with workers not similarly exposed (Ogunranti, 1989).

In Sweden, mortality was investigated among 176 male workers employed at a municipal waste incinerator for at least one year between 1920 and 1985 (Gustavsson, 1989). An excess of deaths from ischemic heart disease was found to be associated with duration of work. Exposure to combustion products and polycyclic aromatic compounds was common.[1] In a study of 14 persons who worked at a liquid hazardous waste incinerator in the United States, investigators found severe neurologic damage (myoclonus and severe tremor) in two workers, and all workers had abnormal psychiatric symptoms, but with different diagnoses. Workers had experienced exposure to solvents during their work (Kawamoto, 1992). In Poland, dermatologic examinations were conducted on 393 workers whose work brought them into contact with waste volatile ashes in electric power stations. Eczema was found in 18% of workers and hypersensitivity to chromium was noted in 21% (Kiec-Swierczynska, 1989). In Japan, an investigation was conducted of workers' exposure to organic solvents in liquid waste at 35 waste water disposal facilities (Ikatsu et al., 1989). Incineration was the method of disposal. Workers' exposure to all solvents combined exceeded the acceptable threshold limit value. Exposures were greatest in the incineration areas. In none of these studies were actual levels of workers' exposure to hazardous substances measured or estimated.

Some investigators have measured biologic levels of hazardous substances or their metabolites in incineration workers. In a joint United States/Germany study, individual blood samples were analyzed for polychlorinated dibenzo-p-dioxins (PCDDs) and polychlorinated dibenzofurans (PCDFs) in two cohorts of German incineration workers (Schecter et al., 1995). One group consisted of 10 workers from an incinerator that lacked adequate pollution control. The second group consisted of 11 workers from a newer incinerator that had contemporary pollution control. The comparison group consisted of 25 persons from the general German population, matched for age, sex, and race. Statistically significant increases were found in blood levels of specific PCDDs and PCDFs in workers from the older incinerator in comparison with the levels in both workers at the new incinerator and the comparison group.

In another study, exposure to lead was assessed in solid waste incinerator workers in New York City. Blood lead and erythrocyte protoporphyrin (EP) levels were measured in 56 incinerator workers and 15 nonincineration heating plant workers who served as a comparison group (Malkin et al., 1992). Blood lead levels were increased in incinerator workers (mean = 11.0 µg/dL, SD = 3.6) in comparison with the heating plant group (mean = 7.4 µg/dL, SD = 1.9). All blood lead levels were well below the OSHA action level of 40 µg/dL for workers exposed to lead in the workplace. Curiously, the mean EP level in the comparison group was higher than that in the incineration workers group. Multiple regression analysis showed the increase in blood lead in incinerator workers was associated with the number of times a worker had cleaned the incinerator's precipitator during the previous year, degree of wearing personal protective equipment, and frequency of cigarette smoking. The authors concluded that workplace lead exposure could be minimized by greater use of personal protective equipment, cessation of smoking, and rotating the workforce to reduce contact with precipitator ash.

The incineration of municipal waste can expose workers to mutagenic compounds in combustion gases and particulates if adequate workplace precautions are not taken (Ma et al., 1992). Although municipal waste incinerators differ in design and operation from hazardous waste incinerators, both can present hazards to incineration workers. Ma et al. used the Ames mutagenicity assay to assess three sets of urine samples collected from 37 municipal waste incinerator workers and from 35 reference workers at

---

[1] See Chapter 5 for a summary of Jansson and Voog (1989), who conducted a study in Sweden to analyze whether the incidence of cleft lip and palate had increased after incineration of refuse began.

a water treatment plant. The three urine samples were collected over a one-week period. The first set of urine samples showed a significantly increased presence of both direct-acting mutagens and promutagens (i.e., substances that when metabolized by enzymes become mutagens) in urine of incinerator workers compared with that of referents. This difference was not seen in the next two urine collections. Investigators explained these results by speculating that because incinerator workers knew they were being monitored for indication of workplace exposure to hazardous substances, they began greater use of protective equipment.

The studies summarized in this section show that waste disposal can bring workers into contact with hazardous substances if adequate safeguards are not in place. Moreover, some workplace conditions that hazardous waste workers face are associated with adverse health effects. Studies that include measured levels of hazardous substances in workers are particularly useful for assessing the degree of health risk associated with waste disposal operations.

Work practices should be used that prevent direct contact with hazardous materials, and protective equipment must be provided and used. Training and education of workers and managers on how to conduct waste management operations is essential. Programs of medical monitoring and health surveillance should be implemented if chronic health effects are to be prevented.

Occupational Safety Risk

Traumatic injuries are also a concern for workers' health. Because site remediation work can involve the extensive use of heavy equipment, manual labor, and movement of materials, these pose potential hazards to workers. The safety implications of site remediation work have led to a limited amount of research on estimating the level of risk. Workers who respond to chemical emergencies are also put at safety risk by thermal stress, falling objects, acute chemical exposure, and transportation mishaps.

Knowledge of the equipment used by site-remediation workers is important because equipment operation can bring workers into contact with devices that have the potential to injure if improperly designed, operated, or maintained. Ruttenberg et al. (1996) cite the EPA data showing the 10 most frequently used pieces of equipment on at least one-third of the 100 hazardous waste sites surveyed. These are listed in Table 6.1 in descending order of use at the sites surveyed. Each kind of equipment brings its own safety hazards, and therefore workers' safety programs must include precautions specific to the equipment being used.

Site Remediation Workers

One group of researchers estimated the risk of occupational fatalities associated with hazardous waste site remediation by assessing acute traumatic fatality risks for workers in three kinds of site remediation alternatives (Hoskin et al., 1994). They were interested in assessing the level of risk of fatalities in comparison with the level of adverse health risks. They note that government regulatory agencies typically use 10^{-6} as an explicit human cancer risk criterion. How does the risk of fatal traumatic injury compare with this level of risk? The researchers argue that knowing this level of safety risk would permit incorporating quantitative risk estimates into site remediation plans, thereby improving the level of protection afforded to cleanup workers.

Table 6.1. Site Remediation Equipment in Descending Order of Use (EPA data cited by Ruttenberg et al., 1996).

Backhoe/excavator
Front-end loader
Lowboy
Bulldozer
Generator
Hand tools
Pressure washer/laser
Diaphragm pump
Air compressor
Tractor

Hoskin et al. compiled occupational employment and fatality statistics from U.S. Bureau of Labor Statistics data for 11 states for the years 1979–1981 and 1983. They analyzed these data for 17 occupations known to be associated with three common remediation alternatives: excavation and landfill, capping, and capping plus slurry wall. The occupations included truck driver, laborer, manager, operating engineer, secretary, and others. Assumptions were made about the extent of remediation required under each of the three remediation alternatives and the hours of labor required for each occupation. They then converted the expected number of fatalities, using the Poisson distribution, to the risk of experiencing at least one fatality during remediation.

The excavation and landfill remediation method was found to be the most labor intensive. The most hazardous occupation was that of truck driver, followed in order by laborer, oiler, and bulldozer operator. The other remediation alternatives, capping only and capping plus slurry wall, were both less labor intensive than the excavation and landfill alternative. For both these alternatives, truck drivers and laborers were the occupations with greatest risk. Overall, the risk of experiencing at least one fatality during remediation was calculated to be 1.49×10^{-1} for excavation and landfill, 1.2×10^{-2} for capping, and 1.4×10^{-2} for capping plus slurry wall. Hoskin et al. observed, "The fatality risks to workers engaged in remediation, 10^{-1} to 10^{-2} as found here, are orders of magnitude greater than the 10^{-6} human cancer risk criterion often used in association with remediation risk discussions."

Hoskin et al. postulate that fatality risks vary according to the method of remediation chosen. Less labor-intensive remedial methods represent smaller fatality risks to workers. These findings are important for factoring human safety and health risk into site remedial designs. The study's findings are tempered by limitations inherent in the databases analyzed. Occupational data from only 11 states were evaluated, and the data were not current-year because of limitations in occupational codes in U.S. Census data.

In another study, Travis et al. (1993) developed a method for estimating workers' risk during remediation of sites contaminated with radioactive materials. This work was conducted for the Department of Energy's (DOE) program that remediates DOE hazardous waste facilities that are part of the nuclear weapons production complex in the United States. The researchers characterized remedial workers' risk associated with remediation activities at 17 DOE sites that contained radiologic as well as chemical contaminants.

The method for estimating risks to remedial workers encompassed two elements: (1) radiologic risk and (2) construction- and transportation-related risk. Exposure to hazardous substances was not included in the risk analysis. Cancer mortality risks were estimated separately for exposures from direct radiation and those from inhalation of radionuclides. Three different remediation options were evaluated for each of the 17 DOE sites. The remediation alternatives differed according to site. Using site-specific data, estimates of radiation exposure rates were calculated and doses to workers were estimated. Bureau of Labor Statistics data on fatalities from construction incidents and transportation mishaps were used for construction and transportation risks. Only fatality rates from off-site transportation were considered in the analysis.

Travis et al. presented their findings as fatality risk for direct radiation exposure (1×10^{-13}–8×10^{-2}), inhalation exposure of radionuclides (2×10^{-12}–8×10^{-2}), general construction work (2×10^{-4}–6×10^{-1}), and off-site transportation (2×10^{-6}–2×10^{-1}). The authors concluded, "Radiation-induced cancer fatalities from direct radiation and inhalation of radionuclides are on the average nine times lower than fatality risks from construction and transportation accidents." They further concluded that radiation doses were high, warranting stringent worker-protection measures. Their data do not permit any generalizations about the kind of remediation alternatives that pose the greatest risk to workers. However, their data do make clear that the risk (radiation and construction/transportation) posed to workers varies according to site conditions and remediation alternative.

The studies of Hoskin et al. and Travis et al. are important because they show that waste site remediation presents quantifiable fatality risks to workers that depend on the choice of site remediation. Construction labor and transportation work are occupations that rank high on risk fatality for remediation workers.

NUMBERS OF HAZARDOUS WASTE WORKERS

One source notes, "Cleaning up the nation's hazardous waste sites is an enormous undertaking, requiring the efforts of millions of workers and hundreds of billions of dollars. On-site remedial action alone during the years 1990–2010 will utilize three million job years, or 4.5 billion hours, of labor. Operations and maintenance work will require another one billion labor hours" (Ruttenberg et al., 1996). Although the precision of these numbers is unknown, remediation of hazardous waste sites and emergency responding will certainly require large numbers of workers through part of the twenty-first century. What are the estimates of the current number of hazardous waste workers and future employment trends?

Site Remediation Workers

Although remediation of hazardous waste sites is a relatively new industry in the United States, it already involves large numbers of workers. The precise number of waste-site remediation workers is unknown, but some rough estimates of the workforce are available. One source used labor–demand models that estimate the kinds of skills needed to implement the various stages of a remedial response (i.e., site characterization, design, and cleanup) (Warhit, 1995). Drawing on DOE and DOD estimates of site-remediation costs, the number of professional workers involved in remediating DOD and DOE uncontrolled waste sites was projected to approximate 86,000 persons in fiscal year 1995 and to increase by 29% to approximately 111,000 workers in fiscal year 1998 (Warhit, 1995). Included in these figures are engineers, scientists, managers, and technicians.

The demand for craft labor at federal sites was estimated to be about 131,000 full-time workers in fiscal year 1995, decreasing to about 92,000 craft laborers in fiscal year 1998 (Warhit, 1995). For nonfederal waste sites, the labor demand was estimated to be about 84,000 craft laborers in fiscal year 1995. Many assumptions underlie these estimates. A key assumption is that current remediation procedures and technologies will characterize those used in the future. However, as cost-containment concerns mount, site remediation methods and technologies will likely change, thus changing the mix and number of remediation workers—both professionals and craft laborers. The number of craft laborers is particularly important for health risk purposes because they have the greatest day-to-day contact with sites and therefore potential exposure to substances found at sites.

In another study, estimates projected through year 2010 (Ruttenberg et al., 1996) of the demand for remediation workers include jobs at NPL, RCRA, DOE, DOD, underground storage tank, and state/private sites. They found that "...remediation job demand is expected to grow by 60%, or almost 300,000 jobs, from the 1990 to 1995 five year period through the five year period 1995 to 2000—from 447,000 to 740,000. Demand for jobs continues to grow by nearly another 300,000 in the 2000 to 2005 time interval. During this peak period nearly two million jobs will require workers. As many as 7.5 million more workers will require training—either basic or refresher." However, as Ruttenberg et al. observe, the actual number of jobs will depend on political and policy decisions concerning the amount of resources dedicated to site remediation.

Emergency Response Workers

The precise number of persons in the occupational workforce who respond to HazMat events is unknown. However, the ATSDR has made some approximations (Barry, 1996).

A community's fire department typically has primary responsibility for HazMat incidents. The National Fire Protection Association (NFPA) estimates 1,073,600 firefighters are in the United States (265,700 career, 807,900 volunteer) (Barry, 1996). Fire, police, and emergency medical service (EMS) workers may encounter emergencies that involve hazardous materials. Hazardous materials first responders in the United States are often fire service personnel who are specially trained for this kind of response.[2] Title III of CERCLA, as amended in 1986, requires all firefighters, police, and emergency medical technicians who might be the first to arrive at the scene of a release of hazardous substances to be trained to the "first responder" level. Fire service personnel are usually employees of counties or municipalities. Fire department hazardous materials teams most often comprise line fire companies whose personnel have received the additional training required by federal regulations, are specially equipped, and perform hazardous materials response in addition to their firefighting and rescue duties. Federal regulations require the employer to maintain health records for the employees' medical monitoring program.

---

[2] Training requirements are found in federal regulations 29 CFR 1910.120 and 40 CFR part 300.

The ATSDR contacted several organizations to seek information on the number of hazardous materials workers and to determine whether a national database exists on these workers' health (Barry, 1996). They contacted the National Fire Protection Association, the International Association of Fire Fighters (IAFF), the Chemical Manufacturers Association, the International City/County Mangers Association, the EPA's Chemical Emergency Preparedness Program, the National Fire Academy, the National Association of SARA[3] Title III Officials, the National Association of Emergency Medical Technicians, NIOSH, OSHA, and the U.S. Fire Administration.

None of the organizations contacted could estimate the number of public service employees or volunteers that have hazardous materials response assignments as a primary duty or are assigned to a hazardous materials response team (HMRT). NIOSH and IAFF advised that no centralized repository exists for health and medical monitoring data for hazardous materials workers. OSHA advised that no information on public employees is collected because the Occupational Health and Safety Act of 1970 excludes public employees from coverage.

The National Association of SARA Title III Officials responded to ATSDR that 3,900 Local Emergency Planning Committees (LEPCs) in the United States are established in accordance with the provisions of Sec. 301 of SARA Title III. Using a conservative estimate of one organized hazardous materials response team per planning district (3,900 planning districts), staffed for three shifts with six persons per shift (Kansas City, Missouri, Fire Department estimate), the ATSDR estimated that 70,200 first line responders have primary duties related to hazardous materials response (Barry, 1996). This figure does not include responders in private industry, police, emergency medicine services, and other firefighters who provide security, medical response, equipment operation or technical support at the scene of a hazardous materials incident.

Other emergency service workers who might respond in some degree to HazMat incidents include 10,000 registered emergency medical technicians (source: National Association of Emergency Medical Technicians), 97,000 registered emergency room nurses (source: Emergency Room Nurses Association), and 26,000 emergency medicine physicians (source: American College of Emergency Physicians).

With the caveats stated previously, the following numbers of first responders and allied workers are estimated to be:

HazMat First Responders = 70,200
HazMat teams in 60 major companies = 1,080[4]
Registered EMTs = 100,000
Registered Emergency Room Nurses = 97,000
Emergency Room Physicians = 26,000

Therefore, the ATSDR estimates the number of first responders to hazardous materials emergencies to approximate 70,000 workers, with another approximately 225,000 per-

---

[3] Superfund Amendments and Reauthorization Act of 1986. This act amended and reauthorized the Comprehensive Environmental Response, Compensation, and Liability Act of 1980.

[4] Assuming 60 companies, 6 persons per team, 3 shifts (Chemical Manufacturers Assoc. CHEMTRAC [Barry, 1996]).

sons potentially involved in health care-related work. As noted, these sets of figures do not include the police and fire personnel who respond to chemical emergencies.

In a separate analysis, the number of emergency responders was derived on the basis of the national number of local fire departments (28,000), in-plant emergency response teams (22,000), commercial hazardous materials response teams (750), and public hazardous materials response teams (200) (cited by Ruttenberg et al., 1996). The same source estimated the average number of workers assumed to be involved in response work was ten each for local fire departments, commercial response teams, and public response teams, and eight workers per in-plant response team. This results in approximately 500,000 workers engaged in response work. This figure is higher than the ATSDR estimate because it includes fire service personnel.

HEALTH AND SAFETY REGULATIONS

Protecting workers who clean up hazardous waste sites and those who respond to chemical emergencies was not included as a requirement in the original CERCLA, as enacted by Congress in 1980. Congress believed that existing provisions of the Occupational Safety and Health Act of 1970 (OSHAct), which created the Occupational Safety and Health Administration (OSHA), would suffice for protecting remediation site workers. The OSHAct is intended to ensure safe and healthful conditions in the American workplace. It requires that employers take steps to protect employees from recognized workplace hazards that can cause illness or injury.

OSHA has developed numerous safety standards to protect the safety and health of workers involved in construction work and general labor that are applicable to activities on hazardous waste sites (Andrews, 1990). However, as Melius (1995) described, labor unions and federal government agencies became aware after 1980 of the need to upgrade protection for hazardous waste workers, largely because OSHA construction work standards were thought to be inadequate for site remediation work. Moreover, major chemical emergencies like the one in Elizabeth City, New Jersey, in 1980 which caused large numbers of emergency responders to become acutely ill following exposure to toxic fumes, identified emergency responders as another occupational group at health risk. In this episode, emergency response workers had been inadequately trained and lacked fully protective equipment. By extrapolation from the conditions faced by emergency responders, site remediation workers were considered to be at similar risks because of potential contact with hazardous substances. It remained for the 1986 amendments to CERCLA to mandate into law workers' health and safety protections for hazardous waste operations.

HAZWOPER Standard

The Superfund Amendments and Reauthorization Act of 1986, Title I, Section 126, directed OSHA to develop regulatory actions (called standards) to cover hazardous waste workers (Moran, 1994). In 1989 OSHA promulgated its final standard for Hazardous Waste Operations and Emergency Response, which became known as the HAZWOPER standard[5] (Moran, 1994; Melius, 1995). The standard went into effect on

---

[5] Found in federal regulations 29 CFR 1910.120.

March 6, 1990, and covers both employers and employees engaged in the following operations:

- Cleanup operations required by a government body (federal, state, local, or other) that are conducted at uncontrolled hazardous waste sites, including initial investigations of government-identified sites conducted before the presence or absence of hazardous substances has been identified;
- Corrective actions involving cleanup operations at sites covered by the Resource Conservation and Recovery Act of 1976 (RCRA), as amended;
- Voluntary cleanup operations at sites recognized by federal, state, local, or other government bodies as uncontrolled hazardous waste sites;
- Operations involving hazardous wastes that are conducted at treatment, storage, and disposal facilities regulated under RCRA;[6] or by agencies under agreement with the EPA to implement RCRA regulations; and
- Emergency response operations for releases of, or substantial threats of, hazardous substances without regard to the location of the hazard.

The HAZWOPER standard therefore addresses three major categories of hazardous waste management: the remediation of uncontrolled hazardous waste sites; emergency response operations in reaction to the release, or likely release, of hazardous substances; and hazardous waste operations at facilities permitted to dispose of hazardous waste under the RCRA statute, as amended. Hazardous waste operations covered by RCRA are not included in this chapter.

In 1989, the EPA extended the HAZWOPER standard to cover state and local government employees engaged in "hazardous waste operations," as defined by HAZWOPER, in states that do not have state plans approved under Section 18 of the OSHAct of 1970 (EPA, 1989).

The HAZWOPER standard is a comprehensive, prescriptive regulation. Although it has been criticized for not being comprehensive enough (Moran, 1994), it sets forth the following elements that cover employers and employees:

1. **Safety and Health Program:** Employers must develop and implement written safety and health programs for their employees engaged in hazardous waste operations. The programs must be designed to identify, evaluate, and control safety and health hazards, and provide for emergency response in hazardous waste operations. The standard specifies what a safety and health program must include.
2. **Site Characterization and Analysis:** Hazardous waste sites must be evaluated to identify site-specific hazards and to determine the appropriate safety and health control procedures needed to protect employees. These must include preliminary evaluations of site characteristics before site entry. Detailed surveys must be performed after preliminary evaluation to further identify any site hazards.
3. **Site Control:** Site control procedures must be implemented before the start of cleanup work to control employees' exposure to hazardous substances.
4. **Training:** "All workers on a site who could be exposed to hazardous substances and the workers' supervisors and managers must receive training. Site workers must receive a minimum of 40 hours of off-site instruction and no fewer than three days of field supervision at the site. Workers who have only occasional contact with the

---

[6] Found in federal regulations 40 CFR Parts 264 and 265.

site must receive at least 24 hours of off-site training and one day of direct field supervision. Eight hours of refresher training is required annually for general workers, managers, and supervisors" (Melius, 1995).

5. **Medical Surveillance:** Hazardous waste workers must receive baseline medical examinations and annual follow-up examinations if they are expected to be exposed to hazardous substances above permissible exposure limits or if they will be wearing respirators 30 or more days annually. As Melius (1995) notes, "The content of the examination is not specified, but there are specific requirements regarding recordkeeping and communication of the results of the examination." Furthermore, the HAZWOPER standard requires an examination as soon as possible of any employee who is injured or becomes ill possibly because of exposure to hazardous substances.

6. **Engineering Controls, Work Practices, and Personal Protective Equipment:** HAZWOPER requires employers to use engineering controls, implement work practices, and provide personal protective equipment to protect employees against exposure to hazardous substances.

7. **Monitoring:** Employers must monitor sites if there is a question of workers' exposure to hazardous substances at concentrations posing potential health hazards. This includes air monitoring on initial site entry to identify any dangerous levels of hazardous substances. Periodic monitoring must be conducted when the possibility exists of exposure levels in excess of permissible exposure limits. Monitoring must be conducted of high-risk employees likely to have the highest exposure to hazardous substances and health hazards.

8. **Informational Programs:** Employers must develop and implement programs, as part of site Safety and Health Programs, to inform employees, contractors, and subcontractors of the nature, level, and degree of exposure likely as a result of hazardous waste operations.

9. **Handling Drums and Containers:** Hazardous substances and contaminated soils, liquids, and other residues must be handled, transported, labeled, and disposed of in accordance with details in the HAZWOPER standard.

10. **Decontamination:** All employees leaving contaminated areas must be decontaminated so that hazardous substances are not carried off-site. Personal protective equipment and work clothing must be cleaned and showers and rooms provided to change clothing when needed.

11. **Emergency Response:** Employers must develop and implement written emergency response plans to handle anticipated emergencies before hazardous waste operations begin. These plans must be available for inspection and copying by employees, their representatives, and relevant government agencies.

12. **Illumination:** Areas accessible to employees must be lighted to meet illumination intensities listed in the standard.

13. **Sanitation at Temporary Facilities:** Potable water must be provided to employees on site and toilet facilities must meet requirements specified in the standard.

14. **New Technology Procedures:** Employers must develop and implement procedures for introducing new technologies and equipment developed for improved protection of employees working with hazardous waste cleanup operations and implement the new technologies as part of site safety and health programs.

In summary, OSHA has developed and promulgated specific regulations for hazardous waste site remediation as mandated in CERCLA, as amended (Moran, 1994). These regulations require employers of remediation workers to develop and implement comprehensive safety and health programs for workers involved in hazardous waste operations. A significant requirement of the regulations is periodic medical surveillance of hazardous waste site workers, although no agreement exists on how to implement this requirement (Melius, 1990). A subsequent section of this chapter discusses medical surveillance of hazardous waste workers.

Coverage of Remediation Workers

After OSHA's HAZWOPER standard was promulgated, its requirements became mandatory at every hazardous waste site and in instances where emergency releases of hazardous substances occur.

Coverage of Emergency Response Workers

The HAZWOPER standard has requirements that pertain to emergency response workers. These workers respond to emergencies caused by uncontrolled release of hazardous substances; for example, release of substances due to transportation mishaps. The standard requires emergency response plans, as well as procedures for identifying hazardous substances released, selection of personal protective equipment, and the use of appropriate decontamination procedures. First responders must have at least eight hours of training. HAZWOPER requires hazardous materials technicians, specialists, and supervisors to have additional training. Only members of designated HazMat teams require medical surveillance. Persons injured while responding must receive medical examination as soon as possible.

Compliance with HAZWOPER

The HAZWOPER standard, like other OSHA regulations, requires employers' compliance under force of law, the OSHAct of 1970. OSHA conducts workplace inspections as part of its compliance program. Since 1992 the agency has conducted several hundred HAZWOPER compliance inspections annually, resulting in about 1,000 to 1,500 citations for violations of the HAZWOPER standard (Melius, 1995). As Melius cautioned, this number of inspections and mix of citations may not accurately represent the level of noncompliance with the standard because it does not represent a statistically drawn sample of work sites or conditions. Nevertheless, the information in the citations accumulated since 1992 is useful for hazard identification and improved worker protection.

Melius evaluated OSHA's HAZWOPER compliance database and found that more than half the citations issued for noncompliance were attributable to two causes. Specifically, 27% of citations involved a deficiency in the employer's written safety and health program plan, and 25% were for various deficiencies in engineering controls and personal protective equipment (Melius, 1995). The categories and distribution of HAZWOPER violations are shown in Figure 6.2. Few data are available on the actual number and severity of injuries and health problems associated with site remediation work, but full compliance with the HAZWOPER standard would help prevent workplace morbidity and mortality.

HEALTH SURVEILLANCE OF HAZARDOUS WASTE WORKERS

The HAZWOPER standard requires employers to provide hazardous waste workers with baseline medical examinations and annual follow-up examinations if they are ex-

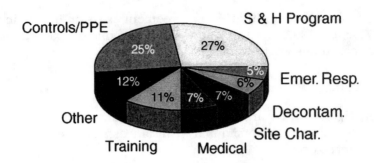

Figure 6.2. HAZWOPER citations for years 1992–1994 (Melius, 1995, with permission of author).

pected to be exposed to hazardous substances above permissible exposure limits or if they will be wearing respirators 30 or more days annually. The employers retain the medical records. Getting an early picture of any adverse health trends among hazardous waste workers requires a health surveillance database. Only one such database appears to exist.

The database was established by the ATSDR and the Laborers' Health and Safety Fund of North America (LHSFNA), who created a health surveillance program for members of the Laborers' International Union of North America (LIUNA) and other construction trade workers who undergo training for hazardous waste work. The LHSFNA is a joint effort between LIUNA and site remediation contractors. LIUNA represents a large number of laborers, construction workers, and others who are engaged in remediation work at hazardous waste sites.

The health surveillance system was created to identify and prospectively follow a cohort of construction trade workers who are trained to carry out hazardous waste cleanup. The surveillance system will be used to detect any trends and clusters in the occurrence of occupational illnesses and injuries associated with hazardous waste cleanup and thereby to identify targets for disease and injury prevention. The primary data for this surveillance system will be acquired by baseline, in-person interviews and annual follow-up telephone interviews. Staff of the LHSFNA track the cohort and conduct the baseline and follow-up interviews. The expected size of cohort enrollment each year will be approximately 2,000 workers.

LHSFNA collaborates with the Laborers–Associated General Contractors Training and Education Fund (Training Fund) to facilitate access to construction trade workers undergoing training for hazardous waste work. The Training Fund provides the HAZWOPER-required, initial (80-hour) training course for hazardous waste workers, offered at 14 centers operated by the Fund. The capacity of this training network is approximately 2,000 trainees per year. The baseline interviews are conducted at the training centers while the workers are attending the initial training course.

Baseline and follow-up interviews use a core questionnaire, developed by the ATSDR and the LHSFNA, which asks about an individual's medical history, reproductive history, personal habits, and occupational history. A supplemental questionnaire is administered to obtain more focused information on occupational and medical history (including neurologic, respiratory, and gastrointestinal symptoms), injuries, heat stress, hearing loss, chemical burns, skin rashes, and poisonings.

Additional sources of data for the surveillance system include death certificates to determine cause of death and medical records to verify information obtained from the baseline questionnaire on morbidity. Data analyses are performed jointly by the ATSDR and the LHSFNA.

SYNTHESIS

Care must be exercised that the workers who remediate hazardous waste sites are not harmed by the conditions of their work. The removal of soil or water contaminated with toxic substances certainly presents the potential for adversely affecting the health of cleanup workers if the necessary precautions are not taken. Work conditions can include exposure to toxic chemicals, biologic hazards, and radiation; strenuous work, and thermal stress. Primary protection measures include personal protective equipment, engineering controls, site monitoring equipment, and training and education of workers. Secondary protection measures include periodic medical evaluations of remediation workers and a program of workers' health surveillance. An adequate program of workers' health protection requires both primary and secondary measures. Similarly, workers who respond to chemical emergencies are put at safety risk because of conditions that can include thermal stress, falling objects, acute chemical exposure, and transportation mishaps. The Superfund Amendments and Reauthorization Act of 1986 directed OSHA to develop regulatory actions (called standards) to cover hazardous waste workers. In 1989, OSHA promulgated its final standard for Hazardous Waste Operations and Emergency Response, which became known as the HAZWOPER standard.

Estimates of the demand for remediation workers were projected by one source through the year 2010. The projections include jobs at NPL, RCRA, DOE, DOD, underground storage tank, and state/private sites. They found "...remediation job demand is expected to grow by 60%, or almost 300,000 jobs, from the 1990 to 1995 five year period through the five year period 1995 to 2000—from 447,000 to 740,000. Demand for jobs continues to grow by nearly another 300,000 in the 2000 to 2005 time interval. During this peak period nearly two million jobs will require workers. As many as 7.5 million more workers will require training—either basic or refresher." However, the source notes that the actual number of jobs will depend on political and policy decisions concerning the amount of resources dedicated to site remediation.

The number of emergency responders has been derived on the basis of the national number of local fire departments, in-plant emergency response teams, commercial hazardous materials response teams, and public hazardous materials response teams. This equates to approximately 500,000 workers engaged in response work. Another 225,000 persons are potentially involved in health care-related work.

Given the relative recency of large-scale site-specific remediation work, there is only a small scientific database on any potential health effects associated with such work. According to one source, clinical and research findings from clinical centers performing health surveillance examinations on hazardous waste workers have not found health abnormalities related to the workers' activities. From a public health perspective, this is a very desirable outcome. However, the investigators caution that the lack of any health problems in hazardous waste workers may be due to the fact that most of the workers examined had performed feasibility (i.e., inspection and assessment) work, not actual remediation of sites.

Traumatic injuries must be a concern in addition to potential adverse effects on workers' health from site remediation work, response to chemical emergencies, and waste disposal operations. Site remediation work can involve the extensive use of heavy equipment, manual labor, and movement of materials. A study of 17 hazardous waste sites showed three crafts accounted for more than 60% of the payroll: operators of equipment, laborers, and truck drivers. The investigators note that the mix of crafts varies with the site being remediated and the stage of remediation. Each craft brings its own level of health and safety risk. One study found that fatality risks vary according to the method of remediation chosen. Less labor-intensive remedial methods represent a smaller fatality risk to workers. These findings are important in terms of factoring human safety and health risk into site remedial designs.

There are few data on the safety risk of persons who respond to emergency chemical events. One set of surveillance system data collected in 1994 by 12 states showed a total of 4,244 events. Of the 4,244 events, there were 20 deaths and 574 events led to evacuation of people. Victims according to population group showed employees (45%), general public (44%), and responders (11%) to be the groups most often injured.

As the nation's hazardous waste is managed through waste site remediation, permitted disposal of materials, and responses to chemical spills, it is likely that a larger workforce will be required. How the work is planned and performed will determine health and safety risk to workers. Occupational health standards, training and education, and surveillance will all be needed for workers' protection.

Chapter Seven

Health Promotion

What is health promotion and how does it relate to the public's concerns about hazardous waste? This chapter contains a definition of health promotion and summarizes the strategies used to improve both personal and community health. Public health practitioners use health promotion strategies, which originated in Europe in the nineteenth century, to prevent or mitigate both general and specific causes of disease. These strategies are adapted here as actions to reduce the human health impacts caused by exposure to hazardous substances.

The subject of health promotion might at first glance seem out of place in a book about hazardous waste. What does health promotion have to do with hazardous substances and waste sites? Moreover, what does "health promotion" mean? This chapter will answer these questions and, it is hoped, contribute to an improved linkage between public health activities and environmental protection programs. An improved linkage is important because human health programs have traditionally been the province of public health agencies, whereas environmental protection has become a regulatory program in separate departments at the federal and many state government levels. This division of interests, resources, and authorities has not always worked well, because opportunities for coordinating and sharing resources, data, and information are sometimes missed due to organizations' myopia and conflicting statutory authorities.

Let's begin with the observation that federal environmental statutes can be thought of as trees with two branches, each associated with an aspect of protection. The two branches are protection of the public's health and protection of the environment. Consider the example of the CERCLA statute. In 1980, the U.S. Congress enacted CERCLA, which was signed into law by President Carter. Predicate events that shaped the formation of this statute included widespread media attention about abandoned hazardous waste dumps leaking into homes in Love Canal, New York, and the discovery of thousands of leaking, buried drums in Kentucky, in the so-called "Valley of the Drums." The public expressed great concern that chemicals released from "toxic dumps" would cause human health problems. Congress acted on this concern to protect human health and prevent damage to natural resources and environmental quality. At Love Canal, health investigators found an elevated incidence of low birth weight babies and children with reduced height born to area residents, supporting the community's fear that adverse health effects were occurring (Vianna and Polan, 1984; Paigen et al., 1987, Chapter 5).

CERCLA has at its core the protection of both human health and the environment. Protection of human health is addressed through various activities mandated of the ATSDR, a federal public health agency, and through various risk management actions (e.g., site remediation) for which the EPA has responsibility. The EPA and states also address protection of the environment through cleanup of waste sites, emergency removal of hazardous substances, and prevention of contaminated sources (e.g., groundwater) from spreading into uncontaminated environmental media. Such actions to protect the environment also benefit the public's health because contact with hazardous substances is decreased or eliminated.

A review of other federal environmental protection statutes shows a commitment to protecting human health as a key goal (Johnson and Weimer, 1995). This is neither surprising nor inappropriate, considering the strong support the American public gives to environmental protection and health-risk reduction. The ensuing sections describe how health agencies address health protection and promotion within a public health context and apply them to hazardous waste issues.

HEALTH PROMOTION STRATEGIES

What is health promotion? Before discussing some history of the term, it is useful to reflect on a few guidelines that persons in industrialized countries now know about enhancing personal health. These include the health hazard of tobacco use, the importance of personal weight control and regular exercise, dietary restriction of certain fats and cholesterol, temperate consumption of alcohol, and regular physical examinations by a physician. These actions, or "lifestyle behaviors," are individual choices in daily activities that can affect one's health.

Information from countries with the economic, scientific, and medical care capacities to develop scientific research provides the fuel to drive a public health engine that promotes health communication messages and actions. For example, epidemiologic studies over many years on the health of cigarette smokers have led to public health campaigns to stop tobacco use. Federal, state, and local health agencies constitute a public health system to inform the public and health care providers about the health hazard of cigarette smoking.

The linkage of science, public health networking, and personalized health risk communication has well served the American public, particularly in situations where the public views the science as robust, and the health consequences derived from the science are clear, convincing, and have personal health implications. Infectious diseases are examples of clear-cut diseases caused by specific organisms, such as the Human Immunodeficiency Virus (HIV), which was found to cause AIDS. Public health campaigns to prevent HIV transmission have followed from these scientific findings.

Direct linkage of specific diseases or illnesses to hazardous substance exposure is not nearly as strong as with infectious diseases. This is partly because environmentally related diseases or other adverse health conditions (e.g., cancer, birth defects, and hypertension) can have other causes besides environmental factors. For example, although considerable scientific evidence associates prenatal exposure to lead with effects on cognitive development and lowered intelligence (e.g., ATSDR, 1993b), genetic disorders and other risk factors can also retard a child's development. A few notable exceptions for which specific environmental hazards are linked with specific diseases include asbestosis, a lung condition unique to asbestos exposure.

As the examples of lead and asbestos suggest, health promotion embraces targeted populations of people (i.e., young children, asbestos workers) as well as individuals within given populations. What then are the historical antecedents of health promotion?

Milton Terris, Editor of the *Journal of Public Health Policy*, has summarized the historical background on what public health agencies now call health promotion. He notes that confusion still exists on what is meant by health promotion (Terris, 1992). To help reduce this confusion, Terris used historical antecedents of social medicine to construct his concept of health promotion. He described events in Europe during 1820 to 1850 that identified poverty and destitution as contributing factors in epidemic disease. Citing the work of Rudolf Virchow in Germany, who in 1847 investigated an epidemic in the industrial districts of Silesia, Terris concluded, "...the causes of the epidemic were as much social and economic as they were physical. The remedy he [Virchow] recommended was prosperity, education, and liberty, which can develop only on the basis of 'complete and unrestricted democracy.'" He observed that this nineteenth century work in Germany and elsewhere in Europe established a link between general factors like social conditions and consequences to the public's health. Social medicine is a term sometimes applied to this body of thought and work.

According to Terris, the first use of the term "health promotion" is attributed to the medical historian Henry E. Sigerist, who in 1945, Terris states, "...defined the four major tasks of medicine as: (1) The promotion of health, (2) The prevention of illness, (3) The restoration of the sick, and (4) Rehabilitation. He [Sigerist] stated that 'Health is promoted by providing a decent standard of living, good labor conditions, education, physical culture, means of rest and recreation.'" In 1986, the World Health Organization, a component of the United Nations, and Canadian health authorities organized the International Conference on Health Promotion, resulting in the Ottawa Charter for Health Promotion, adopted by the 38 countries in attendance. The Ottawa Charter states, "The fundamental conditions and resources for health are peace, shelter, education, food, income, a stable ecosystem, sustainable resources, social justice and equity," and "...health promotion demands coordinated action by all concerned: by governments, by health and other social and economic sectors, by nongovernmental and voluntary organizations, by local authorities, by industry, and by the media." The Charter remains the definitive international statement on health promotion.

According to Sigerist, the promotion of health and the prevention of illness are separable tasks in the practice of medicine. Public health practitioners have adopted the prevention of disease as their major objective (IOM, 1988). One model of disease prevention stipulates five elements: identification (or surveillance) of disease, evaluation of causal factors of the identified disease, control of the causal factors, dissemination of findings and disease prevention recommendations, and infrastructure to support disease prevention efforts. In the example of AIDS, disease surveillance conducted by health departments identified an unexpected high rate of a rare cancer (Kaposi's sarcoma); epidemiologic and toxicologic studies to evaluate possible causes of the disease identified what is now called the Human Immunodeficiency Virus as the cause; control efforts were implemented through programs of public education, virology, and vaccine research; results were disseminated worldwide; and resources and programs were instituted to assure an infrastructure to support disease prevention efforts.

In contrast to Sigerist's separation of health promotion and illness prevention, one can view illness prevention as an element of health promotion. In preventing exposure of young children to lead in the environment, health promotion embraces targeted populations as well as individuals within these populations. Illness prevention through periodic blood lead screening alone provides no assurance that the deleterious effects of lead in young children will be averted. Instead, response occurs when blood lead screening is combined with community health education about lead, technical assistance in environmental medicine, follow-up of individual children with elevated blood lead levels, and elimination of sources of children's exposure to lead.

A person's or a community's health is obviously enhanced when actions to prevent illness have been taken. However, more than preventing illness is required if an individual's complete health status is to be promoted. The view of illness prevention taken in this book is that it is a component—a vital component—of a larger system of actions that promote an individual's or community's overall health status.

On the definition of health promotion, Terris observes, "The Ottawa Charter defines health promotion broadly, as 'the process of enabling people to increase control over, and to improve, their health.'" He notes that health promotion goes beyond healthy lifestyles to include fundamental social and economic conditions (i.e., peace, shelter, education, stable ecosystem, food, income, social justice, and equity). The Ottawa definition of health promotion led Terris to define three interlocking components of the health promotion strategy [with emphasis added for the purposes of this chapter]:

1. *Intersectoral action* to achieve healthy public policy as well as public health policy. [The coordinated action by many groups for purpose of health promotion is an intersectoral arrangement, that is, it cuts across sociopolitical sectors.]
2. Affirmation of the active role of the public in using health knowledge to make choices conducive to health and to increase *control over their own health and over their environments.*
3. *Community action* by people at the local level. Strengthening public participation and public direction of health matters is at the heart of the health promotion strategy" (Terris, 1992).

The broadening of the concept and practice of health promotion, based on social medicine tenets as outlined by Terris, has been called the "new health promotion movement" by some public health specialists. For example, Robertson and Minkler (1994) note that the new health promotion movement is also known as community health promotion and refer to the movement as being like other social movements. In particular, they observe that the new health promotion paradigm has moved past the former, more narrow approach of emphasizing health promotion primarily through changes in an individual's lifestyle and professionally-based interventions. In their view, "...health is seen as instrumental, a means rather than an end. Health is what one must have to accomplish other things in one's life." They identify four elements that they identify as prominent features of the new health promotion movement (Robertson and Minkler, 1994):

- Broadening the definition of health and its determinants to include the social and economic context within which health—or, more precisely, non-health—is produced.
- Going beyond the earlier emphasis on individual lifestyle strategies to promote health through broader social and political strategies.
- Embracing the concept of empowerment—individual and collective (i.e., community)—as a key health promotion strategy.
- Advocating the participation of the community in identifying health problems and strategies for addressing the problems.

Robertson and Minkler emphasize that two of the key concepts in the new health promotion movement are empowerment and community involvement. An emphasis on empowerment becomes a primary health promotion strategy in this new paradigm. They observe that at the core of empowerment is the concept of power, defined as the ability to control the factors that determine one's life. Empowerment is the process by which persons and communities are enabled to take power and act effectively in changing their lives and their social and physical environments.

Robertson and Minkler (1994) assert,[1] "...health promotion practitioners who truly facilitate empowerment do so by assisting individuals and communities in articulating both their health problems and the solutions to address those problems. **By providing access to information, supporting indigenous community leadership, and assisting the community in overcoming bureaucratic obstacles to action, such practitioners may contribute to a process whereby communities increase their own problem-solving abilities** (emphasis added)..." Empowerment then becomes a matter of health promotion practitioners' sharing information, consulting with communities, and advocating ways whereby problems are solved by the community. An example of community empowerment is found in the Stratford, Connecticut, case study described later in this chapter.

Community involvement (or community participation) is the other key feature of the new health promotion movement (Robertson and Minkler, 1994). Inherent in community involvement is the need to define "community." This is a subject of considerable challenge to health promotion practitioners—who is the community whose health needs promoting? Robertson and Minkler refer to community as a social space that has connectedness. Connectedness means a group of persons who share the same basic values and organization. A goal of the new health promotion movement is to bring communities into full, equal partnership with health professionals to set a particular health promotion agenda.

The role of health professionals will change, as Robertson and Minkler observe, if empowerment and community participation are integrated into programs of health promotion. **In the new health promotion movement, health professionals become consultants to the community, rather than being experts who merely define the community's needs and provide solutions through traditional public health interventions.** "Rather than service provider and client, the community and the professionals are equal partners in setting the health agenda for the community" (Robertson and Minkler, 1994). This change in government's role from expert to consultant has significant implications for health promotion efforts, as will be described.

In reflecting on the work of Terris (1992) and Robertson and Minkler (1994), the definition of health promotion articulated in the Ottawa Charter is adopted for the purposes of this chapter. Health promotion is the process of enabling people to increase control over, and to improve, their health (Ottawa Charter, cited by Terris, 1992). Empowerment of the individual and the community, combined with community participation in health promotion efforts, are key components of contemporary health promotion. Furthermore, Terris's three components of the health promotion strategy are paraphrased, as shown in the inset box, and adopted for use in this book. The following sections relate the elements of health promotion to hazardous waste concerns.

---

[1] Robertson, A. and M. Minkler (1994). *Health Education Quarterly* 21:295-312. Copyright Sage Publications, Inc. Excerpted by permission of Sage Publications, Inc.

HAZARDOUS WASTE AND HEALTH PROMOTION

Health promotion efforts must begin with a science base and proceed to empowering individuals to make decisions that improve their health. Terris's three components of health promotion lie between the start and end of any health promotion effort. Previous parts of this book summarized key scientific findings that associate uncontrolled releases of hazardous substances with contamination of environmental media and human health impacts. What does this science portend for health promotion?

In a health promotion sense, what actions should be undertaken when a health study or public health assessment indicates that a person or population is at elevated risk of exposure and potential adverse health effects? And who should take the actions? These are not rhetorical questions. They go to the heart of the practical concern expressed by some community groups and grassroots environmentalists who eschew health studies in favor of health care. As an example of this philosophy, one grassroots organization testified to Congress, "...It has been very difficult for ATSDR to understand that to Superfund communities 'medical assistance' and 'health studies' do not mean endless rounds of inconclusive studies, a few workshops with local doctors, or poster contests. Communities urgently need and want health protection and treatment for very real physical damage caused by exposure to toxic substances; a far better use of ATSDR's budget would be for community-based and -directed health care" (Dunham, 1995).

This concern by some environmentalists is understandable in the context of their feelings of inadequate response by government agencies to community health concerns, unsatisfactory dialogue between government agencies and local residents, limited access to health care by community residents, and frustration over perceived lengthy response time by government agencies to remediate sites. Furthermore, communities can become suspicious of more health studies when inconsequential or no actions followed the completion of past health studies. However, the position of "no more health studies" is in fundamental conflict with public health agencies' experience and with medical practice. Without knowing the nature and extent of an individual's or a community's health problem, it is impossible to implement a medical response plan, much less a health promotion effort. What is needed is an evident (i.e., apparent to the affected community) linkage between a health investigation and specific, identifiable health promotion efforts when such efforts are merited. What health promotion element's efforts should be pursued when there is a science-based reason for action?

Health Promotion

Health promotion is the process of enabling people to increase control over, and to improve, their health (Ottawa Charter, quoted by Terris, 1992).

The Elements of Health Promotion (paraphrased from Terris, 1992):

- Intersectoral action
- Enabling personal control
- Empowering community action

The sections that follow will relate Terris's three elements of health promotion to hazardous waste and public health issues.

Intersectoral Action

A sector is defined as a distinct part of society, economy, group, or entity. Intersectoral health promotion actions therefore refer to efforts taken between distinct parts of a sociopolitical structure that can influence human health. The parts include community groups, government, nongovernment organizations, private industry, academic institutions, and others. Intersectoral action translates simply to forming partnerships to solve social problems. This is much easier said than done. Each organization, including government, brings certain biases, restrictions, and limitations to a partnership. Moving past these built-in conditions in ways that join common purposes is the essence of partnerships, and therefore intersectoral action.

In the context of hazardous waste concerns, what kinds of intersectoral actions are possible? There are many possibilities, but three seem of primary importance. They include (1) hazardous waste management, (2) science that supports hazardous waste management and impacts of hazardous waste on human and ecologic health, and (3) public health interventions in situations where exposure to hazardous substances is of concern. To be successful, each of these three actions must involve activities, resources, and commitment that cut across government agencies and the private sector. This is elaborated in what follows.

The proper management of hazardous waste is surely an intersectoral effort. At the federal government level, legislation crafted by Congress should result in intersectoral action that joins the breadth of stakeholders involved with hazardous waste generation, transportation, and disposal. A similar intersectoral effort should join stakeholders at the state and local level when legislation, regulations, and ordinances are developed on hazardous waste management.

Intersectoral actions to reduce the generation of hazardous waste are particularly germane for mitigation of human health impacts. These efforts should involve the coordinated effort of federal, state, and local government, private industry, and community representatives. The federal CERCLA program to remediate uncontrolled hazardous waste sites is coordinated with state government agencies and engages communities as remedial options are debated, developed, and implemented. However, some community groups have voiced their belief that more local involvement is necessary, especially where environmental justice[2] concerns are thought to be operative (e.g., Robinson, 1996). Voluntary action by private industry to reduce the generation of hazardous waste is occurring more often. This is an intersectoral action because it contributes to a lower possibility of environmental degradation, which has social benefits.

Science is the second area of intersectoral action of relevance to hazardous waste concerns. Decisions on how to remediate hazardous waste sites, protect human and ecologic health, and when to undertake public health interventions like medical monitoring of persons impacted by hazardous substances should all be based on the best contemporary environmental and biomedical science. CERCLA provides for programs

---

[2] Environmental justice is broadly defined as the unfair imposition of environmental hazards on communities because of race, ethnicity, or low income. See Chapter 9.

ATSDR's Site-Specific Intersectoral Health Promotion Strategy

- Develop a health promotion plan
- Interdict any ongoing exposure
- Ascertain any adverse health impacts
- Conduct community health education
- Facilitate technical assistance in environmental medicine
- Conduct medical monitoring, if indicated
- Facilitate the referral of patients
- Evaluate each site's health promotion plan

---

in both basic and applied toxicologic studies and epidemiologic investigations (Chapters 4 and 5), all of which are intended to produce information useful for preventing or lessening the impacts of hazardous waste sites. A particularly important outcome of CERCLA-relevant research has been the identification of populations at special risk of exposure to substances released from waste sites. These groups include children, pregnant women, minorities, and persons with disabilities.

The third major component of intersectoral health promotion actions of relevance to hazardous waste issues is the implementation of public health interventions. The health interventions target communities and groups at elevated risk of exposure to hazardous substances and adverse health effects. In this context, the ATSDR has structured an intersectoral health promotion strategy to address the human health consequences of hazardous waste sites. The strategy relies on the integration of science and public health practice. The eight components of ATSDR's strategy are given in the inset box. The components consist of:

- Develop a site-specific health promotion plan that is based on the findings from a public health assessment, epidemiologic investigation, or some other kind of health-based evaluation of communities. The plan is developed by the ATSDR, in coordination impacted communities, the EPA or other federal agencies, and state/local health agencies. At present, the Public Health Action Plan is always part of public health assessment reports (Chapter 2). This plan outlines specific actions to be taken to interdict exposure to hazardous substances and prevent adverse health effects. The Public Health Action Plan is a key building block in health promotion plans developed for specific populations or communities that are at health risk.
- Interdict any ongoing exposure to hazardous substances. As described in Chapter 2, waste sites that have completed exposure pathways pose the greatest hazard. Of the 1,450 NPL sites, 530 have one or more completed exposure pathways. This means about 36% of all sites assessed by the ATSDR have one or more completed exposure pathways, although the percentage varies from year to year. For instance, in fiscal years 1993 to 1994, 60% of sites had one or more completed exposure pathways (Johnson, 1995), a figure that grew to 80% for sites with public health assessments conducted from 1993 through 1996. The increase in percentage of sites with completed exposure pathways is attributed by the ATSDR to improvements in how NPL sites are ranked by the EPA and an earlier presence by the ATSDR in conducting its health assessments, leading to assessment of sites before exposure pathways might have been interdicted.

- Ascertain any adverse health effects of current or prior exposure to hazardous substances released from the waste site of concern. This site-specific effort requires methods to screen persons who are, or were in the past, at greatest risk of exposure and possible adverse health effects. Where exposure poses an imminent health threat, persons should be referred for medical examination. A primary aim of screening is to detect any subtle adverse health effects before they manifest as more serious disease. Screening is performed in conjunction with community groups, the EPA, and state and local health departments.
- Conduct community health education in partnership with communities to share information on toxicity of substances, exposure pathways, potential health effects, and strengths and limitations of science and statutory authorities. From community health education can come information used by individuals to promote their personal health; for example, ways to reduce exposure to hazardous substances. This point is further elaborated in a following section.
- Facilitate technical assistance in environmental medicine, targeting the needs of local physicians and public health officials. The purpose is to increase knowledge and expertise in environmental medicine of locally practicing physicians in the diagnosis and treatment of environmentally-relevant health problems. This assistance is often requested by physicians and public health officials in communities impacted by hazardous substances because primary care providers are usually not well versed in toxicology and environmental hazards.

Building capacity in environmental medicine and health programs for locally practicing primary care physicians, nurses, sanitarians, and other health professionals—also described as infrastructure development—is integrated in the environmental medicine program. This is accomplished through the creation of public health partnerships (e.g., national medical associations, local medical associations) that engage health education and medical specialty referral networks. In several communities affected by hazardous waste, the ATSDR has integrated the elements of clinical evaluation of exposure and health effects, community education, education and training for health professionals, and infrastructure development into a comprehensive public health response. This response—the ATSDR's medical intervention program—allows for a proactive response to a community's health concerns. Active participation by communities throughout the program is essential.

- Conduct, when indicated, site-specific medical monitoring, which is a program to assess and follow the health status of individuals with documented exposure to substances at levels of health concern. Medical monitoring is described in detail in a subsequent section.
- Facilitate, in conjunction with local health authorities, the referral of persons for specialty medical care when indicated by findings from clinical evaluation of individuals or medical monitoring programs. No provision is made under CERCLA's current authorities to provide for patients' medical treatment and care. The expense of this care would therefore have to be borne by a person's health insurance, health maintenance organization, or public assistance programs of health care.
- Evaluate each site's health promotion intersectoral plan on its outcome, efficiency, and impact. As an example, the ATSDR tracks outcomes of recommendations in its public health assessments. More than 95% of the ATSDR's health recommendations to interdict releases from hazardous waste sites are implemented by the EPA and state and local governments (Johnson, 1995; Johnson, 1997). A database will be developed on what works best by evaluating each site's health promotion plan.

These eight components of the ATSDR's intersectoral health promotion strategy are implemented on the basis of community need and CERCLA relevance. Not all eight components are necessary at each CERCLA site. However, any health promotion program must integrate the elements of enabling personal control and empowering community action.

Enabling Personal Control

The second of Terris's elements of health promotion is paraphrased here as Enabling Personal Control. This is defined as the actions that an individual can take to enhance one's personal health. While there are many possible actions, three are discussed here: reduction of exposure to hazardous substances, civic duty, and lifestyle choices.

Individuals who have been exposed, or are currently being exposed, to hazardous substances can empower themselves with information and take control of individual actions that will reduce the risk of adverse health effects. This, of course, assumes that individuals know of their exposure situation. Knowledge of one's exposure to hazardous substances can come from personal observation or sense of smell (e.g., a chemical spill that releases a visible or smelly substance), from the news media (e.g., television news about a waste site), from local health or environmental agencies, or from results of public health assessments or epidemiologic studies. Knowledge of one's exposure history is the single most important information for personal health promotion empowerment.

Reduction of Exposure to Hazardous Substances

In situations where acute, potentially high exposure levels are occurring, interdiction of the exposure must be the first action. If a chemical spill or similar uncontrolled acute release of hazardous substances poses an imminent threat to human health, evacuation from the contaminated area is often the action of choice. Local hazardous materials teams manage these kinds of emergencies and determine when evacuation of a population at health risk is warranted. These teams can mobilize medical support when persons become ill from the exposure. An individual's actions should be consonant with directives from the emergency response team.

When ongoing exposure is chronic and is not acutely toxic, a person's best course of action is to interrupt the exposure pathway. Relocation or evacuation from one's residence may not be required. For example, if private well water is contaminated, an alternative source of water must be obtained. If the contamination is from an NPL site, alternative drinking water may be funded by the federal or state CERCLA program. Connecting the household to an uncontaminated municipal water supply is the best long-term solution.

For many CERCLA sites, contact with soil contaminated with hazardous substances is controlled by restricting access to the site. This usually takes the form of fences surrounding the site and warnings posted on the fences and elsewhere. Personal health promotion should include avoidance of the contaminated sites, particularly by children who may wish to use the site for play or other recreational purposes. Parents should ensure that their children know about a site's hazards and take responsibility for keeping children away from contaminated sites. Where fences or warnings are inadequate to restrict access to a site, local authorities should be notified by community residents.

In addition to interdicting any currently occurring exposure to hazardous substances, there are situations where persons have been exposed to hazardous substances in the past, but for whom exposure has ceased. The public health concern then becomes one of minimizing any latent health effects. For example, for some waste sites, substances infiltrated groundwater and thence into private wells used as drinking water and for other purposes. Individuals can reduce their exposure by converting to bottled water and seeking connection to municipal water supplies. Moreover, persons who have had long-standing exposure to contaminated water or contaminants in air, food, or soil should inform their health care providers. Health care providers can monitor their patients for any early signs of illness or disease associated with hazardous substances. For sites that meet health surveillance or medical monitoring criteria, health programs under CERCLA may be available to assist individuals.

Because knowledge is empowering, persons are advised to avail themselves of whatever information is available on the nature, extent, and implications of past exposure to hazardous substances released from a waste site or similar source. Obtaining a copy of the public health assessment and asking the sponsoring agency for information is advisable. This information can be shared with local health care providers; for example, personal physicians. For individuals who do not have direct access to health care providers, contact with their local health department is recommended. There are health centers in many communities that also provide health services to persons in need of health care and who lack financial resources.

For some CERCLA sites, past exposure of populations to hazardous substances may have been sufficiently great to warrant federal or state health agencies' conducting health surveillance, adding persons to an exposure registry, or medical monitoring. These activities, while each is different in purpose and conduct, all have the potential to identify adverse health conditions early enough to alert both persons at risk and their health care providers. It is therefore advisable for individuals to participate in such surveillance or monitoring programs when they are available. In a personal health promotion context, persons in such programs should ensure that they know their individual data (e.g., results from any biomedical test) and what the data mean.

Any exposures that could occur from site remediation activities should be part of an individual's health promotion effort. While site remediation methods are selected and implemented in ways that minimize releases of hazardous substances, individuals should nonetheless be aware of the possibility of exposure to site substances and take actions to prevent contact with any released substances. Personal awareness and knowledge of a waste site's remediation status is all the more important when remediation occurs over a long time, thereby increasing the possibility of substances being released from the site. Participation in local community groups that monitor site remediation is therefore advisable for purposes of personal health promotion.

An example follows of the effectiveness of individuals' actions to reduce children's exposure to lead at an NPL site. This example exemplifies the positive impact that individuals can have on reducing exposure to hazardous substances. However, individuals' actions are considered secondary prevention actions by public health officials. Primary prevention, which is the elimination of hazards (i.e., hazardous substance), is preferable to secondary prevention measures.

Reduction of Exposure to Lead: Reduction of children's exposure to lead in the environment was achieved in a community that had a defunct lead smelter, which had become an NPL site. Moreover, lead in the paint of older housing in the community was also a significant source of lead (Kimbrough et al., 1994). A study was conducted of the effectiveness of education and counseling provided to children with elevated blood lead levels and their parents. Health officials visited all households in which at least one child resided with a blood lead level of 10 µg/dL or greater.[3] Sources of lead exposure in each household were identified and training and counseling provided. This consisted of providing literature on lead's toxicity and exposure reduction methods. Counseling was given on personal behavioral factors that can reduce exposure to lead. The factors included washing children's hands before eating and before bedtime, keeping children's fingernails clipped short, and providing a well-balanced family diet. Children were counseled not to put their fingers and nonfood items into their mouth. When indicated, counseling was provided on the safe removal of lead-based, peeling paint. When children were retested post-intervention, reductions had occurred from a mean blood lead value of 15 µg/dL to 4 months later a mean of 8 µg/dL and a plateau mean of 9 µg/dL at 12 months. The investigators concluded that education and counseling of parents and their children was an effective intervention for reducing children's exposure to lead.

Civic Duty

The nineteenth century physicians who realized that social conditions affect human health and the spread of epidemics stressed the importance of good education, lack of poverty, and social justice. They saw these conditions as products of a healthy democracy. For democracies to be healthy, each citizen should be involved through various civic duties, especially by voting in elections.

Civic duty can contribute to a person's health promotion initiative. Informed actions by citizens are the strength of democracy. Regarding hazardous waste, individuals should inform themselves of the nature and long-term direction of decisions on hazardous waste management in their community. For example, individuals should expect local government officials to have a comprehensive waste management plan for their community. There are national environmental groups (e.g., Sierra Club) that have local chapters that can become focal points for discussion and action on the adequacy of community waste management plans. Participation in these groups can be viewed as supportive of health promotion for the community, with positive implications for one's personal health.

Lifestyle Choices

Personal lifestyle choices can affect the toxicity of substances to which an individual is exposed. As a general rule of thumb, actions that promote good health are important for moderating the effects of exposure to hazardous substances. For instance, cigarette smoking and tobacco smoke in ambient air are to be avoided, as a general rule of good health. When exposure to hazardous substances is occurring, or has occurred, avoidance of cigarette smoking is recommended. This is because the toxicants in tobacco smoke will add to other substances in the body that may have been absorbed from other sources of exposure, including exposure to substances released from a hazardous waste site.

---

[3] Health officials consider a blood lead level of 10 µg/dL in young children to be indicative of excess exposure to lead.

In a similar vein, adequate nutrition is always in one's best health interests. Diets deficient in key nutrients can exacerbate the effects of certain hazardous substances. As an example, diets deficient in zinc and calcium can enhance the toxicity of lead (ATSDR, 1993b). Similarly, higher dietary iron intake is associated with lower blood lead levels (Hammad et al., 1996). Because exposure to a mix of hazardous substances is considered to be additive in terms of the toxicity of individual substances unless there is evidence to the contrary, individuals exposed to toxicants in the outdoor environment can promote their health by reducing their exposure to toxicants in the home or workplace.

Some other health-promoting lifestyles include parents' monitoring of their children's play activities so that any contact with hazardous substances in the environment is avoided. For example, in areas that have soils bearing known contaminants, young children should avoid "playing in the dirt" because of the very real possibility of ingesting contaminated soil. Parents should ensure that children who play outdoors should wash their hands and face when they enter the home and replace soiled clothing. Similarly, vegetables grown in areas with contaminated soil and air should be washed prior to food preparation. As another way to reduce children's exposure to hazardous substances, materials brought into the home should be nontoxic. If hazardous substances, like solvents and pesticides, are brought into the home, they should be safely stored away from children and only used according to the manufacturer's instructions.

Empowering Community Action

The third of Terris's health promotion elements is paraphrased here as empowering community action. This is defined as the actions that a community can take to enhance its members' health. While there are many possible actions, three are discussed here: community health education, technical assistance in environmental medicine, and medical monitoring.

Robertson and Minkler (1994) refer to community as a social space that has connectedness. Connectedness means a group of persons who share the same basic values and organization. A community is also a group of persons who are aggregated for some common purpose. For many people, community will be interpreted as the geographic area where they live or a sociopolitical structure like a county, city, or census tract. However, community can also mean a group of persons with no geographic connection; for example, a religious or cultural group that has common characteristics or interests. Because communities are aggregates of individuals, much of what was just stated for individuals' health promotion applies to a community's health promotion efforts. For the purposes of this chapter, community refers to individuals and groups of persons exposed to hazardous substances released from hazardous waste sites or other sources of uncontrolled releases.

Communities can do more than individuals, because communities have the potential for joining resources in ways that an individual cannot. For instance, harkening back to the social medicine roots of health promotion, poverty, inadequate food, social injustice, and poor public education remain as precursors of disease epidemics and poor quality community health. Efforts by communities to reduce poverty, provide social justice, and instill quality education are therefore consistent with health promotion goals.

There are specific community empowerment actions that can address hazardous waste concerns of community residents. Regarding hazardous waste issues, although federal and state laws gird many actions on the generation, transportation, and disposal of hazardous waste, counties and cities ultimately become the locales where issues are joined. Informed community leadership and empowerment of local residents are prerequisites for civic decisions on hazardous waste. In particular, local authorities are advised to address any environmental equity (Chapter 9) concerns over siting, transporting, or disposing of hazardous waste in socioeconomically-disadvantaged communities.

Community Health Education

Knowledge is empowering for communities as well as for individuals. Health education experienced by a community can contribute to improved health promotion actions. Community health education is therefore defined here as the mutual development, dissemination, and using of information and data of relevance to a group of people with shared interests.

Community leaders and organizations concerned about hazardous waste issues are advised to inform themselves about the technical and social implications of hazardous waste. Federal and state agencies, universities, national environmental groups, and private industries are all sources of environmental data and information of relevance to community residents' health concerns. On issues of health, the city or county health department can be a valuable source of environmental health education, working with state and federal environmental health specialists. (Environmental health is used here to denote the area of public health that is concerned with effects on human health of socioeconomic conditions and physical factors in the environment.)

The ATSDR provides a wide variety of services to educate health care professionals and communities about health effects of hazardous substances. Activities in a community around a hazardous waste site can include conducting grand rounds for health care providers about the effects of specific hazardous substances, providing fact sheets on chemicals, conducting workshops on environmental disease, and distributing case studies in environmental medicine.

There has been very little research in communities to evaluate the longevity of awareness about the health implications of hazardous waste sites. In one study, investigators conducted a follow-up survey of residents who lived near the McColl hazardous waste disposal site in Fullerton Hills, California (Lipscomb et al., 1992). The site had previously been the subject of a health study conducted seven years earlier by state health officials. About 36% of the residents surveyed said that they had heard of the health study, which was a low percentage given the prior communications effort. Factors found to be associated with residents' recall of the health study included: age 55 years or greater, some college education, and residence in the area prior to the year of the health study. The investigators advocated focusing on individuals with less education and the use of creative public participation and media relations as ways to enhance community education. An example of a more comprehensive community health education effort follows.

Community health education case example: Stratford, Connecticut—Stratford is a city of approximately 50,000 persons located on Long Island Sound. It is a city of diverse population with a mixed economic base (Chapter 8). In 1919 a plant opened in Stratford that manufactured brake parts and other products that contained asbestos. Waste from the plant was placed in lagoons, but as space became exhausted, the lagoons were dredged and their sediment was spread on the factory's property. The dredged waste and subsequent industrial waste were made available to the public as filler material and thereby found their way into areas throughout the city.

The waste material was later found to contain asbestos, lead, PCBs, dioxins/furans, phenolic resins, adhesives, and solvents. In 1982 a citizens group, the Milford/Stratford Citizens Against Pollution, was formed in response to complaints about odors emanating from the brake manufacturing plant. In 1989 the company ceased all operations and, at the urging of the city, the EPA conducted environmental sampling in areas of the city thought most likely to have received waste from the facility. Initial testing of 15 known waste disposal areas in the city found lead, asbestos, and PCBs at very elevated levels. Lead in soil was found at some sites in concentrations up to 150,000 ppm, asbestos at levels up to 900,000 ppm, and PCBs up to 160 ppm; dioxin contamination was found only at a few commercial properties (cited by Cole, 1996). More than 500 residential, public, and commercial properties were tested for lead, PCBs, asbestos, and other substances. Ultimately, more than 50 properties were remediated in less than two years (cited by Cole, 1996). The ATSDR advised the city, based on the EPA's environmental contamination data, that the sampled sites represented a hazard to the public's health.

In May 1993, the ATSDR issued a Public Health Advisory that warned the Stratford community of a substantial and imminent threat posed by hazardous substances found in soil samples taken from various city locations. A public health advisory is the highest level of concern that is issued by the ATSDR under its CERCLA authorities (Chapter 2). Because of the very high levels of lead in some soils sampled, an early health concern of health authorities was young children's exposure to lead. Asbestos was also of major interest because of its high concentration in soil samples. The public also expressed concern that PCBs and dioxin contamination might cause cancer.

Given the widespread extent of community contamination because of the many years of using hazardous waste from the industrial facility, together with the toxicity of the principal substances (lead, asbestos, dioxin, PCBs), the local health department director took extraordinary measures to establish and maintain communications with the Stratford community. The local health department served as the source for information on health and related concerns, and it helped in the creation of the Stratford Citizens Advisory Council.

The Council, in turn, quickly became a vital link between the Stratford community, the news media, and various government agencies (NACCHO, 1995). Noteworthy was the local health department's coordination of communications among more than 40 local, state, and federal health and environmental agencies.

The Stratford Health Department developed a comprehensive public information and outreach program. Their efforts included: (1) a centralized information source, (2) a full-time community liaison/advocate who responded to telephone inquiries and met often with concerned citizens, (3) public meetings and forums, (4) a quarterly newsletter, (5) letters to Stratford residents impacted by hazardous waste on their property, and (6) a health booth that was operated by the city health department and the Stratford Citizens Advisory Council. Weekly meetings were held with the various community and government stakeholders, and news media contact was regular and factually-based. This coordination, although difficult, was crucial for effective communication. The result was an informed public that actively participated in decisions on human health protection and environmental cleanup.

**Stratford, Connecticut
Community Health Education**

Waste from a company in Stratford that manufactured brake parts was used for decades as filler material for yards, playgrounds, and commercial properties. In 1989, soil testing revealed high levels of lead, PCBs, and dioxins in many areas of the city. The ATSDR issued a public health advisory alerting the city to potential adverse health impacts from contact with the contaminated soil.

Because of widespread contamination in Stratford, the city's health department director led a community health education effort to provide credible information on the toxicity of PCBs and other contaminants, ways to prevent exposure to soil contaminants, and findings from exposure investigations. The news media, local physicians, and environmental groups were partners in providing education to the Stratford community at large. Soil samples from areas where waste had been used as fill, together with measurement of toxicants in persons at presumed greatest risk of exposure, helped characterize the geographic areas of greatest public health concern. Prompt action by government agencies led to an accelerated remediation of contaminated areas in the city. On November 3, 1997, a community event was held to mark the completion of remediation of the Raymark Industries CERCLA site.

As the Stratford, Connecticut case illustrates, local health departments can and should play a key role in communities concerned about hazardous waste problems. However, many local health departments are not adequately prepared to play this role. For this reason, the National Association of County and City Health Officials (NACCHO) initiated training and educational efforts to enhance local health departments' capacity to involve communities on matters of hazardous waste and other environmental issues.

In one study, NACCHO developed a self-assessment guide for local health departments to use in enhancing their service to communities (NACCHO, 1997a). The guide targets local health departments in communities that have hazardous waste sites. Evident in the guide is the philosophy that community collaboration is a social change process of building relationships and sharing decision-making authority. The guide states, "The collaboration process requires that the community be understood in its totality, including its cultural, political, social, and economic dimensions. Based on this philosophy, the guidelines outline a set of assumptions and principles. Examples of the assumptions include statements about the community's right to participate, the importance of the community's knowledge and skills, potential for leadership by the health department, and the implications of the agency's political authority." Adoption of the guide by local health departments should enhance their role when empowering communities in health promotion efforts.

Knowing the needs of communities impacted by hazardous waste sites is integral to efforts to empower communities. NACCHO developed an educational tool to assist local health departments when working with communities to identify environmental health education needs and action plans (NACCHO, 1997b). The tool is called a community needs assessment. The assessment is a structured approach that leads to specific health education programs for targeted communities.

Environmental Medicine Education

Contemporary training in environmental medicine and health is often requested by health care providers in communities that have hazardous waste sites and other environmental hazards. Unfortunately, studies by the Institute of Medicine (IOM) have found that physicians and nurses both require more formal education about environmental hazards. The IOM noted that primary care physicians are often the first contact for patients with environmentally-related illnesses (Pope and Rall, 1995). The IOM opined that graduating medical students would be better prepared to care for patients with environmental illnesses if environmental and occupational health content were integrated into undergraduate medical education.

In a parallel study, the IOM noted that the 2.2 million nurses are the largest group of professional health care providers in the United States (Pope et al., 1995). In both urban and rural settings, nurses are often the initial, and sometimes the only, point of contact with persons seeking medical care. However, the IOM concluded that the vast majority of nurses have had no formal training in occupational or environmental health. The IOM developed a series of recommendations on how to improve nurses' education, practice, and research in environmental health and medicine. For nursing education, they advocate integration of environmental health concepts into all levels of nursing education and in nurses' licensure and certification examinations.

The infusion of environmental health into the curricula of medical and nursing schools is a long-term effort. In the near term, enhanced community medical services are required in some communities with documented exposure to hazardous substances at levels that portend adverse health effects. In such situations, the local health care system may not be able to respond to the challenging, and sometimes unique, health implications of a person's exposure to specific hazardous substances. For instance, primary care physicians practicing in such communities might not have current medical information on the toxicologic and human health implications of the substances of concern to residents. In such situations technical assistance in environmental medicine can be sought through public health resources, as described in the following section.

Technical Assistance in Environmental Medicine

Hazardous substance issues such as contamination of local drinking water supplies involve complex questions of toxicology, exposure assessment, and human health interpretation. In turn, environmental contamination is often the subject of individuals' health concerns. Community health promotion is enhanced by utilization of local resources in environmental health and medicine. Local medical societies, for example, are an excellent opportunity for a community's physicians to learn of health implications of a chemical spill or hazardous waste site. Under its CERCLA authorities, the ATSDR has facilitated technical assistance to local medical societies to update them on current toxicologic information and community-specific health issues.

Some community groups concerned about hazardous waste sites in their midst have advocated that medical services be provided. For example, a representative who spoke for a coalition of grassroots environmental groups testified to Congress, "...Citizens continue to pay the cost of Superfund with loss of property values, degradation of their quality of life, and most costly of all, loss of their health, and loss of beloved friends, neighbors, and family members. Furthermore, in many of the communities most affected by Superfund sites, there is inadequate or no access to proper medical attention, proper housing, good nutrition, and other social and educational factors, all of which exacerbate the exposure to toxic chemicals. Consequently, we advocate the establishment of specifically designated medical units for each Superfund Community that will monitor, diagnose, and treat the citizens of that community in the hopes of achieving better health outcomes" (Robinson, 1996).

No data were given in the testimony about the number and nature of the medically underserved communities. However, there is no doubt that access to health care is inadequate in many urban, rural, and tribal areas. Recall that the historical roots of health promotion are embedded in the knowledge that poverty, inadequate food, poor education, and social injustice are major contributors to poor health. Where health care in a community is inadequate, there are federal and state government programs that can be brought to bear. For example, the Health Resources and Services Administration of the Public Health Service provides grants to health centers in communities and rural areas. Rather than structuring "specifically designated medical units for each Superfund Community," a better strategy would be to enhance environmental medicine capacity in existing health centers, rather than forming new medical units.

An example follows of a community with an NPL site that required environmental medicine assistance. This example is noteworthy because of linkages established between the community, public health agencies, and clinical resources.

Case example: Del Amo/Montrose NPL sites—The Del Amo/Montrose NPL sites are located in Torrance, California. From 1943 through the mid- to late-1960s, a chemical plant manufactured synthetic rubber at the Del Amo site. In 1991, the EPA proposed that the 280 acre site be listed on the NPL. Contaminants in the waste disposal area included benzene, ethylbenzene, and polycyclic aromatic hydrocarbons (PAHs). The Montrose Chemical NPL site is located adjacent to the Del Amo site. A chemical plant on the Montrose site had manufactured DDT. Testing of off-site soils and biota resulted in detectable levels of DDT and related compounds. The soil contamination findings triggered a community health concern that in turn led to intersectoral technical assistance in environmental medicine.

An existing clinic in the Del Amo community was funded by the ATSDR to add an environmental medicine specialist and laboratory services to the clinic. This occurred in response to the community's health concerns. Environmental medicine specialists at the University of California at Irvine, University of California at Los Angeles, and California Department of Health Services collaborated to provide technical assistance to the clinic.

The purpose of providing technical assistance in environmental medicine was to assess the health status of persons living near the Del Amo/Montrose sites and to effectuate public health interventions. The Del Amo/Montrose technical assistance effort has four components:

1. *Community outreach and health education.* Community outreach was conducted in the Del Amo and Montrose communities to increase public awareness of environmental medicine resources available to the community. The outreach included a commu-

> **Case Example:**
> **Del Amo/Montrose Medical Assistance**
>
> Residents in an area of Torrance, California, were at risk of exposure to DDT, PAHs, benzene, and ethylbenzene associated with two former chemical production facilities. Discovery of clumps of DDT in residential soil led to a community-based medical intervention effort, coordinated by the ATSDR, the EPA, state/local health agencies, and community organizations.
>
> A local health clinic was staffed with an environmental medicine specialist, and laboratory services were enhanced. The clinic provides local residents with exposure assessments, clinical evaluations, and laboratory tests. Persons found to have specific health disorders are referred to local health care providers for follow-up.

nity health fair in the Montrose neighborhood, bilingual (English and Spanish) materials distributed door to door that described environmental contamination in the area, and information about available clinical services. Meetings with community representatives were held on a regular basis.

2. *Environmental health education for local health care providers.* Grand rounds have been conducted for family practice residents at a medical clinic near the Del Amo/Montrose community on topics in environmental medicine pertinent to the hazardous waste sites. A course on risk communication was also provided to the clinic's physicians.

3. *Clinical evaluation.* Persons who reside in areas near the Del Amo/Montrose waste sites are eligible for clinical evaluation. The clinical evaluation consists of an exposure history questionnaire, a physical examination, and laboratory tests to measure levels of contaminants in biologic specimens. If the findings of the clinical evaluation are normal, the participant is counseled on his/her results, and no further follow-up occurs. If findings warrant follow-up, the patient receives further diagnostic testing and referral to other medical specialists. Approximately 500 community members, of which half are children, participated in clinical evaluations through December 1996.

4. *Infrastructure building in environmental medicine.* The ATSDR's intersectoral health promotion strategy includes an evaluation component. For the Del Amo/Montrose sites, an independent evaluation of the technical assistance in environmental medicine was conducted by a clinical association not affiliated with the clinic (Paranzino et al., 1996). The evaluators found the Del Amo/Montrose environmental medicine technical assistance to have generally met the goals of the program, but recommended changes in record keeping, community outreach, and improved clarity of the purpose of the technical assistance.

Through December 1996, approximately 2,000 persons from the Del Amo/Montrose community have benefited through improved community education about hazardous substances, clinical examinations, exposure assessment, and medical referrals for some individuals.

Medical Monitoring

CERCLA requires the Administrator of the ATSDR to initiate a health surveillance program for populations at significantly increased risk of adverse health effects as a result of exposure to hazardous substances released from a facility. Specifically, CERCLA Section 104(i)(9) states the following:

"Where the Administrator of ATSDR has determined that there is a significant increased risk of adverse health effects in humans from exposure to hazardous substances based on the results of a health assessment conducted under paragraph (6) [Section 104(i) of CERCLA], an epidemiologic study conducted under paragraph (7), or an exposure registry that has been established under paragraph (8), and the Administrator of ATSDR has determined that such exposure is the result of a release from a facility, the Administrator of ATSDR shall initiate a health surveillance program for such population. This program shall include but not be limited to: (A) periodic medical testing, where appropriate, of population subgroups to screen for diseases for which the population or subgroup is at significant increased risk; and (B) a mechanism to refer for treatment those individuals within such population who are screened positive for such diseases."

A program included under health surveillance is referred to as "Medical Monitoring or Screening" by the ATSDR and is defined in CERCLA as "...the periodic medical testing to screen persons at significant increased risk for disease." Medical monitoring is one of the elements of the ATSDR's health promotion strategy. When there is no known association between the exposure and specific adverse health effects, medical monitoring is not an appropriate public health activity under the authorities of CERCLA. Where there is sufficient information on a specific health effect's relationship to an exposure, then options such as exposure registries, disease and symptom prevalence studies, or epidemiologic investigations are more appropriate (Chapter 5). When adequate information associates exposure to a hazardous substance with a specific adverse health effect, further consideration must be given to medical monitoring of the exposed population.

Medical monitoring should be directed toward a target community known to be at "significant increased risk for disease" on the basis of exposure to specific hazardous substances. Significant increased risk will vary for particular hazardous waste sites depending on such factors as the underlying risk of the selected health outcome, the risk attributable to the exposure, and the presence of sensitive subpopulations (e.g., children). These factors are considered when evaluating the appropriateness of medical monitoring in a community. CERCLA's medical monitoring mandate requires that a mechanism be in place to refer persons for medical care who are screened positive for the selected health outcomes.

Therefore, a mechanism to refer people for diagnosis, interventions, or treatment must be in place prior to the initiation of a medical monitoring program. Authority to provide or fund long-term health care or medical treatment is not authorized by CERCLA.

The primary purpose of a medical monitoring program is not considered to be a research activity that further investigates cause-effect relationships between exposure and outcome. Rather, the purpose of a medical monitoring program is to identify persons with early signs of illness or disease. Within this framework, medical monitoring includes testing both for any early biologic effect and an assessment of exposure using biologic specimens (for example, blood or urine), when appropriate.

ATSDR's Criteria for Medical Monitoring (ATSDR, 1995c)

Medical monitoring is periodic medical testing to screen persons at significant increased risk for disease.

- Evidence of exposure is documented, and
- A well-defined population is at risk, and
- A scientific basis exists for health concern, and
- Medical monitoring tests are consonant with health concerns, and
- Medical screening requirements must be met, and
- Accepted methods of treatment and/or interventions must exist, and
- Logistics must be resolved prior to the program.

Criteria for Considering Medical Monitoring—The seven criteria listed in the inset box are used by the ATSDR to determine the appropriateness of conducting medical monitoring in a community. The criteria are applied in a phased approach (ATSDR, 1995c). Phase I, conducted by the ATSDR, evaluates exposure and health outcome criteria. Phase II consists of an evaluation of the system criteria. This is conducted with the input of community representatives and state and local health officials. At the end of Phase II, a detailed medical monitoring plan is developed. All seven criteria must be met in order for a medical monitoring program to be established for communities impacted by hazardous waste sites. In addition, resources must be available to initiate and sustain the medical monitoring program.

Phase I: Exposure and health outcome assessment. Phase I of the program is conducted by the ATSDR for the purpose of determining if exposure data associated with a particular waste site are at sufficient levels to warrant health concern. A description of the ATSDR's seven criteria for determining whether a community qualifies for medical monitoring under CERCLA's provisions follows. Commentary is shown, indented, for each criterion.

Criterion 1. There should be evidence of contaminant levels in environmental media that suggest the high likelihood of environmental exposure to a hazardous substance and subsequent adverse health outcomes.

> The primary criteria for medical monitoring should be documented evidence of exposure of a population to a hazardous substance in the environment. An exposure is considered to be at a sufficient level if there is documentation of an increased opportunity for exposure to a level that meets or exceeds some health-based comparison value (such as the ATSDR's Minimal Risk Levels [MRLs] or the EPA's Reference Doses [RfDs]) or that meets or exceeds a level reported in the peer-reviewed scientific literature to result in some adverse health effect.

Criterion 2. There should be a well-defined, identifiable target population of concern in which exposure has occurred to a hazardous substance at a sufficient level of health concern.

Initially, the target population of concern is defined geographically on the basis of exposure. In addition, all populations considered are assessed for the presence of any subpopulation at increased risk of the adverse health effects associated with the exposures. The target population of concern is the population in which there is documented exposure at a sufficient level to place the individuals in that population at significant increased risk for developing one or more specific adverse health effect.

Criterion 3. There should be documented human health findings that demonstrate a scientific basis for a reasonable association between an exposure to a hazardous substance and a specific adverse health effect (such as an illness or change in a biologic marker of effect).

Studies on human populations must have demonstrated a reasonable association between a particular exposure and an adverse health effect. To make that association, consideration should be given to the strength, specificity, and consistency of the association among the identified studies.

Criterion 4. The monitoring should be directed at detecting adverse health effects that are consistent with the existing body of knowledge and amenable to prevention or intervention measures.

Medical monitoring should be established for specific adverse health effects. The specific adverse health effect being monitored should be a result of the possible exposure consistent with the existing body of knowledge. An adverse health effect is consistent with the existing body of knowledge if it has been described in the scientific literature as caused by that agent or by similar agents, taking into account structure-activity relations.

Phase II: System criteria. Phase II of the program is conducted by the ATSDR with input from the community. When the ATSDR has determined that exposure from a site has met the exposure and outcome criteria, a site-specific panel is formed based on recommendations from the community and the state and/or local health departments. The panel reviews the system criteria and assists in the development of a site-specific medical monitoring plan. The site panel includes representatives from the community, state or local health departments, local medical societies, and subject experts (e.g., radiation) as necessary. The site panel is responsible for assessing the available community health resources and determining the feasibility and extent of the screening program for the community. If the panel determines that a screening program is feasible in the community and the ATSDR concurs with that decision, the ATSDR develops a site-specific monitoring plan. That plan is presented to the site panel for review and concurrence. After the plan has been developed and has undergone peer review, it is presented to the community at large for their input prior to establishing the medical monitoring program.

Criterion 5. The general requirements for a medical screening program should be satisfied.

The monitoring aspect of a health surveillance program consists of periodic medical testing to screen individuals who are at increased risk of disease. Monitoring serves to identify those individuals that have an unrecognized adverse health effect. This is consistent with one definition of screening as[4] "...the presumptive identification of unrecognized disease or defect by the application of tests, examinations, or other procedures which can be applied rapidly. Screening tests sort out apparently well persons who probably have a disease from those who probably do not. A screening test is not intended to be diagnostic. Persons with positive or suspicious findings must be referred to their physicians for diagnosis and necessary treatment" (CCI, 1957).

Criterion 6. There must exist an accepted treatment, intervention, or both, for the condition (outcome or marker of exposure), and a referral system should be in place prior to the initiation of a medical monitoring program.

There should be established criteria for determining who would receive referral for intervention or treatment. These criteria are based on the selected effect being screened for and the screening test being used. Results are evaluated by the ATSDR longitudinally and cross-sectionally to identify any changes in the system or screening tools that require follow-up (Gochfeld, 1990). A referral mechanism should exist so that those who are eligible for the intervention can be referred to a qualified health care provider for further diagnosis, treatment, or intervention. Under CERCLA, the referral must be for treatment or intervention that is standard practice and not experimental in nature. The medical monitoring (screening) program is not responsible for the cost of the referral, the intervention, or the treatment of individuals participating in the program.

Criterion 7. The logistics of the system must be resolved before the program can be initiated.

After medical monitoring has been determined to be appropriate for a site, the specifics of the monitoring system are detailed in a site-specific medical monitoring plan. The site panel works with the ATSDR to develop and review the site-specific medical monitoring plan. The specifics of the medical monitoring system will vary for each site. The monitoring plan is the protocol for the specific program to be proposed for a community. The plan outlines the target community, the types of health outcomes to be screened for, the participants in the referral system, and the program reports.

Monitoring for Evidence of Continuing Exposure—At sites with exposure occurring in the community, the monitoring program could include biologic markers of exposure. Such sites would be those where exposure is known to have a variety of adverse health effects, but for which no tests are available to detect the effects at a time when intervention could affect the course of the disease process. In those instances, the primary intervention is to remove the individual from the exposure. This allows the medical monitoring system to recommend referral for medical evaluation prior to the onset of detectable, adverse health effects. A monitoring system that includes biomarkers of exposure is similar to medical surveillance of hazardous waste workers where changes indicative of increasing or continued exposures occur sufficiently early so that the exposure can be curtailed and the risk for disease reduced (Gochfeld, 1990).

Because of the need and time taken to establish medical monitoring criteria, the establishment of a medical monitoring for a specific CERCLA community only began

---

[4] Excerpted with permission from *Chronic Illness in the United States, Vol. 1*. Copyright 1957 Harvard University Press, Cambridge, Massachusetts.

in 1996. The ATSDR determined in 1996 that medical monitoring for thyroid cancer should be initiated for a segment of the population exposed to radioactive iodine (^{131}I) released in the late 1940s from the Department of Energy's Hanford facility (Spengler, 1997). Radiation exposure occurred during the period 1945 to 1951. The most important pathway of human exposure was the consumption of contaminated milk produced by cows grazing on pasture located downwind of Hanford. Children were estimated to have received the highest doses because of their consumption of milk and other dairy products. When the medical monitoring program is established, approximately 14,000 individuals with significant radiation doses to the thyroid gland will be offered medical evaluations to detect thyroid neoplasms and other thyroid and parathyroid conditions. The medical monitoring program is expected to detect 36 cases of thyroid cancer in a population of 6,000 persons; 12 cases would be expected in a population of comparable age. Early detection of precancerous thyroid conditions permits earlier medical treatment and increased likelihood of survival.

Medical Care Referral

Persons who are identified in medical monitoring programs or from clinical evaluation in need of medical care are referred to private or public systems of health care. For example, persons found by the Del Amo/Montrose clinic to have specific chronic diseases were referred to patients' primary care physicians or to the California Medicare system. In general, a referral system will consist of the review of the screening results and then a referral to appropriate health care providers. The specific mechanisms for determining who needs referral and for selecting the health care providers in the referral pool must be in place prior to the initiation of a medical monitoring program.

Once the patient has been referred to health care providers, those providers are responsible for any subsequent diagnosis, treatment, and care. The health promotion strategies of the ATSDR are based on the agency's authorities under CERCLA. No authority to provide or fund long-term health care or medical treatment is authorized in CERCLA's medical monitoring provisions.[5]

SYNTHESIS

Health promotion is the process of enabling people to increase control over, and to improve, their health (Ottawa Charter, quoted by Terris, 1992). According to Terris, the first use of the term "health promotion" is attributed to the medical historian Henry E. Sigerist, who in 1945, "...defined the four major tasks of medicine as: (1) The promotion of health, (2) The prevention of illness, (3) The restoration of the sick, and (4) Rehabilitation. He [Sigerist] stated that 'Health is promoted by providing a decent standard of living, good labor conditions, education, physical culture, means of rest and recreation.'" Paraphrasing the work of Terris, there are three elements of health promotion: intersectoral actions, enabling individuals' control, and empowering community action. These three elements must be integrated into a cohesive effort if individuals in a

---

[5] Regarding medical care, CERCLA sections 104(i)(1)(D) and (E) refer to "in cases of public health emergencies" for which medical care is to be provided. This authority has not been utilized by the ATSDR, owing to uncertainty about Congressional intent.

CERCLA community are to be impacted. This is most effectively achieved when environmental and biomedical science is brought to bear on community health concerns, and through an integrated, partnership effort involving the community, government, and nongovernment entities.

Concerning Terris's first element of health promotion, a sector is defined as a distinct part of society or of an economy, group, or entity. Intersectoral health promotion actions therefore refer to efforts taken between distinct parts of any sociopolitical structure that can influence human health. These parts include community groups, government, nongovernment organizations, private industry, and academic institutions. Intersectoral action translates to forming partnerships to solve social problems. This is easier said than done. Each organization, including government, brings to a partnership certain values, biases, restrictions, and limitations. Moving past these built-in conditions in ways that join common purposes is the essence of partnerships, and therefore intersectoral action.

In the context of hazardous waste concerns, what kinds of intersectoral actions are possible? There are many possibilities, but three seem of primary importance. They include (1) hazardous waste management, (2) science that supports hazardous waste management and impacts of hazardous waste on human and ecologic health, and (3) public health interventions in situations where exposure to hazardous substances is of concern. To be successful, each of these three actions must involve activities, resources, and commitment that cut across government agencies and the private sector.

The second of Terris's elements of health promotion was paraphrased in this chapter as enabling personal control. By this is meant the actions that an individual can take to enhance one's personal health. While there are many possible actions, three were discussed: reduction of exposure to hazardous substances, civic duty, and lifestyle choices. Terris's third element of health promotion was paraphrased as empowering community action. By this is meant the actions that a community can take to enhance its members' health. Three were discussed: community health education, technical assistance in environmental medicine, and medical monitoring.

The broadening of the concept and practice of health promotion has been called the "new health promotion movement" by some public health specialists (Robertson and Minkler, 1994). They assert the new health promotion paradigm has moved past the former, more narrow approach of emphasizing health promotion primarily through an individual's lifestyle changes and professionally based interventions. In their view, "...health is seen as instrumental, a means rather than an end. Health is what one must have to accomplish other things in one's life."

Robertson and Minkler assert, "...health promotion practitioners who truly facilitate empowerment do so by assisting individuals and communities in articulating both their health problems and the solutions to address those problems. **By providing access to information, supporting indigenous community leadership, and assisting the community in overcoming bureaucratic obstacles to action, such practitioners may contribute to a process whereby communities increase their own problem-solving abilities** (emphasis added)..." Empowerment then becomes a matter of health promotion practitioners' sharing information, consulting with communities, and advocating ways whereby problems are solved by the community. The role of health professionals will change if empowerment and community participation are integrated into programs of health promotion. **In the new health promotion movement, health professionals become consultants to the community, rather**

than being experts who merely define the community's needs and provide solutions through traditional public health interventions. Robertson and Minkler conclude, "Rather than service provider and client, the community and the professionals are equal partners in setting the health agenda for the community."

Regarding communities, CERCLA and other federal environmental programs that involve risk management actions are increasingly involving communities in these actions. Health promotion, which empowers communities, is surely in the spirit of greater community involvement and empowerment on matters of risk management. Based on need, providing communities with technical assistance in environmental medicine, community health education, and medical monitoring are key components of any health promotion effort.

Chapter Eight

Risk Communication

How is information about risks to personal and community health communicated? In the current context, how are the health risks of hazardous waste communicated with concerned communities, local health care providers, and others? How does the public perceive risk? What constitutes a program of participatory risk communication, and what are the principles of risk communication? This chapter addresses these questions and provides case studies of both effective and ineffective risk communications regarding hazardous waste issues.

Government agencies at all levels have struggled with improving communication with the public. Moreover, the public expects to be more involved in decisions on policies, regulations, and practices that relate to environmental hazards like hazardous waste. Community involvement in communications is an important element in avoiding gridlock. For example, Chess et al. (1995) note that government agencies have been prodded to improve communication with the public about environmental problems that require individuals to change lifestyle behaviors such as waste recycling, decreased automobile use, and testing for radon in homes.

It is increasingly evident that all the thorough planning, skillful conduct, and thoughtful analysis of elegant biomedical research will go for naught if the personal health implications of research are ineffectively communicated with target audiences. Target audiences for specific risk communications are almost always multiple, and include local community groups, local health and environmental agencies, health care providers, state government, and private industry. Whatever the nature and number of audiences, participatory (i.e., two-way) communication is essential.

Risk communicators are confronted with many challenges. Even under optimal conditions, segments of the American public are skeptical of actions by, and messages from, government and private industry. Communicating the risk of environmental hazards is especially challenging. Incomplete science and uncertainties in environmental data and biomedical science will often encumber particular communications. Yet communication is essential for communities at potential risk of adverse health effects because hazardous substances were released from waste sites. What to do and how to do it become the challenge.

Some terms need clarification before proceeding. The National Research Council (NRC, 1989c) defined **risk communication** as an interactive process of exchanging information and opinion among individuals, groups, and institutions. Risk communication flows from the NRC's risk assessment and risk management paradigm developed in 1983 (NRC, 1983). In a narrow sense, risk communication should be communication about the probability of occurrence (i.e., risk) of an adverse outcome associated with a hazard, but is seldom used this way.

The term **hazard communication** predates the term risk communication. Hazard communication informs a target group that contact with a particular environmental agent could be harmful. For example, the U.S. Occupational Safety and Health Administration (OSHA) promulgated a comprehensive regulation about communicating the hazards of chemicals in the workplace to employees (Waldo and Hinds, 1993). The OSHA standard requires that manufacturers, importers, and distributors provide a Material Safety Data Sheet with the first shipment of a hazardous chemical to commercial customers. Firms that ship, store, sell, or use chemicals must prepare and implement comprehensive hazard communication programs for their employees.

Health warnings are messages from public health officials and health care providers to alert the public about threats to their health. Perhaps the most familiar health warning is found on cigarette packages and advertisements warning smokers that cigarette smoking will injure their health.

The single most influential report on risk communication, *Improving Risk Communication*, was prepared by the Committee on Risk Perception and Communication, convened by the NRC (1989c). In retrospect, the Committee produced a remarkably visionary document by deliberately taking a different approach to risk communication. The Committee noted, "...We found a focus on one-way messages too limiting, however. Instead, we make a crucial distinction between risk messages and the risk communication process. We see risk communication as an interactive process of exchange of information and opinion among individuals, groups, and institutions." Framed in this context, risk communication is a political process that engages democratic traditions.

The NRC's definition of risk communication as an interactive process of exchange involving[1] "...multiple messages about the nature of risk and other messages, not strictly about risk, that express concerns, opinions, or reactions to risk messages or to legal and institutional arrangements for risk management" (NRC, 1989c) seems broad enough to encompass hazard communication and health warnings. It is therefore adopted here (see inset box).

Risk assessment and risk management have emerged as methods that government regulatory agencies use to control environmental hazards. Risk communication has become part of this process and as a result risk communication, hazard communication, and health warning have largely become synonymous terms. Therefore, risk communication is used in this broad communications context throughout this chapter.

PUBLIC HEALTH AND RISK COMMUNICATION

Public health practitioners have a long-standing tradition of communicating the health hazards of infectious agents. Seventy years ago, the American public readily understood a sign nailed on a house warning QUARANTINE, "If you enter this house, you risk contracting a life-threatening disease." As a health warning, a quarantine message was

---

[1] Excerpted with permission from *Improving Risk Communication*, pp. 20–21. Copyright 1989 by the National Academy of Sciences. Courtesy National Academy Press, Washington, DC.

> ### Definition of Risk Communication
>
> "Risk communication is an interactive process of exchange of infor-
> mation and opinion among individuals, groups, and institutions. It pro-
> vides multiple messages about the nature of risk and other messages,
> not strictly about risk, that express concerns, opinions, or reactions to
> risk messages or to legal and institutional arrangements for risk man-
> agement" (NRC, 1989c).

simple and effective because it came from a trusted source—a local physician or a public health official. It was also effective because the disease was known and feared and scientific data and experience supported the gravity of the message. In addition, it had societal, not just individual import, because if the quarantine were breached, the community was also at risk of disease.

Another illustration of successful risk communication, using an environmental haz-ard as the example, is the effort by public health officials to prevent the adverse conse-quences of tobacco use. Substantial scientific evidence associates cigarette smoking and environmental tobacco smoke with lung cancer and heart disease. This evidence has shaped antismoking health warning messages. Surveys show the American public is quite aware of tobacco's consequences to personal health. Physicians, public health officials, health care providers, and environmental officials continue to communicate the health hazards of cigarette smoking through targeted media campaigns by govern-ment, the Surgeon General, school communications, physician interventions, news media reports, and the efforts of public interest groups like the American Lung Associa-tion and the American Cancer Society. The result has been a steady decrease in the percentage of Americans who smoke cigarettes, although the decrease seems to have leveled off or reversed for some groups, particularly adolescents.

Like health messages conveyed by quarantine notices, risk communications about cigarette smoking are effective for many of the same reasons. The messages are based on compelling science and emanate from trusted sources (i.e., physicians, public health authorities, and public interest groups). Both personal, "My smoking can harm my health," and societal, "My smoking can affect your health," messages are present. In contrast, public meetings convened by governmental agencies to discuss particular CERCLA sites have occasionally become raucous and confrontational. Government officials are sometimes confronted by disbelieving communities that discredit findings from health investigations or take issue with proposed site remediation actions. An example of the latter comes from communities where on-site incineration of hazardous materials has been recommended as the government's technologically preferred site remediation method. The public associates incineration of hazardous waste with re-leases of dioxins and furans in air emissions from incinerators. The news media cap-ture such confrontations on videotape and in print, and the risk communication intended by government quickly becomes obscured and lost. Making public meetings more effective should be a goal in any effort to improve risk communication.

Although the preceding scenario of difficult public meetings seems to be occurring less often, hazardous waste issues will nevertheless remain a challenge for risk communicators. Two reasons support that assertion. First, the CERCLA statute, as enacted into law in 1980 and reauthorized in 1986, contains "the polluter pays" principle. This principle forges confrontation and fosters litigation. Attitudes harden, communities form lines of opinion, and communications can become difficult. Irrespective of how CERCLA is ultimately reauthorized by Congress,[2] there will remain a legacy of confrontation among government, parties found financially responsible for costs of site remediation, and communities impacted by hazardous waste sites. This legacy will ensure that risk communication remains a challenge. Second, toxicologic and epidemiologic data on many hazardous substances will remain incomplete for the foreseeable future (Chapters 4, 5). These data gaps lead to uncertainty in the public's mind about possible adverse health effects and thereby compound the difficulty of effective risk communication.

The foregoing may seem gloomy, and communicating health risks about hazardous waste will remain challenging; however, there are success stories to share and principles of risk communication to state. In fact, as described in this chapter, much has been learned about enhancing the effectiveness of risk communication for environmental hazards in general and for hazardous waste sites in particular.

This chapter shares four case studies specific to hazardous waste sites. These examples illustrate and illuminate current principles and practices in risk communication. Three studies illustrate effective risk communications with communities; the other describes less effective communication. The case studies will set the agenda for this chapter. Following the first case study, background material is presented on the evolution of risk communication, followed by a summary of key research findings on how the public perceives risk and the essential involvement of community groups in risk communication.

Case Study: Chattanooga Creek, Tennessee

A successful risk communication effort, as ascribed by government officials and community environmentalists, was achieved in Chattanooga, Tennessee, in an area bordering Chattanooga Creek. This effort, like that described later for Stratford, Connecticut, involved collaboration of several government groups, working with local citizen groups. As in Stratford, local leadership was crucial to success.

Background—Chattanooga Creek is approximately 23.5 miles long. A 7.5-mile stretch of the creek passes through predominantly African-American neighborhoods in the city of Chattanooga. The community has a large youth population, and an estimated 10,000 persons live in areas along the creek.

Environmental Data—Over the years, Chattanooga Creek was polluted by waste from industries that produced coke, organic chemicals, textiles, bricks, pharmaceuticals, and leather and tanning products. Other operations included wood preservation and metallurgic and foundry work (Tinker et al., 1995, 1996a). Both the creek's floodplain and the creek itself were used as dumping sites for some municipal and household wastes. Of the 42 hazardous waste sites located near Chattanooga Creek, 13 are being managed under Tennessee's hazardous waste site program. Large coal-tar deposits, some of which are 6 feet deep, are located along the creek, posing safety hazards to children who play in and along the creek.

---

[2] CERCLA is required to be reauthorized by Congress every five years. It was reauthorized in 1986 and extended "as is" in 1991.

Health Issues—Hazardous substances found in Chattanooga Creek and along its course are at concentrations of concern to health agencies. Community residents complained over the years of increased rates of cancer, miscarriages, respiratory problems, headaches, and eye and skin irritation (Tinker et al., 1995). In 1992, a member of the community group Stop Toxic Pollution (STOP) petitioned the ATSDR to evaluate the site's impact on their community's health.[3] The ATSDR's public health assessment found that local residents used the creek for fishing, swimming, and other recreational activities. The ATSDR classified the Chattanooga Creek site as a human health hazard because of past, current, and potential future exposure of south Chattanooga residents to hazardous substances. Three completed exposure pathways were attributed to Chattanooga Creek: sediments, surface water, and fish. PCBs, dieldrin, 4,4-DDT, PAHs, and bacteria were found in one or more of the exposure pathways.

Communication Issues—The ATSDR, the EPA, State of Tennessee agencies, and local authorities initiated comprehensive community education and risk communication efforts. The effort was headed by STOP, and its leader became the community's focal resource and most credible source of information. The ATSDR provided technical information and personnel in support of STOP's effort. Because children who swam in, fished in, and played along the banks of the creek were of special concern, education programs about the creek's chemical and biologic contamination were given in local elementary schools. For example, one elementary school conducted an art contest to help students express their concerns about the creek's hazards. The winning poster was entitled "Don't be a creek geek" (Tinker et al., 1995). Furthermore, local health care providers participated in medical rounds conducted by environmental medicine specialists on hazards posed by the creek. Local media played an important role in communicating hazards of the creek.

Outcome—As the result of the ATSDR public health assessment, the EPA placed the Chattanooga Creek site on the NPL, which will eventually result in the site's cleanup. STOP has continued to communicate the hazards posed by the creek by collaborating with local health and environmental agencies, medical and nursing schools, and private industry. Although interviews with community residents, children, and local health authorities report a high awareness of the creek's contamination and the implications for health, a formal evaluation of the risk communication effort in the Chattanooga Creek community has yet to be completed.

Lessons Learned—Residents in the Chattanooga Creek area say that effective risk communication occurred in their community because of a coordinated effort among Chattanooga community groups and government agencies. Several lessons seem evident. First, local leadership provided by STOP was crucial to building trust. STOP built this trust through small community meetings and local networking. Another key factor was prompt responses by local, state, and federal health and environmental authorities. Government was seen as a technical resource to the community, having the resources to effectuate risk reduction; that is, remediate the area and warn residents of the health hazards of swimming and fishing in Chattanooga Creek. Another positive element was the informed involvement of local news media, which accurately depicted health hazards along and in the creek.

---

[3] CERCLA, as amended, permits an individual or licensed physician to petition the ATSDR to conduct a public health assessment of a site of concern to the petitioner (Chapter 2).

EVOLUTION OF RISK COMMUNICATION

Case studies portray some of the challenges risk communicators face and provide valuable lessons on how to make risk communication more effective. However, this kind of practical experience and lore should be complemented with findings and practices developed by persons who have researched health risk issues. This section will include some of the roots of risk communication. In particular, an awareness of how the public perceives risk is essential for any risk communicator. Furthermore, distillation of research findings on risk communication has led to the development of "cardinal rules of communication" and other advice. This section provides sufficient detail on how risk communication has evolved to help persons who communicate environmental risks to the public.

Some academicians attribute the modern age of environmental risk communication to William Ruckelshaus, who was at the time serving his second term as Administrator of the EPA (e.g., Chess et al., 1995; Covello and Peters, 1996). Ruckelshaus promoted the goals of informing and involving the public as foundation principles in environmental risk management. He advocated "participatory democracy" as a means of resolving environmental policies (Chess et al., 1995). Similarly, the National Research Council (NRC, 1989c) observed,[4] "...Citizens of a democracy expect to participate in debate about controversial political issues and about the institutional mechanisms to which they sometimes delegate decision-making power. A problem formulation that appears to substitute technical analysis for political debate, or to disenfranchise people who lack technical training, or to treat technical analysis as more important to decision-making than the clash of values and interests is bound to elicit resentment from a democratic citizenry." Both Ruckelshaus and the NRC therefore advocate democratic ideals as a key in risk communication.

Although neither Ruckelshaus nor the risk communicators who proceeded from his ideas discussed the notion of "participatory democracy," it is consonant with practices of the new health promotion movement discussed in Chapter 7. As described there, health promotion has at its core the ideas of community involvement and empowerment, ideas that comport readily with participatory democracy.

The Public's Perception of Risk

How the public perceives risk will influence the effectiveness of any risk communication because public perceptions may differ from those of government and private sector agents who conduct risk assessments of environmental hazards and prepare attendant risk communications. Government agencies, in particular, have recognized the need to improve the effectiveness of risk communications (Covello et al., 1987). This has led to risk perception becoming an area of research that now involves the social, physical, and biologic sciences.

Considerable research on risk perception now exists (e.g., Covello et al., 1983; Fischhoff et al., 1978, 1984; Krimsky and Golding, 1992; NRC, 1989c, 1996; Slovic, 1992, 1993). This

---

[4] Excerpted with permission from *Improving Risk Communication*, pp. 20–21. Copyright 1989 by the National Academy of Sciences. Courtesy of the National Academy Press, Washington, DC.

work was stimulated in large measure because of the need to better understand the public's perception of risk as it relates to changes in technology. Slovic describes the influence of new biologic, chemical, and nuclear technologies on the public's risk perception and acceptance of hazards (Slovic, 1987). The proliferation of news media and other information sources has been simultaneous with the advance of new technologies introduced into commerce.

People in the industrialized nations, in particular, have access to cable television, computer information networks, and a plethora of materials prepared by journalists. Disasters that occur anywhere in the world are almost instantly communicated. The public's attention is caught by technology disasters like the Chernobyl, Ukraine, nuclear reactor meltdown and the Bhopal, India, chemical plant leak of a very toxic substance. Similarly, the discovery of hazardous substances leaking into the Love Canal, New York, community elicited tremendous public interest, concern, and reaction. The public perceived these events as failures of technology, and suspicion of new technologies increased. One evident result was sustained opposition in the United States to the siting of new nuclear power generation plants.

Specific environmental hazards have become the subject of individual political exercises in the United States. Various advocacy groups on the American landscape oppose the introduction or enlargement of technologies perceived to be hazardous to human health and the environment. Environmental groups that oppose the use of high technology incinerators for waste disposal are an example. As part of their advocacy, these groups work actively to influence public opinion—and in the process they, perhaps unwittingly, also shape the public's broader perception of technologic risk. A primary benefit of such problem-specific advocacy is a public more aware of a specific hazard. However, when the public is confronted with multiple expressions of hazard, each one being presented as a serious threat to human health, this can result in an attitude of "zero risk" as an expectation. Thus, the risk communicator must work within a system of public bias and preformed attitudes to initiate a specific risk communication.

In the mid-1970s, because of shifting public attitudes toward risks posed by technology, researchers began working to better understand how individuals make personal decisions about risk. Researchers have developed techniques to assess the complex and subtle nature of people as they perceive and act upon risk information. As Slovic (1987) stated, "The basic assumption underlying these efforts is that those who promote and regulate health and safety need to understand the ways in which people think about and respond to risk." Slovic further notes that research on risk perception should help policy makers improve communication with the public and forecast the public's response to new technology.

In a key area of risk perception research, a psychometric paradigm with a taxonomy of hazards allowed the public to express its reactions (Slovic, 1987; Fischhoff et al., 1978). The psychometric paradigm uses psychophysics, scaling, and multivariate analysis techniques to produce quantitative representations of risk attitudes and perceptions (Slovic, 1987). Thus, psychometrics is an approach psychologists developed to quantitatively measure and depict persons' feelings and attitudes. Psychometricians use structured questionnaires to assess a person's attitudes about conditions of interest to both the psychologist and the person being assessed. Factor analysis and other statistical methods seek common factors that may account for how people perceive risk.

Findings from psychometric research on risk perception reveal a complex set of factors that influence how people react to technologic hazards. Studies have shown that

Table 8.1. Characteristics of Risk that Affect the Public's Perception of Risk Level (Fischhoff et al., 1981; Lum and Tinker, 1994).

Risks perceived as being...	are more readily accepted than risks perceived as being...
Voluntary	Imposed
Under an individual's control	Controlled by others
With clear benefits	With little or no benefit
Fairly distributed	Unfairly distributed
Natural	Created by humans
Statistical	Catastrophic
Generated by a trusted source	Generated by an untrusted source
Familiar	Exotic
Harmful to adults	Harmful to children

risk acceptance by the public involves characteristics of individual hazards. These characteristics include voluntariness of exposure, familiarity (of the hazard), degree of control, catastrophic potential, equity, and level of knowledge about the hazard (Slovic, 1987). The most important characteristics that affect how the public perceives risk are listed in Table 8.1. For instance, the public more readily accepts risks that have clear benefits to them over risks for which the benefits are thought to be few. As will be described subsequently, differences according to gender and race influence the public's perception of risk.

Factor-analytic analysis of risk characteristics like those listed in Table 8.1 have identified three core factors that influence risk perception (Slovic et al., 1979; Slovic, 1987). This "factor space" has been replicated across groups of lay persons and technology experts who were asked to judge large and diverse sets of hazards. According to Slovic, Factor 1, which has been labeled "dread risk," is defined at its high end by hazards judged by respondents to be unobservable, dreaded, have catastrophic potential, present fatal consequences, and to have benefits that are inequitably distributed. Nuclear weapons and incidents at nuclear power reactors are examples of dread risk. Factor 2, which has been labeled "unknown risk," is defined at its high end by hazards judged to be unobservable, unknown, new, and delayed in their effect. Hazards involved with chemical technologies fit into Factor 2. Slovic notes, "A third factor, reflecting the number of people exposed to the risk, has been obtained in several studies." The third factor is not designated by name or label.

Of the three factors, "dread risk" is the most important in shaping lay persons' perception of risk (Slovic, 1987). From their psychometric research, Slovic and associates have developed graphs that place individual hazards on a map according to Factor 1 (dread) and Factor 2 (familiarity). A risk perception map constructed along these two dimensions for selected hazards is shown in Figure 8.1.[5] (Additional hazards are found in Slovic, 1987.) In the figure, hazards that rate highest on dread (i.e., to the far right on the horizontal axis) are those for which lay people want stringent control regulations

---

[5] Excerpted from Slovic, P., Fischhoff, B., and Lichtenstein, S., in *Perilous Progress: Managing the Hazards of Technology*, Kates, R.W., Hoheriemser, J., and Kasperson, J.X., Eds. Copyright 1985 by Westview Press. Reprinted with permission of Westview Press.

Gauging the Dread Factor

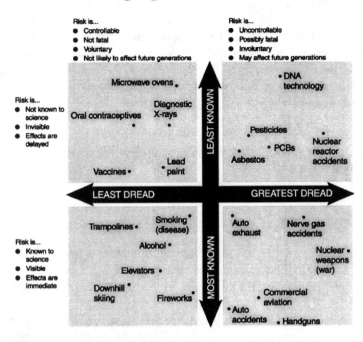

Figure 8.1. Lay persons' perception of risk (adapted from Slovic, 1987).

(Slovic, 1987). Incidents involving nerve gas illustrate a hazard greatly dreaded by lay persons. By contrast, trampolines present much lesser dread to lay persons, even though the likelihood of injury from using trampolines is much greater because more trampolines exist than do occurrences of nerve gas releases.

The lay public's perception of the risk from uncontrolled hazardous waste sites is not shown in Figure 8.1, because this hazard was not included in Slovic's psychometric analysis. However, by analogy with pesticides, PCBs, and nuclear waste, it is reasonable to assume that uncontrolled hazardous waste sites would rank high on dread risk.

By contrast with lay persons' perceptions of risk, experts' perceptions are not closely related to any of the various risk characteristics ascribed to the lay public (Slovic, 1987). Experts appear to judge risk on the basis of expected annual mortality, according to Slovic. This obviously puts experts at variance with the lay public and can lead to misperceptions between the two groups. Slovic (1987) has written sagely about the differences in risk perception between lay people and experts,[6] "Lay people sometimes lack certain information about hazards. However, their basic conceptualization of risk is much richer than that of the experts and reflects legitimate concerns that are typically omitted from experts' risk assessments. As a result, **risk communication and risk management efforts are destined to fail unless they are structured as a two-way process** (emphasis added). Each side, expert and public, has something valid to

---

[6] Excerpted with permission from Perception of Risk, *Science* 236:280–285. Copyright 1987, American Association for the Advancement of Science.

Two-Way Communication

"...risk communication and risk management efforts are destined to fail unless they are structured as a two-way process. Each side, expert and public, has something valid to contribute. Each side must respect the insights and intelligence of the other" (Slovic, 1987).

---

contribute. Each side must respect the insights and intelligence of the other." This advice about the importance of participatory communication and mutual respect is a succinct recipe for more effective risk communication.

Gender, Race, and Ethnic Differences

As described previously, a person's perception of risk is influenced by certain characteristics of the hazard. To complement research on the qualities of health hazards, investigators have examined the role that gender, race, and ethnicity play in shaping a lay person's perception of risk.

Flynn et al. (1994) conducted a national telephone survey of a random sample of 1,512 English-speaking persons in the United States. More than 200 persons of color participated in the survey. The objective of the survey was to assess attitudes, perceptions, values, knowledge, and beliefs about specific hazards. Respondents were asked to rate the risk they associated with each of 25 hazards. Hazards on the list were wide-ranging and included AIDS, chemical pollution, ozone depletion, climate change, radon in homes, medical X-rays, and motor vehicle crashes. The investigators constructed a hazard index by averaging the respondents' ratings across the 25 hazards. Respondents who identified themselves as "Hispanic, black, Asian, or American Indian were included in the nonwhite category."

Several important findings came from the survey. Flynn et al. replicated the consistent difference between risk perceptions of men and women that other investigators had reported; namely, the percentage of high-risk responses was greater for women on every hazard item. Furthermore, nonwhite respondents had consistently higher mean ratings of perceived risk than did white respondents. Most strikingly, white males gave mean risk perception ratings that were consistently much lower than those of white females, nonwhite females, and nonwhite males. The investigators examined whether the observed associations among race, gender, and risk perception could be explained by other social and demographic variables. Stepwise multiple regression analyses gave little support to the role of factors such as income, education, and age in explaining the race and gender associations with risk perception.

Flynn et al. concluded that gender and race (which includes ethnicity) are major determinants of how a person perceives risk. They recommended further research to clarify any biologic and sociopolitical factors that contribute to gender and race differences.

Environmental justice has emerged as a very significant sociopolitical development (Chapter 9) and thus has great import for risk communication. Vaughan (1995) suggests that communication and participatory strategies will be successful only if diverse communities are engaged as partners in the policy process. She notes, however, that incorporating sociocultural perspective into risk communication has been rare in the environmental policy literature. Vaughan observes that great disparity exists in the social histories and cultural characteristics of many communities selected for participatory risk communication. She further notes that adopting a justice framework to define the terms of risk debates is becoming more frequent. This shifts a risk management debate from a technical, risk-assessment basis to a justice-centered basis. This can lead to risk communications becoming discussions about racial or ethnic equity, participatory democracy, empowerment, and community control over risk management decisions and actions.

The best advice for risk communicators is to involve not only the community but specialists in environmental justice debates and communications. As an example, Clark Atlanta University in Atlanta, Georgia, has established an environmental justice center that specializes in issues of community involvement and risk management.

Risk Communication: Myths and Actions

Conducting a public health assessment or an epidemiologic investigation of a community potentially affected by exposure to hazardous substances must involve the community at risk. Involvement is a key element of successful communication. As the Chattanooga Creek case study shows, public health interventions (e.g., changing children's play activities) involve much planning and attention to detail—especially on issues of risk communication.

Many health and environmental officials acknowledge the importance of effective risk communication and seek ways to improve it. But that has not always been the case, and contemporary examples of ineffective risk communication still exist. Part of this failure to practice effective risk communication may stem from myths about communication with the public. Chess et al. (1988), augmented by Lum and Tinker (1994), listed several key myths that have contributed to the poor communication performance of some government officials. Selected key myths and counteractions to the myths are listed in Table 8.2. Myths can become excuses or other impediments to developing a risk communications program within an agency or for initiating particular risk communication efforts. Counteracting myths like those in Table 8.2 is therefore important.

Both government and private sector organizations have recognized the importance of making risk communication more effective and getting the public more involved. Research on risk communication and risk perception, together with considerable experience accrued by government agencies that work with community groups, provide practical guidance for moving past myths like those in Table 8.2 and toward more effective communication.

Interacting with the Community

Effective risk communication must begin and end with the community at risk, that is, involve the community throughout the communication process. In the context of hazardous waste, the community will be persons at increased risk of adverse health effects from exposure to hazardous substances. Within that community will be persons of special concern, for example, young children and persons with existing disabilities whose health status may be exacerbated by exposure to toxicants. Interacting with the

Table 8.2. Some Key Myths about Risk Communication and Their Counteractions (Lum and Tinker, 1994).

Myth	Counteraction
There aren't enough resources for a risk communication program.	Effective communication is less costly in the long run.
Telling the public about a risk is more likely to unduly alarm people than keeping quiet.	Decrease the potential for alarm by encouraging people to express their concerns. Address each concern and respond proactively.
Communication is less important than education. If people knew the true risks, they would accept them.	Pay as much attention to the process of working with people as is given to explaining the data.
Health and environmental issues are too difficult for the public to understand.	Separate public disagreement with your policies from misunderstanding of the highly technical issues.
Technical decisions should be left in the hands of technical people.	Provide the public with information; listen to community concerns; involve the public in planning, implementing, and evaluating specific actions.
If the public is listened to, scarce resources will be devoted to issues not relevant to the public's health.	Listen early to avoid controversy and to hear the potential for disproportionate attention to issues not related to the health problems at hand; redirect nonhealth concerns to appropriate sources for follow-up.

community is *the* most important element of any risk communication effort. Chess et al. (1988) and Lum and Tinker (1994) have developed useful recommendations that should be considered by persons who communicate health risk messages. The following guidance is adapted from those sources.

Recognize the importance of community engagement. Community involvement is essential because (1) people are entitled to make decisions about issues that directly affect their lives, (2) advice from the community can help an agency make better decisions, (3) involvement in the communication process leads to better understanding of—and more appropriate reaction to—a particular task, (4) persons affected by a problem bring different variables to the problem-solving equation, and (5) cooperation increases credibility. Moreover, when community involvement is not sought, conflict and contentiousness can sometimes result.

Involve the community in the decision-making process to the extent possible. Government agencies have been slow to recognize the importance of involving communities in actions designed to benefit those same communities. Several reasons for this reluctance are evident. Some agencies, like individuals, have practiced avoidance behavior for difficult encounters. Occasionally, this has occurred in public meetings when the impacts of environmental hazards were in dispute. In some government agencies, including public health agencies, a "command and control" approach has sometimes prevailed, in effect telling the community what to do, instead of working with them to shape a response or action.

Conditions that mitigated against the government involving communities have begun to change because of community opposition to government programs; government awareness of ineffective decision-making; and the public's desire to have government become more effective, efficient, and accountable. Chess et al. (1988) have offered practical advice on how to involve communities in decisions that affect them. They advise, **"Involve the community at the earliest stage possible. Clarify the public's role from the start. Find out from the community what kind of involvement is being sought. Acknowledge situations where the agency can give the community only limited power in decision-making."**

Identify and respond to the needs of different audiences. Any community has multiple audiences with which an environmental or health agency should interact. One important challenge for the risk communicator will be to determine those audiences. Each community is different, but some common themes for audience identification have been cited (Chess et al., 1988). Agencies should try as early as possible to identify the various audiences for a particular risk communication. This occurs by networking within communities. Local health and environmental agencies and the health care community are often the first points of contact. For some communities, the formation of a citizen advisory group can be an essential resource for developing and disseminating risk messages. With every audience, the communicator must act with candor, honesty, and fairness.

Smaller, more informal meetings are more effective than large public meetings. Persons who have followed the story of communities concerned about hazardous waste sites in their midst are familiar with the dynamics of large public meetings. Large public meetings are often the first contact between government agencies and communities concerned about hazardous waste sites. These meetings must be planned in advance, with considerable input from community groups and local officials. However, even the most thorough planning and involvement of communities will not guarantee effective public meetings. Some large public meetings have become confrontational and fodder for television newscasts. Chess et al. (1988) have offered the following practical advice. If a large public meeting is held, the logistics should enable both the agency and the community to be treated fairly. Consider breaking large groups into smaller ones. Be clear about the goals for the meeting. In certain situations, one-on-one communication may work best.

Chess et al. (1988) provide some additional practical advice to government agencies and others who engage communities' concerns over environmental issues: "Provide a forum for people to air their feelings. Listen to people when they express their values and feelings. Acknowledge persons' feelings about an issue. When people are speaking emotionally, respond compassionately. Do not merely react with data and statistical jargon. Show respect by developing a system to respond promptly to calls and inquiries from communities. Recognize and be honest about the values incorporated in agency decisions. And as a risk communicator, be aware of your own values and feelings about issues of concern to a community."

Accept the public's values, feelings, and concerns as genuine. Government agencies have not always been attentive to communities' concerns and involved them in decisions. This failure of government derives in part from how some areas of government have worked. Some agencies practiced command and control approaches without involving the affected public. Secrecy and protection of national security data were necessary for some federal agencies, but these became impediments to more open communication with the public. A lack of basic experience for engaging the public explained some agencies' reluctance to engage. As greater expectations of the public and legislative

Table 8.3. Seven Cardinal Rules of Risk Communication (Covello and Allen, 1988; Lum and Tinker, 1994).

1. **Accept and involve the public as a partner.**
   The goal is to achieve an informed public, not to defuse public concerns or replace actions.
2. **Plan carefully and evaluate your efforts.**
   Different goals, audiences, and media require different actions.
3. **Listen to the public's specific concerns.**
   People often care more about trust, credibility, competence, fairness, and empathy than about statistics and details.
4. **Be honest, frank, and open.**
   Trust and credibility are difficult to obtain; once lost, they are almost impossible to regain.
5. **Coordinate and collaborate with other credible sources.**
   Conflicts and disagreements among organizations make communication with the public much more difficult.
6. **Meet the needs of the media.**
   The media are usually more interested in politics than risk, simplicity than complexity, danger than safety.
7. **Speak clearly and with compassion.**
   Never let your efforts prevent your acknowledging the tragedy of an illness, injury, or death. People can understand risk information, but they may still not agree with you; some people will not be satisfied.

direction like the Government Performance and Results Act[7] make government more customer oriented, changes will increasingly occur in how government programs and actions incorporate the public's values, feelings, and concerns.

Several risk communication researchers have developed principles, guidelines, and rules for "effective" risk communication, although what makes an "effective" communication is not always defined. Covello and Allen (1988) developed one set of such rules for the EPA, called the Seven Cardinal Rules of Communication (Table 8.3). The Public Health Service's Environmental Health Policy Committee (Tinker, 1996b), which evaluated how agencies of the Public Health Service conducted risk communication, endorsed these rules. Although these rules were written for a federal government agency, their use by other branches of government, private industry, and environmental groups seems appropriate.

Preparing the Communication

Discussion to this point has emphasized the essential involvement of the audience (i.e., communities at risk) in shaping and evaluating risk communications, bearing in mind the Seven Cardinal Rules of Risk Communication (Table 8.3). To actually prepare a risk communication, a risk communicator can find manuals (e.g., Lum and Tinker, 1994) that provide useful tips on how to prepare for a public meeting and work with news media. Some of these tips are summarized in this section, but risk communica-

---

[7] The Government Performance and Results Act (GPRA), Public Law 103-62, was enacted in 1993, in part, to "improve Federal program effectiveness and public accountability by promoting a new focus on results, service quality, and customer satisfaction" (U.S. Congress, 1993).

Table 8.4. Communicating with the Public: Ten Questions to Ask (Chess and Hance, 1994).

1. Why are we communicating?
2. Who is our audience?
3. What do our audiences want to know?
4. What do we want to get across?
5. How will we communicate?
6. How will we listen?
7. How well did we respond?
8. Who will carry out the plans? When?
9. What problems or barriers have we planned for?
10. Have we succeeded?

tors are advised to seek more complete information when they plan a risk communication effort. The NRC (1989c) report on risk communication is still an excellent resource.

After deciding to initiate a risk communication with an at-risk group, the communication must be planned. Ten questions listed in Table 8.4 can be used to plan public meetings, communications with news media, and interagency planning sessions.

Public meetings present a special challenge to risk communicators. The meetings can involve hundreds of persons and are sometimes held in a climate of tension and uncertainty about government agencies' health findings and remediation proposals. Agencies must therefore provide their staff with risk communication training. Materials are available from several sources for risk communication training (e.g., Lum and Tinker, 1994). Some general tips drawn from several sources include preparing thoroughly before the public meeting, using minimum scientific jargon, restricting the key messages to three or fewer, encouraging questions from the audience, answering questions in simple, clear terms, keeping promises made during the public meeting, and showing honest personal concern about the waste site under discussion.

Evaluating Risk Communications

This chapter began with the assertion that risk communication evolved in part because of government's need to improve communication with communities impacted by agencies' programs. Improvement connotes the concepts of reference, measurement, and evaluation. How can we know whether improvement has occurred if no point of reference existed when the communication effort began? How will we measure a communication's "success"? What mechanisms are available for evaluating the effect of a risk communication? Although answers to these basic questions are only now emerging from risk communication research, and consensus agreement on the "right" answers is still fluid, some advice can be given.

One could begin with what represents "success" in risk communication. The committee that prepared the NRC report, *Improving Risk Communication* (1989c), struggled with the concept of successful risk communication. They concluded, "...risk communication is successful only to the extent that it raises the level of understanding of relevant issues or actions and satisfies those involved that they are adequately informed within the limits of available knowledge." This definition has some drawbacks because it begs the questions of what is "relevant" and what constitutes "satisfaction." Because it may not be possible to develop and make operational a generic definition of successful risk communications, communicators may wish to develop a statement of success between themselves and the intended audience before mutually developing the communication. This would result in agreed-upon measures to determine the outcome and impact of a particular risk communication. In other words, success would be defined one risk communication at a time.

Literature has developed on the principles and techniques for evaluating health risk communications. Evaluation of communication efforts is essential because it provides vital information about whether risk messages were received, understood, and acted upon by the targeted audience. Without evaluation, risk communicators cannot know which messages and processes are most effective and are left with only subjective impressions of what works "best."

Three kinds of evaluation have been adapted to assess the effectiveness of risk communication programs (DHHS, 1994).

Formative: Evaluation during the formative stages of a risk communication effort assesses the strengths and weaknesses of materials or message campaign strategies before they are implemented. Formative evaluation permits revisions before the risk communication effort begins. Among other things, risk communication materials can be tested for clarity, tone, and comprehensiveness.

Process: Process evaluation examines the procedures and tasks involved in implementing a risk communication activity. Process evaluation can collect and assess the importance of administrative and organizational aspects of a risk communication. Such data could include number of staff, schedule of activities, number of materials distributed, attendance at meetings, number of calls to a hotline, number of facsimile messages received, number of public inquiries, number of newscasts, and number of articles printed.

Outcome: Outcome evaluation involves collecting information to judge risk communication efforts and their effectiveness in meeting stated communication objectives. Outcome evaluation is best pursued if the risk communication effort or program has clear and measurable goals and consistent replicable materials, organization, and activities. The following kinds of information are needed for an outcome evaluation: changes in knowledge and attitudes, expressed intentions of the target audience, and changes in behavior.

Eight elements in any evaluation design, whether formative, process, or outcome evaluation are listed in Table 8.5. These elements require careful planning and thoughtful implementation when applied to a specific risk communication.

A crucial step in assessing the effectiveness of risk communication efforts is to determine what message ideas or concepts have the best chance of "connecting" with the target audience and promoting their health (Roper, 1993). This process begins with formative research and evaluation.

Literature reviews, in-depth interviews, and focus groups are examples of formative research tools. They can help determine whether one concept is more salient to an audience segment than another, and what concepts should eventually be developed into specific risk messages. The general approach to pretesting message concepts is to share them with members of the target audience, gauge their reactions, and solicit their participation. Pretesting is conducted while materials are in draft form to permit adjustments before they are used in an actual risk communication effort. Pretesting methods include focus groups and intercept interviews (e.g., interviews of persons randomly selected at a shopping mall) with members of the target audience. Materials are avail-

Table 8.5. Elements of Design for a Communication Evaluation (NCI, 1992).

1. A statement of communication objectives
2. Definition of data to be collected
3. Methodology (i.e., study design)
4. Data collection instruments
5. Data collection
6. Data processing
7. Data analysis
8. Reporting

able that give good advice on whom to include in reviewing and pretesting risk messages, methods for structuring print materials, guidelines for analyzing and using pretest findings (e.g., Roper, 1993).

Case Study: Saltville, Virginia, NPL Site

One component of successful risk communication is presented in Figure 8.2. This shows pages of a pamphlet that the ATSDR risk communicators and a community group in Saltville, Virginia, developed jointly (ATSDR, 1996j). As background, the city of Saltville is located in the Appalachian Mountains in Virginia. The Saltville Waste Disposal site is an NPL site along the North Fork of the Holston River between the town of Saltville and the community of Allison Gap. Waste ponds cover about 120 acres and extend along the river for more than a mile.

Industrial operations at the site began in 1894 and produced soda ash, chlorine, caustic soda, dry ice, liquid carbon dioxide, and hydrazine. Waste products were discharged into large settling ponds, which leaked into the North Fork of the Holston River. Mercury, used in producing chlorine gas, was one waste generated. Mercury has since been detected as a contaminant in on-site soil, groundwater, surface water, fish tissues, and river sediment. Saltville residents report that waste material was used as fill in different areas of the town.

The river runs by an elementary school and is a favorite place to fish. Annual sampling of fish tissue and sediment in the North Fork of the Holston River has been conducted since 1970. Mercury levels in fish tissue have exceeded the Federal Drug Administration's (FDA) recommended action level. Local health authorities disseminated fish consumption advisories within the Saltville community. The health risk of consuming mercury-contaminated fish had to be communicated.

Community residents and the ATSDR risk communicators worked together to prepare several community education efforts warning of mercury's toxicity. One risk communication focused on elementary school children (ATSDR, 1996j). Figure 8.2 indicates the use of nontechnical language, a clear statement of the population at risk, and purposeful statements of what to do. Using the materials in Figure 8.2, a Saltville community group sponsored poster contests and meetings in schools. Community and government risk communicators incorporated pretest and posttest analyses into the community education project. Findings from focus groups indicated that the community gave highest value to ideas or strategies in which they were actively involved in the decision-making process, but government agencies valued actions that were most directly related to their statutory mandates.

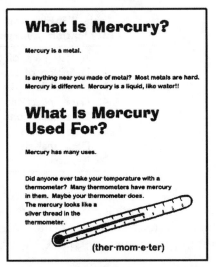

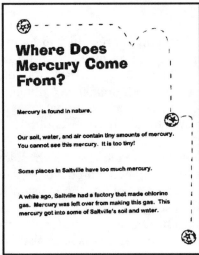

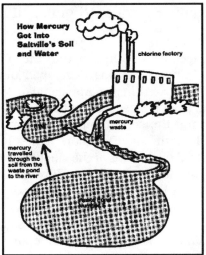

Figure 8.2. Saltville, Virginia, Risk Communication Pamphlet.

Although the Saltville example does not provide a complete answer about the effectiveness of a risk communication effort, focus groups in the community gave strong endorsement to being actively involved in designing and implementing the risk communication. Moreover, a post-test evaluation of children conducted five months after the community education campaign found that they were still able to describe the "catch and release [fish]" public health message, the symptoms of mercury poisoning, and how fish become contaminated with mercury.

Case Study: Escambia Wood–Pensacola NPL Site, Pensacola, Florida

A community adjacent to an uncontrolled hazardous waste site in Pensacola, Florida, provides a contrast to the successful risk communications achieved in Chattanooga and Saltville.

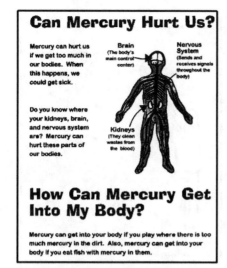

Can Mercury Hurt Us?

Mercury can hurt us if we get too much in our bodies. When this happens, we could get sick.

Do you know where your kidneys, brain, and nervous system are? Mercury can hurt these parts of our bodies.

Brain (The body's main control center)

Nervous System (Sends and receives signals throughout the body)

Kidneys (They clean wastes from the blood)

How Can Mercury Get Into My Body?

Mercury can get into your body if you play where there is too much mercury in the dirt. Also, mercury can get into your body if you eat fish with mercury in them.

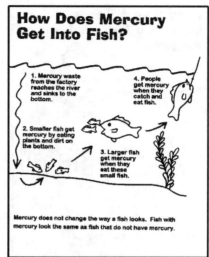

How Does Mercury Get Into Fish?

1. Mercury waste from the factory reaches the river and sinks to the bottom.

4. People get mercury when they catch and eat fish.

2. Smaller fish get mercury by eating plants and dirt on the bottom.

3. Larger fish get mercury when they eat these small fish.

Mercury does not change the way a fish looks. Fish with mercury look the same as fish that do not have mercury.

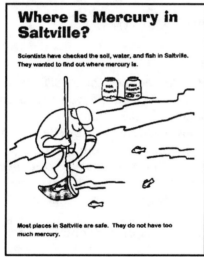

Where Is Mercury in Saltville?

Scientists have checked the soil, water, and fish in Saltville. They wanted to find out where mercury is.

Most places in Saltville are safe. They do not have too much mercury.

Two places in Saltville do have too much mercury. One place is called Waste Pond Number 5.

The other place is the North Fork of the Holston River. The North Fork runs through Saltville down into Tennessee. There is too much mercury in fish in the river from Saltville all the way to Tennessee.

Do you know where Waste Pond Number 5 and the North Fork of the Holston River are in Saltville? If you do not, you can ask your parents or teacher to tell you.

Map of Saltville

Rt. 613 — Waste Pond Number 5 — Saltville Elementary School — Perryville Road — Downtown Saltville — To Tennessee — North Fork of the Holston River — Northwood High School

Figure 8.2. Saltville, Virginia, Risk Communication Pamphlet (continued).

Background—The Escambia Wood–Pensacola wood-treatment facility (EWP) operated from the 1940s to the early 1980s (FDHRS, 1996). The facility treated utility poles and wooden foundation pilings with creosote and pentachlorophenol (PCP) dissolved in diesel fuel from 1945 to 1981. Substantial concentrations of creosote and PCP were found in soil at the facility's site. The EPA started excavating contaminated soil in April 1991 under CERCLA's emergency removal authority. Approximately 250,000 cubic yards of waste sludge and soil containing pentachlorophenol (PCP), creosote, and dioxin compounds were excavated, placed on-site in mounds, and covered with large sheets of plastic. Excavation was completed in 1993, and the site was secured by a chain link fence that had metal warning signs attached. The largest mound is 60 feet high and is called "Mt. Dioxin" by local residents and the news media. A local group, Citizens Against Toxic Exposure (CATE), was formed as a forum to express environmental and health concerns and facilitate community actions.

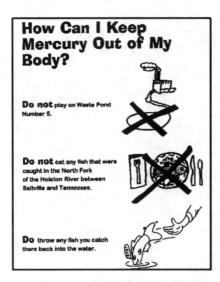

Figure 8.2. Saltville, Virginia, Risk Communication Pamphlet (continued).

Environmental Data—In July 1992, at the EPA's request, the ATSDR reviewed air and soil contamination data, an air sampling plan, and cleanup levels proposed for the EWP site. Following the review, the ATSDR concluded that potential exposures to hazardous substances associated with the EPA's removal activities at the site were unlikely to pose an imminent health threat. However, the ATSDR recommended that (1) health evaluations be conducted of residents living near the site because of concern that exposure to hazardous substances from the site may have occurred, and (2) an environmental health education program be initiated for Escambia residents and Pensacola health care providers.

Health Issues—The Escambia community consists of residents in single family dwellings and persons living in a public housing complex. The community feared that exposure had occurred to creosote compounds, PCP, and dioxin/furan compounds. They expressed their fears to local, state, and federal health officials about perceived elevated rates of skin diseases, respiratory problems, birth defects, and cancer in the neighborhood adjacent to the EWP site. Some residents attributed these health problems to dust released from Mt. Dioxin.

In June 1992, the ATSDR received the first of 25 letters petitioning the agency to conduct a public health assessment of the EWP site. The Florida Department of Health and Rehabilitative Services (FDHRS) conducted the public health assessment through an agreement with the ATSDR. At the same time, CATE asked the ATSDR to conduct a health study of its community. In response, the ATSDR developed a draft cross-sectional study design consisting of (1) a symptom-and-disease prevalence questionnaire survey, (2) biomarker tests to evaluate organ damage and dysfunction in immune, renal, hepatic, and hematopoietic systems, and (3) measurement of contaminants of concern in biologic media.

The ATSDR and the FDHRS sponsored environmental health training and education for 180 physicians in Pensacola in 1993 to 1994. Occupational physicians from a Florida university provided the training. Health education sessions focused on the toxicity and adverse health effects of contaminants found at the EWP site. In August 1994, the EPA proposed the site for listing on the NPL. In May 1995, the FDHRS released its public health assessment of the EWP site, which concluded that the site constituted a human health hazard and recommended that the EPA conduct further environmental sampling on and near the site (FDHRS, 1996). The assessment also recommended that a comprehensive health study be conducted of local residents and former workers of the Escambia Wood–Pensacola wood-treatment facility.

Communication Issues—The draft protocol for the community health study was presented at a public meeting in September 1993. The ATSDR proposed the formation of a broad-based Community Assistance Panel that would have included CATE spokespersons, local physicians, local government officials, and other area residents. The Escambia community did not accept the proposal. Because the broad-based panel was rejected, CATE by default became the spokesgroup for the Escambia community. Comments received from CATE during or after the meeting focused on study design, participant selection criteria, comparison group selection, review of birth and death records, cancer prevalence, and a request for testing of dioxin in biologic samples. A revised draft protocol was presented to CATE in March 1994.

CATE referred the revised study protocol to several reviewers. One reviewer, an epidemiologist at a southern university, recommended that the ATSDR revise its study protocol and conduct a retrospective cohort study of current and past residents to assess several health conditions that may have occurred over a period of 40 years. The other reviewers were in general agreement with the ATSDR's study protocol. The dissenting reviewer became the principal technical adviser to CATE, serving as a volunteer. At a November 1994 meeting with the ATSDR staff, CATE representatives and their principal technical adviser met in Pensacola and agreed on a study design. After this meeting, the ATSDR staff visited the Pensacola community to assess available resources for helping identify former residents and explore the possibility of acquiring Escambia residents' medical records for evaluation.

At a public meeting in February 1995 in Pensacola, the ATSDR staff summarized the revised study protocol and the feasibility of conducting a retrospective cohort study, including exposure characterization. Residents attending the meeting made it very clear that they wanted no health studies until after they had been permanently relocated away from the site. Even after the ATSDR staff explained that a study might provide important information on the current health of the Escambia community, the response remained that relocation must come before any studies. The ATSDR later learned that two weeks before the public meeting, a national environmental grassroots organization had visited the community and advised CATE and others not to accept the ATSDR health study proposal because it would be "inconclusive by design."

In June 1995, the ATSDR staff briefed CATE's principal technical advisor and an attorney retained by the Escambia community on the status of the proposed health study. At this meeting, CATE's lack of response to the proposed health study was discussed. A spokesperson for CATE stated that their first priority was permanent relocation. No further contact between the ATSDR and the Escambia community has occurred regarding the proposed health study.

Outcome—In April 1996, the EPA stated their willingness to permanently relocate a portion of the Escambia community, based on the presence of dioxin in some soil samples. After the EPA proposal, the ATSDR suggested to the community that biologic exposure assessment be considered for determining individuals' exposure to dioxin and other hazardous substances.

In October 1996, the EPA announced its plan to spend about $18 million to relocate residents from the 158 houses and 200 apartments that constitute the whole Escambia community (Nossiter, 1996). An EPA press release stated, "The U.S. Environmental Protection Agency has decided to relocate all of the residents living around the Escambia Treating Company Superfund site in Pensacola. This decision is based on the best scientific information available about this community's unique environmental, health and safety factors. It also is based on the community's concerns as well as our goal to protect the neighborhood's health, welfare, and well-being" (EPA, 1996d).

The EPA's decision followed by two days a half-page advertisement that Lois Gibbs, a former resident of Love Canal, who directs a national grassroots environmental organization, placed in the October 1, 1996, Florida edition of *USA Today*. The advertisement, which was addressed to President Clinton, reminded him of a campaign speech in which he said that no child should have to live near a hazardous waste site (Satchell, 1997); the newspaper advertisement preceded by two days a debate in Florida between the vice-presidential candidates in the 1996 election.

The basis for the EPA's decision to relocate the Escambia community is found in Section 5 of the Record of Decision for the EWP site (EPA, 1998b), which states, "The residents living in the area will be relocated because of the cumulative impacts of the following factors, not because of any single factor: 1. EPA has identified health risks, above the 10^{-4} risk level, due to the presence of contamination from dioxin and BaP [benzo(a)pyrene] in portions of the Relocation Area. 2. The adverse impacts on the residents of the Relocation Area from fear stemming from uncertainty relative to health impacts, loss of property values, and psychological stress, are difficult to quantify but are very real considerations.... 4. Removing the residents in the Relocation Area will provide greater flexibility for final remedy selection with a significant potential for lowering overall project costs...."

Lessons Learned—It is doubtful that anyone associated with the Escambia Wood–Pensacola site and the Escambia community would consider risk communications to have been a success. Several lessons are apparent. The EPA's initial removal actions, which led to a huge pile of dioxin-contaminated soil left on the EWP site, created a daily, visible reminder to the Escambia community of government inertia. Contrast this with the quick, focused health and remediation actions that occurred in Chattanooga, Tennessee, and Stratford, Connecticut (described later). In contrast to Chattanooga and Stratford, local, state, and federal government agencies did not have an integrated plan for responding to the legitimate health and environmental concerns of the Escambia community.

Local media were generally involved only when public meetings or some controversy occurred. The ATSDR and other health agencies probably erred by simply accepting the extant community grassroots organization rather than establishing a more broadly based community advisory group. Moreover, the community, which had the ultimate objective of being bought out and relocated, provided inconsistent positions about their desire for "health investigations." This inconsistency prevented any agreement on what kind of health study to conduct.

One journalist concluded, "And if Escambia is any model, the local politics will be as important in determining which communities move as the science" (Satchell, 1997).

The ATSDR's offer to measure residents' exposure to dioxin and other toxicants remains unaccepted by the Escambia community.

COMMUNITY ADVISORY GROUPS

Lynn and Busenberg (1995) noted that the concept of citizen advisory committees (CACs) on environmental issues has had more than 30 years application in the United States. By their definition, "A CAC is a relatively small group of people who are convened by a sponsor for an extended period of time to represent the ideas and attitudes of various groups and/or communities for the purpose of examining a proposal, issue, or set of issues." They observe that CACs have been involved, particularly by state and local government, in shaping the development of legislation and regulatory standards, land-use decisions, and the planning, permitting, and remediation of large-scale projects that include waste facilities, dams, and water resources.

The investigators associated several benefits with the CAC process. By their reckoning, CACs have: (1) helped educate their sponsors (e.g., federal government agencies) on community attitudes, (2) helped inform the community, (3) provided a forum for citizen involvement in decision-making, (4) improved the public's support for local decisions, and (5) allowed government and industry to work with a focused part of the community, rather than with the whole community. However, as will be noted later in this chapter, the effectiveness of CACs in helping shape local environmental decisions and in communicating risk is open to question.

The philosophy and mechanics are still evolving about how government and other agents share health information with communities affected by environmental hazards. Some government agencies have implemented the principles of risk communication by focusing on community groups. These groups are called by different, but similar names: Community Advisory Groups, Community Assistance Panels, and Community Working Groups. The creation of such groups has by necessity become a component of the federal CERCLA program, given that statute's troubled history of difficult communications with community groups. Both the ATSDR and the EPA have created site-specific Community Assistance Groups as a means for communicating with communities. The following sections will summarize some salient results and the impacts of the two agencies' experiences.

ATSDR's Community Assistance Panels

In 1992, grassroots environmental groups advised ATSDR to enhance community involvement in the agency's public health programs at CERCLA sites. In response, the agency announced guidelines for creating Community Assistance Panels (CAPs) in communities impacted by hazardous substances (ATSDR, 1992f). These panels were designed to improve communication between the ATSDR and the public, establish an avenue for the agency to communicate site-specific scientific findings, and provide a means for communities to participate in the ATSDR's activities. In its announcement, the ATSDR stated that CAPs would consist of persons, generally 12 to 15 in number, from concerned communities. Serving on the CAP would be a voluntary, unpaid activity.

If an ATSDR public health assessment of a hazardous waste site or facility indicates the need for public health interventions (e.g., an exposure study or epidemiologic in-

The CAP for the Brio NPL Site

The Brio NPL site, located in Houston, Texas, was a former oil processing facility (Chapter 5). Local residents became concerned when they perceived a pattern of birth defects and other health problems. A public health assessment conducted by the ATSDR in 1989 found no likely pathway of exposure to hazardous substances from the site. (However, data collected in 1995 suggest soil gas may have migrated from the waste site into the adjacent residential area.) In the mid-1980s, the EPA and potentially responsible parties (PRPs) had entered into a consent decree that was legally binding. This agreement led to PRPs' conducting the environmental sampling and providing the results to the EPA. Some community residents questioned the credibility of the environmental contamination data. Later, some community residents disputed the EPA's proposed use of incineration to remediate the site.

In 1992, in response to community concerns about residents' health and the proposed method of site remediation, the ATSDR created a Community Assistance Panel (CAP) for the Brio site. The CAP consisted of 15–18 members, including a local newspaper publisher (who chaired the CAP) and the mayor of the community. The CAP met 13 times through May 1996 and played the key role in keeping lines of communication open between the Brio community, government agencies, and PRPs. It provided the ATSDR with community-collected health data and reviewed health study protocols and findings that the ATSDR developed. The CAP actively debated the EPA on the health implications of incineration technology, with the result that incineration was discarded as the remediation method in 1996. Alternative remediation methods are under review by the EPA and the community.

vestigation), a CAP may be deemed advisable. When considering the formation of a CAP, the ATSDR works with local media and other local sources to announce its intent. Written nominations are sought for 30 days. The ATSDR gives preference to persons nominated for CAP membership who are at least 18 years old, reflect area residents' various viewpoints; have knowledge of the site; the hazardous substances of concern, and/or the community's health concerns pertaining to the site, and are permanent residents of the community. Members of a CAP are selected by the ATSDR.

All CAP meetings are announced in advance and held in public. The issues discussed by CAPs vary with each site, but generally consist of the following activities: (1) reviewing findings from public health assessments, (2) elaborating communities' health concerns, (3) reviewing study protocols when health studies are to be pursued, (4) reviewing data-collection procedures and interim findings, (5) reviewing draft study reports, and (6) serving as a sounding board for community reactions and residual concerns, including as a forum for discussing and reacting to remediation options proposed by government agencies.

EPA's Community Advisory Groups

As stated previously, the need to enhance community involvement in CERCLA decisions has driven much of federal agencies' efforts to improve their work with communities. The EPA's work with community advisory groups is noteworthy because of its outreach to minority and low-income populations. In their policy document on community involvement, the EPA states, "Community Advisory Groups (CAGs) are another mechanism designed to enhance community involvement in the CERCLA process. CAGs respond to a growing awareness within EPA and throughout the Federal government that particular populations who are at special risk from environmental threats—such as minority and low-income populations—may have been overlooked in past efforts to encourage public participation" (EPA, 1995).

The EPA's guidance document on Community Advisory Groups contains practical advice on the purpose, creation, operation, and impact of CAGs (EPA, 1995). The EPA says the purpose of CAGs is to serve as public forums for government representatives and diverse community interests to present and discuss their needs and concerns related to the CERCLA decision-making process. However, the EPA states, as do other federal agencies, that it cannot abrogate its responsibility to make final decisions on actions to be taken at CERCLA sites.

The EPA guidance does not provide criteria or guidelines for when to form a CAG in a specific community. Under current federal government policies, a CAG can be a voluntary activity, that is, persons volunteer their time to serve on the CAG. However, for some communities, CAGs might be chartered under the Federal Advisory Committee Act (FACA). FACA-chartered committees must announce all meetings in advance and conduct them in a public setting; CAG members are paid for their service.

Concerning membership on a CAG, the EPA recommends, "To the extent possible, membership in the CAG should reflect the composition of the community near the site and the diversity of local interests, including the racial, ethnic, and economic diversity present in the community—the CAG should be as inclusive as possible" (EPA, 1995). The EPA recommends that CAG membership should be drawn from the following groups: (1) residents or owners of residential property near the site and those who may be directly affected by the site, (2) those who may be potentially affected by releases from the site, even if they do not live or own property near the site, (3) local medical professionals practicing in the community, (4) Native American tribes and communities, (5) minority and low-income groups, (6) citizens, environmental, or public interest group members living in the community, (7) recipients of CERCLA Technical Assistance Grants (TAG), if a TAG[8] has been awarded by the EPA to the community, (8) local governments, including pertinent city or county governments, and government units that regulate land use in the vicinity of the site, (9) the local labor community, (10) facility owners and other significant Potentially Responsible Parties, (11) the local business community, and (12) other local, interested individuals.

The number of members in a CAG will vary from site to site, based on local interest, but typically 15–20 members have constituted the committees. Each CAG is advised to

---

[8] Technical Assistance Grants (TAGs) are authorized under CERCLA, as amended. TAGs are three-year grants that do not exceed $50,000 per grant. The EPA awards them to community groups to help the groups participate in Superfund processes.

select its own chairperson, establish its purpose, undergo training, and establish a clear set of administrative procedures (EPA, 1995).

EFFECTIVENESS OF COMMUNITY ASSISTANCE COMMITTEES

Community Assistance Committees (CACs) that have been engaged in local, community-based environmental issues have become the subject of research. Such scholarly evaluation is important because of increased attention and commitment of resources by government and private industry in working with CACs. In particular, agencies need to know the most effective principles and practices when working with CACs.

Lynn and Busenberg (1995) reviewed 14 empirical studies of CACs that had advised government bodies on environmental issues. The reviewed studies spanned the years 1976 to 1993. The review was to identify (1) how CACs (or the researchers studying them) defined and measured success, and (2) the characteristics of CACs and their sponsors that were associated with success. The reviewers noted that definitions of success used in the studies they reviewed included the implementation by the sponsor of CAC recommendations, the representativeness of the CAC, the involvement of citizen participants in decision-making, and enhanced communication between the stakeholders.

The investigators drew several conclusions about CACs from the 14 studies they reviewed. They found a varied picture of impacts of CACs on public outcomes. Impacts were highly contingent on the intentions and expectations of the institutions or sponsors being advised by the CACs. Lynn and Busenberg found few examples of studies where researchers had examined process variables; that is, the steps and procedures that an agency had taken when forming a CAC and assessing its outcomes and impact. In fact, few examples were found of agencies or sponsors that had built formal evaluation procedures into their relationship with CACs.

Although Lynn and Busenberg drew no conclusions about which process variables are associated with successful CACs, they mention several process variables that other researchers felt were important. Until further research is available, these variables would seem quite useful for others to consider when forming CACs. **The characteristics of successful CACs might include: (1) being representative of the community, (2) having a clear definition of tasks or expectations, (3) having agreed-upon rules of conduct, (4) having mechanisms for accountability to the larger community, (5) using a professional facilitator, (6) using sources of information other than the sponsor, and (7) maintaining independence of the CAC from its sponsor.**

Having now summarized some key research findings on risk communication principles and concomitant community involvement, a case study is presented that incorporates many of the previously described recommendations on risk communication.

Case Study: Stratford, Connecticut

Events and outcomes in Stratford, Connecticut (NACCHO, 1995) provide an example of an effective risk communication about hazardous waste. The assertion here of effectiveness is predicated on an evaluation by a reviewer who examined several community involvement efforts that the ATSDR had undertaken (Cole, 1996). Any risk communication with the public should involve an evaluation element that is designed before the communication of the message.

The author has embedded in this case study references to each of the Cardinal Rules of Risk Communication listed in Table 8.3. Whether these rules were knowingly implemented in the planning and conduct of the Stratford risk communication and risk management is not known, but the seven rules were clearly operational in Stratford.

Background—Stratford is a city of approximately 50,000 persons located on Long Island Sound. The diverse population has a mixed economic base. In 1919, a plant opened in Stratford that manufactured brake parts and other products containing asbestos. Waste from the plant was first placed in lagoons, but as space was exhausted, the lagoons were dredged and their sediment was spread on the factory's property. Dredged wastes and other wastes generated by the company were made available as filler material to the public and found their way into wetland areas throughout the city.

Environmental Data—The waste material was later found to contain asbestos, lead, PCBs, dioxins/furans, phenolic resins, adhesives, solvents, and phenolic mixtures. Use of the company's waste as fill material in wetland areas was thought to have begun in the 1930s and continued for decades. In the 1970s, asbestos-containing material was discovered in many areas throughout the city. Temporary covers were placed on areas frequented by the public, but the covers eroded with the passage of time. In the 1980s, the city tried to monitor and regulate land alterations in known waste sites. In 1982, a citizens group, the Milford/Stratford Citizens Against Pollution, was formed in response to complaints about odors emanating from the plant. *[Rule 1: Accept and involve the public as a partner.]* In 1989, the company ceased operations in Stratford. Not long thereafter, the city realized it had a serious environmental and public health problem. A disaster was in the offing if the situation was not managed correctly, and plans were undertaken to deal with a major environmental contamination problem in the city's midst. *[Rule 2: Plan carefully and evaluate your efforts.]*

The company began work with the EPA to remediate hazards and environmental pollution at the plant site. At the urging of the city, the EPA conducted environmental sampling in areas of the city thought most likely to have received waste fill from the facility. *[Rule 5: Work with credible sources.]* Initial testing of 15 known waste disposal areas in the city found lead, asbestos, and PCBs at quite elevated levels. Lead was found at some sites in concentrations up to 150,000 ppm, asbestos at levels up to 900,000 ppm, and PCBs up to 160 ppm; dioxin contamination was found only at a few commercial properties (Cole, 1996). More than 500 residential, public, and commercial properties were tested for lead, PCBs, asbestos, and other substances. Ultimately, more than 50 properties were remediated in less than two years (Cole, 1996). Using the EPA's environmental contamination data, the ATSDR advised the city that the sampled sites were a potential hazard to the public's health. *[Rule 5: Work with credible sources.]*

Health Issues—In May 1993, the ATSDR issued a Public Health Advisory that warned the Stratford community of a substantial and imminent threat posed by hazardous waste attributable to the industrial facility. A public health advisory is the highest level of concern that the ATSDR issues under its CERCLA authorities (Chapter 2). Because of very high lead levels in some soil samples, an early concern of health authorities was children's exposure to lead. Asbestos was of major interest because of its high concentration in soil samples. The public also expressed concern that PCBs and dioxin might cause cancer rates to increase. *[Rule 3: Listen to the public's specific concerns.]*

Communication Issues—Because of both the widespread environmental contamination following many years of using waste from the industrial facility, and the known toxicity of the principal hazardous substances (lead, asbestos, dioxin, PCBs), the local health department director took extraordinary measures to establish and maintain communications with the Stratford community. The local health department was the clearinghouse for information on health and related concerns. It helped create the Stratford Citizens Advisory Council. The Council, in turn, quickly became a vital link between the Stratford community, the news media, and various government agencies (NACCHO, 1995). Coordination by the local health department of communications among more than 40 local, state, and federal health and environmental agencies was noteworthy.

The Stratford Health Department developed a comprehensive public information and outreach program. *[Rule 4: Be honest, frank, and open.]* Their efforts included: (1) public meetings and forums, (2) a centralized information source, (3) full-time community liaison/advocate for the community who responded to telephone inquiries and met often with concerned citizens, (4) a quarterly newsletter, (5) letters sent to Stratford residents impacted by hazardous waste on their property, and (6) a health booth operated by the city health department and the Stratford Citizens Advisory Council. Weekly meetings were held with the various community and government stakeholders, and news media contact was regular and factually based. *[Rule 6: Meet the needs of the media.]*

Outcome—Quick and resolute response by the city health department director was the essential ingredient in what became a successful health communication program (NACCHO, 1995; Cole, 1996). *[Rule 7: Speak clearly and with compassion.]* The community developed trust with the local health department, and the formation of a local citizens' group was an essential bridge between the concerned community and various government agencies. Moreover, prompt action by the EPA and state agencies to quickly remove hazardous substances from several contaminated properties help reduce the public's health concerns and built confidence in the community that something was being done. Blood lead screening of children revealed a small number of children who had elevated blood lead levels, and efforts were taken to prevent their further exposure to lead. Other health investigations for cancer incidence are under review and would be of a long-term nature. One source estimates that more than $100 million will be required to remediate the hazardous waste contamination in the Stratford area (Cole, 1996).

Lessons Learned—The risk communication and health and environmental actions in Stratford were effective in addressing community concerns and evidencing appropriate actions to the community. Several lessons can be gleaned from the Stratford experience. First, and foremost, the timely, sympathetic, professional response by the city health director was essential for everything else that followed. Whether the director recognized it or not, the risk communications principles practiced by the city health department were consistent with the Cardinal Rules of Risk Communication (Table 8.3). Another important development was creation of a community-based advisory group, which provided oversight of government actions and linked community concerns to government actions. The positive, informed reporting by local news media of the Stratford environmental problems was another important contribution to the whole risk communication effort. An extraordinary level of cooperation and coordination across several lines of government facilitated a positive outcome. However, local leadership by the city health department was the essential glue that ensured inter-government and community cooperation.

SYNTHESIS

The assessment and management of environmental hazards has become the staple of government agencies. With this has come risk communication, which was defined by the National Research Council in 1989 as "...an interactive process of exchange of information and opinion among individuals, groups, and institutions." This definition and the ensuing evolution of risk assessment and management have led to what is called participatory communication, in which the risk communicator and the audience for the communication share a joint responsibility for shaping, disseminating, and evaluating the communication effort.

An important body of experience and research has accrued on how to communicate risks. As proposed at the start of this chapter, communication is at the heart of public health practice. This assertion applies irrespective of the nature of the public health hazard. Effective communication of health risks that result from uncontrolled hazardous waste sites and emergency releases of substances is therefore integral to preventing or mitigating health effects from such releases. However, to achieve effective communication, the risk communicator must integrate the experience and recommendations from research on risk perception, from community involvement, and from risk management of hazards.

Risk communicators and risk managers need to know how the public perceives risk and then involve the public in their actions. Dread of and familiarity with the hazard are the two key characteristics that shape lay persons' risk perception and therefore the two characteristics of risk perception that risk communicators should recognize. The hazards most dreaded by lay persons are also the hazards the public wants to have regulated and acted on to reduce the risk. Because uncontrolled hazardous waste sites have characteristics that suggest they are dreaded by the lay public, risk communicators should factor this knowledge into their health risk messages and risk management actions.

Through case studies, this chapter has outlined the importance of planning and evaluating risk communications and has provided tips and advice to risk communicators on how to plan their efforts, evaluate their communications, and conduct public meetings and efforts with the news media. It specifically advocates use of the Cardinal Principles of Risk Communication (Table 8.3).

The role of Community Advisory Committees (or groups by similar names) in decision-making on hazardous waste issues will likely continue to increase.[9] This will include their involvement in land use, site remediation, and health-based actions. However, the record of CACs's involvement in such decisions is mixed, as has been the role of government agencies. Available research findings suggest that characteristics of successful CACs include:

- being representative of the community,
- having a clear definition of tasks or expectations,
- having agreed-upon rules of conduct,
- having mechanisms for accountability to the larger community,
- using a professional facilitator,

---

[9] Both the 103rd and 104th Congresses drafted legislation to reauthorize CERCLA that included language to create Community Working Groups.

- using sources of information other than the sponsor, and
- maintaining independence of the CAC from its sponsor.

This chapter closes the way it began. Risk communication must be viewed and practiced in the context of public health. Solutions to the public's health concerns require communication with, and involvement of, the impacted public. Experience has shown that solutions must include the practice of contemporary risk communication, community involvement, and health education. Honesty, candor, and fairness form a powerful glue that can keep risk communications intact and effective.

Chapter Nine

Environmental Equity

Environmental equity is a subject of great contemporary importance. Although the terms 'environmental equity' and 'environmental justice' are sometimes used interchangeably, important differences exist in definition and intent. What is the origin of environmental equity? This chapter summarizes the history, key issues, concepts, findings, and social and governmental responses bearing on inequities faced by persons of color and of low income because they live near hazardous waste sites and waste disposal facilities.

Several environmental issues have contemporary ethical importance. One such issue for government's and private industry's environmental policies and practices is called environmental equity. Other terms that predate environmental equity are 'environmental racism' and 'environmental inequity' (Sexton et al., 1993). The term 'environmental justice' appeared somewhat later and is favored by many grassroots groups, some government agencies, and elected officials.

As a prelude to comments on the environment's impact on persons of color and on communities challenged by economic and social conditions, the generally inferior health status (compared with Caucasians) of persons of color, particularly those of African-American descent, should be borne in mind. One way of depicting the disparity is to calculate the excess deaths experienced by African-Americans beyond those for white Americans (Warren, 1993). In 1990, 75,000 excess deaths occurred among African-Americans. The primary causes for the excess are heart disease and stroke, cancer, cirrhosis, diabetes, homicides, unintentional injuries, infant mortality, and AIDS (Warren, 1993). This same source comments, "Although many associations with these excess deaths, such as income, education, access to care, and preventive practices, have been determined, the basic causal factors associated with ill health and race remain unanswered."

The consensus view is that much of the excess morbidity in persons of color is still unexplained. This assertion includes the uncertain degree to which environmental hazards contribute to excess morbidity in these populations. What is known, as will be subsequently described, is the fact that in the United States some environmental hazards are experienced more often by minority populations than by white populations.

As the title of this chapter implies, this author prefers the term environmental equity rather than environmental justice. Fairness is the core concern of both environmental equity and environmental justice. (As a consequence of individuals' differences, a healthy democracy that treats people fairly may not always be able to treat them equally.) Although equity and justice are both rooted in a concern for fair treatment, equity seems more directly synonymous with fairness than does justice. Justice has a more litigious image than does equity, and litigation does not always result in fair outcomes. Moreover, equity seems to connote a more prospective approach—actions that will guard against inequities in how environmental hazards are shared. Justice seems more retrospective—actions that give emphasis to redressing past actions that imposed disproportionate shares of environmental hazards.

The EPA gave a different argument in favor of the term environmental equity in their report on reducing risk for all communities (EPA, 1992a): "EPA chose the term environmental equity because it most readily lends itself to scientific risk analysis. The distribution of environmental risks is often measurable and can be quantified. The Agency can act on inequities based on scientific data. Evaluating the existence of injustices and racism is more difficult because they take into account socioeconomic factors in addition to the distribution of environmental benefits that are beyond the scope of this report. Furthermore, environmental equity, in contrast to environmental racism, includes the disproportionate risk burden placed on any population group, as defined by gender, age, income, as well as race." In sum, the EPA's preference in 1992 for the term environmental equity was based on their belief that measurability (i.e., of equity) was important.

The State of Washington, as did the EPA in 1992, prefers the term environmental equity "...based upon the connotation that the word 'equity' better relates to something measured, as opposed to 'justice',..." (WDOE, 1995). They define **environmental equity** as, **"...the proportionate and equitable distribution of environmental benefits and risks among diverse economic and cultural communities** (emphasis added). It ensures that policies, activities, and the responses of government entities do not differentially impact diverse social and economic groups. Environmental equity promotes a safe and healthy environment for all people."

However, in 1994, the Protocol Committee that organized a national conference on environmental justice in Washington, D.C. asserted, "Environmental justice encompasses more than equal protection under environmental laws (environmental equity). It upholds those cultural norms and values, rules, regulations, and policies or decisions to support sustainable communities, where people can interact with confidence that their environment is safe, nurturing, and productive. Environmental justice is served when people can realize their highest potential, without experiencing sexism, racism and class bias. Environmental justice is supported by clean air, water and soil; sufficient, diverse and nutritious food; decent paying and safe jobs; quality schools and recreation; decent housing and adequate health care. Environmental justice is supported by democratic decision-making and personal empowerment; and communities free of violence, drugs, and poverty and where both cultural and biological diversity are respected" (Protocol Committee, 1994). This definition of environmental justice is more expansive and idealistic than that used to define environmental equity. The Protocol Committee's concept of environmental justice does not seem as amenable to measurement as do the definitions of environmental equity.

An environmental justice conference held in 1994 had considerable impact on the EPA, leading it to move away from the term environmental equity in favor of environmental justice. As a consequence, the EPA developed working definitions for "environmental justice" and "fair treatment" (ATSDR, 1995d):

Case Example: Navajo Nation CERCLA Sites

The Navajo Nation and the indigenous people of America hold a deep reverence for the earth and the environment (Antonio et al., 1992). The Navajo Nation consists of 25,000 square miles of tribal land in Arizona, New Mexico, and Utah. Because of uranium mining, the Navajo Nation contains several hundred abandoned uranium mines. As a matter of environmental inequity, inadequate protection of tribal land occurred as mines were developed and later abandoned after mining operations ceased. Mining wastes, uranium tailings, and other hazards were left as a legacy to the Navajo Nation.

The abandoned mines pose severe threats to human health and the environment (Antonio et al., 1992). Uranium mining on Navajo lands occurred from the 1940s to the early 1990s. Open-pit surface mining was common, leading to abandoned pits. Many pits are immense, and some are used as trash dumps, swimming holes, and livestock water ponds.

The Navajo Superfund Program has evaluated the hazards posed by these abandoned uranium mines. They include: exposure to ionizing radiation, inhalation of radon gas, ingestion of contaminated ground and surface water, direct ingestion of site contaminants, ingestion of contaminated food in contact with releases from sites, and physical hazards. Efforts are ongoing to remediate abandoned mines and remove hazardous waste.

- Environmental Justice—the fair treatment and meaningful involvement of all people, regardless of race, ethnicity, culture, income, or education level with respect to the development, implementation, and enforcement of environmental laws, regulations, and policies.
- Fair Treatment—no population, due to political or economic disempowerment, is forced to shoulder the negative human health and environmental impacts of pollution or other environmental hazards.

The EPA's current definition of environmental justice, which is found on their website,[1] is "Environmental justice is the fair treatment for people of all races, cultures, and incomes, regarding the development of environmental laws, regulations, and issues." This is a refinement of their 1995 working definition, stated above. It is curious why "meaningful involvement" is omitted in the current EPA definition.

The language of environmental equity and justice is apparently still evolving. For example, the Presidential/Congressional Commission on Risk Assessment and Risk Management defined environmental justice as, "Concern about the disproportionate occurrence of pollution and potential pollution-related health effects affecting low-income, cultural, and ethnic populations and lesser clean-up efforts in their communities" (CRARM, 1997). This definition seems directed to CERCLA purposes, given the mention of cleanup efforts, and therefore would be more limited in application.

---

[1] The website address is http://www.epa.gov.

The preferred nomenclature seems dependent on the ethos and motives of the group using the term. Although this chapter is titled "Environmental Equity," much of what is summarized here and expressed elsewhere in reports and publications uses the term environmental justice. Thus, both terms will be used throughout this chapter. However, the definition of environmental equity developed by the State of Washington will be adopted.

Environmental equity, as the term suggests, addresses the question of whether environmental risk factors are shared equally by all segments of a population. More specifically, are minority groups and persons of low income disproportionately exposed to environmental risks because of racist policies and actions by government and private sector agents (e.g., Bullard, 1990)? Some history is instructive.

HISTORY

Pinpointing any one event that triggered what became the environmental equity movement is difficult. As seen in much of the American civil rights movement and social justice crusade, societal change occurs as the result of many events over time. A societal movement occurs if sufficient supportive public opinion develops and enough supporters are drawn to the cause. The movement then mobilizes its energy to prevail upon institutions, often governmental bodies, that can effectuate change. Federal voting rights legislation and ordinances that protect against sexual discrimination are examples of such changes.

The American environmental equity movement is like a tree with many roots and branches. It is not yet a mature tree and its survival will depend on many factors. Certainly one factor will be public perception. Will the general public support environmental equity actions and at what cost? The tree has not grown enough to provide an answer. What are the seeds of our metaphoric tree called environmental equity?

Warren County, North Carolina, Protest

Several key events have shaped the environmental equity movement. If any one event can be termed "the" event, it occurred in Warren County, North Carolina, in 1982 (Foreman, 1996). Local opposition arose when the state announced its proposal to locate a hazardous waste facility in this county, which had a large African-American population. The landfill was targeted to receive PCB-contaminated waste (Bullard, 1995) at a site near the community of Afton, which was 84% African-American (Bullard, 1990). When protests very similar to those of the American civil rights movement of the 1960s resulted, and more than 500 persons were arrested, the protests and arrests caught the attention of national news media. Despite the protests and media attention, delivery of hazardous waste to the landfill began.

The final attempt to stop the landfill occurred in July 1982. The local chapter of the National Association for the Advancement of Colored People sought a preliminary injunction in federal court to prohibit placement of PCBs in the Warren County landfill. The court denied the request, stating that race was not an issue in siting the landfill because race was never mentioned as a motivating factor throughout all federal and state hearings and private party suits (GAO, 1983). Although community opposition, national media attention, and legal proceedings did not halt construction and use of the landfill, the civil rights demonstrations and their aftermath mobilized attention on a new issue, which some called environmental racism (Bullard, 1992).

The events in North Carolina led Walter Fauntroy, District of Columbia Delegate to Congress, and Congressman James Florio of New Jersey to ask the General Accounting Office (GAO) in 1982 to assess the racial implications of facilities in the southern states that received hazardous waste. The GAO found African-Americans were the predominate population living near three of the four largest facilities in the South (GAO, 1983). That gave weight to the belief that landfills were deliberately being targeted for location in minority communities.

Issues that came from the Warren County, North Carolina, PCB landfill siting set into motion a series of events that have shaped the current environmental equity/justice movement. Eight key events that shaped the movement are summarized in Table 9.1. Subsequent sections in this chapter describe the nature and import of the events listed, but for historical perspective, one report and four events merit elaboration here.

Bullard's Book and Thesis

The events in Warren County, North Carolina, gave impetus to research that Robert D. Bullard, a sociologist, published as *Dumping in Dixie: Race, Class, and Environmental Quality* (1990, 1994). The book, which quickly became a cardinal, seminal work within the civil rights movement, focused on five African-American communities struggling with environmental problems: Houston, Texas; Dallas, Texas; Institute, West Virginia; Alsen, Louisiana; and Emelle, Alabama. Bullard described in detail the concerns of local residents and the potential health risk ascribed by them to the presence of hazardous waste landfills or operating chemical plants.

Bullard's book contains methods of dispute resolution and grassroots strategies that can be used to counter environmental inequities. This work places him at the center of scholars who developed the intellectual framework that constitutes environmental equity (Bullard, 1990, 1992, 1994, 1995) and has contributed much to improving social and environmental policies intended to prevent environmentally discriminatory actions.

Three Key Conferences

Three conferences have had great impact on the evolution of environmental equity by helping identify disparities according to race, income, or culture. Each influenced subsequent actions that advanced environmental equity and justice policies, particularly those of federal government agencies.

The debate over environmental equity had advanced sufficiently by 1990 to warrant a conference on race and environmental hazards. In January of that year, the University of Michigan's School of Natural Resources convened scholar-activists in a national conference to address the distribution and management of environmental risk (Bryant and Mohai, 1992; Reilly, 1992). Nine of 12 scholars who presented papers were minorities, marking the first environmental equity conference where the majority of presenters of scholarly papers were persons of color (Bryant and Mohai, 1992). The Michigan conference resulted in a compilation of papers that advanced the debate about race and environmental equity.

However, the most important outcome of the Michigan conference was the creation of what became known as the Michigan Coalition, a subgroup of conferees who com-

Table 9.1. Eight Key Events that Shaped the Environmental Equity/Justice Movement.

Event/Year	Impact of Event
1. Warren County, North Carolina, civil rights opposition to a proposed landfill/1982	Brought national attention via news media and made environmental concerns a matter of civil rights
2. GAO Study of 4 Hazardous Waste Landfills/1983	Elevated to congressional attention the potential inequity of placing hazardous waste facilities in areas that have large minority populations
3. Release of Commission on Social Justice Study of Minorities and Waste Facilities/1987	Report gave credence to concerns that minorities were overrepresented in areas around waste sites; became seminal document within civil rights and social justice movements
4. University of Michigan Natural Resources Conference and Formation of the Michigan Coalition/1990	Michigan Coalition had major impact on the EPA's recognition of environmental equity as a concern (Reilly, 1992)
5. Publication of the book *Dumping in Dixie*/1990	First academically based report of patterns of environmental inequities according to race
6. National Conference on Minorities and Environmental Pollution/1990	First national conference to focus on research findings and gaps in knowledge about health effects of environmental hazards on minorities
7. National Conference on Environmental Justice/1994	First national conference specific to environmental justice; developed strategies for environmental justice pursuits
8. Issuance of Presidential Executive Order on Environmental Justice/1994	Provides resources, authority, and imprimatur of federal government

posed an agenda for environmental equity and conducted a series of meetings with senior federal government officials to present their agenda. According to EPA Administrator William Reilly, "It was the arguments of this group that prompted me to create the Environmental Equity Workgroup" (Reilly, 1992). In turn, the EPA Equity Workgroup evaluated environmental risk and race data and produced the report *Environmental Equity: Reducing Risk for All Communities*. This EPA report, which is described below, was the federal government's first official expression on environmental equity and became a primary resource for environmental equity advocates who lobbied for government support and action.

In 1990, the ATSDR organized the first national conference on minority health and environmental contamination (Johnson et al., 1992). The 400 participants were primarily researchers and investigators from government agencies and universities. The conference concentrated on adverse health effects of hazardous substances in minority populations, educational needs of low-income communities, and improvements needed in risk assessment to account for potential disproportionate impact of hazardous substances on minorities. The meeting resulted in agreement that minorities were at increased health risk from various environmental hazards, that risk assessments should integrate concern for minorities and susceptible populations (e.g., children), and that additional research and data collection were warranted.

The ATSDR meeting led to a much larger conference in 1994 that brought together 1,100 environmental justice advocates, state and federal government representatives, university researchers, and others. The meeting was preceded by issuance of a set of 10 review papers that helped shape dialogue during the conference (Sexton and Anderson,

Case Example: Mojave Desert Hispanic Community

Health studies of minority communities at potential health risk from exposure to hazardous waste must be conducted in ways that respect cultural sensibilities. An example is a study conducted by the California Department of Health Services of a small Hispanic community in the Mojave Desert (Teran et al., 1994). The Department had found quite high levels of dioxin in the ash from a junkyard smelter where insulated wire was burned to reclaim the copper. This led to concerns that exposure to dioxin had occurred among the residents. A protocol was designed to collect blood samples for dioxin analysis.

The state health department prepared all materials in both English and Spanish. A door-to-door survey was conducted bilingually. Small meetings were held with local residents to explain all blood sampling procedures. Special efforts were taken to overcome mistrust of government agencies. However, despite all these efforts by the study team, only 3 of 41 persons eligible for measurement of blood dioxin levels participated. Investigators believed that the large amount of blood, 1 pint, was a deterrent. Nonetheless, the culturally sensitive efforts of the research team were commendable.

1993). The conduct of the meeting ranged from confrontation to conciliation. From the meeting came agreement on a set of five recommendations designed to impact government actions on environmental justice (Protocol Committee, 1994).

The protocol committee consisted of 23 persons, consisting in approximately equal numbers of grassroots environmentalists, federal government officials, and academic representatives. They developed the following recommendations:

I. **Conduct meaningful health research in support of people of color and low-income communities.** Preventing disease in all communities and providing universal access to health care are major goals of health care reform. Effective preventive measures cannot be equitably implemented in the absence of a targeted process that addresses the environmental health research needs of high risk workers and communities, especially communities of color.

II. **Promote disease prevention and pollution prevention strategies.** Although treating disease and cleaning up environmental hazards are essential, long-term solutions must rely upon truly preventive approaches.

III. **Promote interagency coordination to ensure environmental justice.** Although at-risk communities and workers are most threatened by occupational and environmental hazards, government agencies (federal, regional, state, local, and tribal) are also important stakeholders. Unfortunately, environmental problems are not organized along departmental lines. Solutions require many agencies to work together effectively and efficiently.

IV. **Provide effective outreach, education, and communications.** Findings of community-based research projects should be produced and shared with community members and workers in ways that are sensitive and respectful to race, ethnicity, gender, language, culture, and in ways that promote public health action.

V. Design legislative and legal remedies. [No narrative accompanied this recommendation.]

These five elements were accompanied by specific strategies and activities that should be pursued by government and private sector entities. The five elements have had substantive impact on the federal government's environmental justice strategies, as described later in this chapter.

During the conference, President Clinton issued an executive order on environmental justice, which is described in a following section. The long-term effect of the executive order remains to be determined, but clearly the national environmental justice conference held in 1994 played a large role in ensuring the issuance of the executive order.

The events in Table 9.1 were seeds that planted environmental justice in the orchard of civil rights. A series of studies followed that tried to better define minority groups and persons of low income who are at risk because of exposure to environmental hazards. Most of these studies concentrated on the demographics of populations living near uncontrolled hazardous waste sites and commercial, operating hazardous waste facilities.

FIVE SEMINAL REPORTS

Concerning environmental equity, are hazardous waste treatment, storage, and disposal facilities (TSDFs) and uncontrolled hazardous waste sites (i.e., CERCLA sites) found more often in minority communities than elsewhere? Moreover, regarding environmental justice, do data support the concern that minority populations and persons of low income have been targeted for placement of TSDFs in their communities?

General Accounting Office (GAO) Study of 1983

In December 1982, following the Warren County episode, District of Columbia Delegate to Congress Walter Fauntroy and Congressman James Florio of New Jersey requested the GAO to "...determine the correlation between the location of hazardous waste landfills and the racial and economic status of the surrounding communities" (GAO, 1983). According to the GAO, agreement with the study's requestors led to examining sites only in the eight southeastern states. The agreement also included examining only off-site landfills, those not contiguous to industrial facilities.

The GAO identified four operating landfills in the Southeast that received hazardous waste: Chemical Waste Management, Sumter County, Alabama; Industrial Chemical Company, Chester County, Alabama; SCA Services, Sumter County, South Carolina; and the Warren County PCB Landfill, North Carolina. For each site, Bureau of Census racial and economic data of 1980 were obtained for census areas in which the landfills were located and the census areas that had borders within about 4 miles of the landfill.

The GAO found Blacks[2] were the majority population in census areas at three of the four sites: Chemical Waste Management (CWM), Industrial Chemical Company (ICC), and Warren County PCB Landfill sites. At all four sites, the Black populations in census

---

[2] The current preferred terminology is African-Americans. Mention of Blacks is a consequence of terminology used by the investigators whose work is described in the text.

Table 9.2. 1980 Census Data for Census Areas Where Landfills Are Located (GAO, 1983).

Landfill	Landfill's Census Area Population (% Black)	Landfill's County Population (% Black)	Mean Family Income for All Races in Landfill's Census Area	Mean Family Income for Blacks in Landfill's Census Area
Chemical Waste Management, Alabama	626 (90%)	16,908 (69%)	$11,198	$10,752
SCA Services, South Carolina	849 (38%)	88,243 (44%)	$16,371	$6,781
Industrial Chemical Co., Alabama	728 (52%)	30,148 (39%)	$18,996	$12,941
Warren Co. PCB Landfill, North Carolina	804 (66%)	16,232 (60%)	$10,367	$9,285

areas containing landfills had mean incomes lower than the mean income for all races combined in the same census area. The findings are summarized in Table 9.2. For example, data in Table 9.2 indicate that the Warren County PCB Landfill was located in a census area with a population of 804 persons, of which 66% were Black. The landfill was located in a county with 16,232 population, of which 60% were Blacks. The mean family income for all races in the landfill's census area was $10,367, compared with $9,285 mean family income for Blacks in the same census area.

The data in Table 9.2 show that percentages of Blacks in census tracts containing landfills generally mirrored the minority population of the counties in which the landfills were located. For all four sites, the mean family income for African-Americans living within the landfills' census area was lower than the mean income for all races in the landfills' census areas. These data suggest that African-Americans living near the four landfills had lower incomes than did other persons in the areas. However, little in the GAO report sheds light directly on what factors led to each site's location. Nonetheless, the GAO's findings added weight to the belief that areas with high percentage of minorities were being targeted for location of hazardous waste sites.

United Church of Christ Report of 1987

Drawing from the GAO report, the United Church of Christ's Commission for Social Justice (CSJ) conducted two studies to determine racial and socioeconomic characteristics of Americans living (1) in residential areas surrounding commercial TSDFs and (2) near uncontrolled toxic waste sites (CSJ, 1987).

The first CSJ study sought to determine whether the variables of race and socioeconomic status played significant roles in the location of commercial TSDFs. The methodologic approach compared geographic characteristics presumed to be relevant to the siting of commercial hazardous waste facilities. The study analyzed five sets of national data: (1) minority percentage of the population, (2) mean household income, (3) mean value of owner-occupied homes, (4) number of uncontrolled toxic waste sites per 1,000 persons, and (5) pounds of hazardous waste generated per person. Racial classifications were taken from the 1980 U.S. Census, and data on TSDFs were obtained from EPA databases.

Minority percentage of the population was used to measure racial composition of communities. Mean household income and mean value of owner-occupied homes were included in the analysis to determine whether socioeconomic factors were more important than race in locating commercial hazardous waste facilities. Existence of uncontrolled waste sites was evaluated to see whether underlying historic or geographic factors were associated with siting of commercial hazardous waste facilities in ways that were not accounted for by other variables used in the analysis.

Results of discriminant analysis showed that minority percentage of the population was statistically significant in relation to the presence of commercial hazardous waste facilities. The percentage of community residents that belonged to particular racial and ethnic groups was a stronger predictor of the level of commercial hazardous waste activity than was household income, the value of homes, the number of uncontrolled toxic waste sites, or the estimated amount of hazardous wastes generated by industry. A key finding was that in ZIP code areas[3] having one commercial TSDF operating in 1986, the percentage minority population, on average, was twice that of areas that did not contain TSDFs.

The second CSJ study was descriptive in nature. Its primary purposes were (1) measure the number of racial and ethnic persons who lived in residential areas where uncontrolled toxic waste sites were located and (2) make comparisons between the extent to which uncontrolled waste sites were located among different racial populations. Investigators used U.S. Census data for 1980 and data in the EPA's national list of uncontrolled hazardous waste sites, which is called the Comprehensive Environmental Response, Compensation, and Liability Information System (CERCLIS). At the time of the study, CERCLIS contained information on 18,164 uncontrolled toxic waste sites. Residential 5-digit ZIP code areas were used to define "communities."

The CSJ's descriptive study found the presence of uncontrolled waste sites to be "highly pervasive." More than half the U.S. population lived in residential ZIP code areas with one or more uncontrolled toxic waste sites. Moreover, three of every five African-American and Hispanics lived in communities with uncontrolled toxic waste sites, which amounted to more than 15 million African-Americans and 8 million Hispanics. The investigators estimated that 2 million Asian/Pacific Islanders and 700,000 Native Americans also lived in such communities.

Mohai and Bryant Study of 1992

Mohai and Bryant (1992) examined the regional demographics of persons living near 16 commercial TSDFs in three counties (Macomb, Oakland, Wayne) in the Detroit, Michigan, area. Data on race and income were obtained from face-to-face interviews of persons in a sample of households selected with equal probability. An additional oversample was drawn of households within 1.5 miles of existing (n = 14) and proposed TSDFs (n = 2). Information about race and household income was obtained for 793 respondents; for analysis, all nonwhites were combined into one category "minority."

---

[3] ZIP code areas are administrative units established by the U.S. Postal Service for the distribution of mail and do not generally respect political or census statistical area boundaries (BOC, 1992). ZIP code areas for 1980 contained, on the average, about twice as many persons (6,500) as did census tracts (3,900) (Anderton et al., 1994b).

Analyses were conducted of respondents living within one mile of a facility; from 1 to 1.5 miles of a facility; and persons living more than 1.5 miles from a facility. Results showed that percentage minority population and percentage below poverty level varied with distance from TSDF facilities. Within one mile, 48% were minority and 29% were below the poverty level; for 1 to 1.5 radial distance the corresponding numbers were 39% and 18%; and for more than 1.5 mile, 18% and 10%. Chi-square tests indicated all these percentage differences were statistically significant.

A second objective of the Mohai and Bryant study was to examine relationships between race and income on the distribution of commercial hazardous waste facilities. Multiple linear regression was used to test the strength of associations. Investigators tested whether race (coded as 1 = white and 0 = minority) and household income (measured in dollars) had independent relationships with the distance of residents from a TSDF. The investigators found that the relationship between race and location of TSDFs in the 3-county area was independent of income. Moreover, race was the stronger predictor of proximity to a TSDF.

Mohai and Bryant concluded,[4] "Review of 15 existing studies plus results of our Detroit area study provide clear and unequivocal evidence that income and racial biases in the distribution of environmental hazards exist. Our findings also appear to support the claims of those who have argued that race is more importantly related to the distribution of these hazards than income."

EPA Study of 1992

At the direction of Administrator William Reilly, an EPA Environmental Equity Workgroup was formed in July 1990 to review evidence that racial minority and low-income communities bear disproportionate burdens of environmental risks. The workgroup conducted a comprehensive evaluation of the scientific literature on environmental equity and related issues, examined environmental and human exposure databases, and reviewed data collected by federal health agencies on the health of minorities. They also reviewed socioeconomic data pertinent to environmental equity concerns.

Six findings came from these evaluations: three are specific to risk communication and government policy issues, the other three are recommendations specific to human health and environmental hazards and are therefore more germane for this chapter. As quoted from the EPA Environmental Equity Workgroup's final report (EPA, 1992a):

1. There are clear differences between racial groups in terms of disease and death rates. There are limited data to explain the environmental contribution to these differences. In fact, there is a general lack of data on environmental health effects by race and income. For diseases that are known to have environmental causes, data are not typically disaggregated by race and socioeconomic group. The notable exception is lead poisoning. A significantly higher percentage of Black children compared with White children have unacceptably high blood lead levels.

2. Racial minority and low-income populations experience disproportionate exposures to selected air pollutants, *hazardous waste facilities* (emphasis added), contaminated fish and agricultural pesticides in the workplace. Exposure does not always result in an immediate or acute health effect. High exposures, and the possibility of chronic effects, are nevertheless a clear cause for health concerns.

---

[4] Excerpted with permission. Copyright 1992 Westview Press, Boulder, Colorado.

3. Environmental and health data are not routinely collected and analyzed by income and race. Nor are data routinely collected on health risks posed by multiple industrial facilities, cumulative and synergistic effects, or multiple and different pathways of exposure. Risk assessment and risk management procedures are not in themselves biased against certain income or racial groups. However, risk assessment and risk management procedures can be improved to better take into account equity considerations.

The EPA found major limitations in the environmental and health databases pertinent to environmental equity issues. However, for three hazards compelling data supported an EPA finding that environmental exposures were disproportionately borne by minorities. The three hazards, air pollution, children's exposure to lead, and hazardous waste sites, and others in the EPA report are discussed in the following sections.

Air Pollution

As the EPA noted (1992a), air pollution is primarily a problem of urban areas, where pollution emission densities are greatest. The EPA noted that the percentages of various populations living in polluted urban areas differ by ethnic category: White (70.3%), Black (86.1%), Hispanic (91.2%), and Other (86.5%). Citing the work of Wernette and Nieves (1991), who analyzed the demographics of areas that the EPA designated as being out of compliance with the Clean Air Act, the EPA concluded that minorities were disproportionately exposed to air pollutants. Table 9.3 contains the data that undergird the EPA's conclusion. These data show the importance to minorities of attaining urban air quality standards under the Clean Air Act.

The impact of air pollution on Hispanics is also a matter of concern. The National Coalition of Hispanic Health and Human Services Organizations has pointed out that reducing exposure to air pollution is a priority issue for Hispanic communities because, in an update of data in Table 9.3, about 80% of Hispanics still live in areas that do not attain air quality standards (COSSMHO, 1996). These locales fail to meet the EPA ambient air quality standards. By updated comparison, about 65% of non-Hispanic Blacks and 57% of non-Hispanic Whites still live in nonattainment areas. The implications for Hispanics is a greater rate of respiratory morbidity and mortality and other adverse health effects than for other groups.

Children's Lead Exposure

The EPA concluded that children's exposure to lead was the environmental hazard for which the strongest evidence supports a disproportionate effect on minority populations. The relevant data are given in Table 9.4, which is drawn from the ATSDR (1988) and the EPA (1992a) documents. Drawing on data assembled by the ATSDR (1988), the EPA concluded that the evidence was unambiguous: children of color have higher blood lead levels than do White children. Moreover, all socioeconomic and racial groups have children with lead in their blood high enough to cause adverse health effects. This was found to be particularly true for African-American children. As evident from data in Table 9.4, lower family income was associated with higher prevalence of elevated blood lead levels in children.

Following the EPA report (1992a), the Centers for Disease Control and Prevention (CDC) updated their data on blood lead levels in young children (CDC, 1997). The CDC's

Table 9.3. Percentages of Populations Living in Air Quality Nonattainment Areas (EPA, 1992a).

Air Pollutants	Whites (70.3% Urban)	Blacks (86.1% Urban)	Hispanics (91.2% Urban)
Particulate matter	14.7%	16.5%	34.0%
Carbon monoxide	33.6%	46.0%	57.1%
Ozone	52.5%	62.2%	71.2%
Sulfur dioxide	7.0%	12.1%	5.7%
Lead	6.0%	9.2%	18.5%

Table 9.4. Estimated Percentages of U.S. Children in 1988 with Blood Lead Levels that Exceeded 15 µg/dL and Family Annual Income (ATSDR, 1988).

Race	Income Less Than $6,000	Income Between $6,000–$15,000	Income More Than $15,000
Black	68%	54%	38%
White	36%	23%	12%

National Health and Nutrition Examination Survey (NHANES) is a population-based, periodic series of national examinations of the health and nutritional status of the civilian, noninstitutionalized U.S. population. Geometric mean blood lead levels in the U.S. population, age one year or older, declined from 12.8 µg/dL in 1976–1980 to 2.3 µg/dL in 1991–1994. This remarkable outcome is attributed largely to removing lead from gasoline, which in turn reduced ambient air levels of lead.

However impressive the decrease in the national mean blood lead level, considerable disparity continues to exist across racial/cultural and income lines. As shown in Table 9.4a, the percentages of children, age one to five years, who have blood lead levels ≥ 10 µg/dL (i.e., CDC's action level) were highest in urban, low income, Black, non-Hispanic children (CDC, 1997). Older housing containing lead-based paint accounts for much of this elevation in blood lead levels.

Waste Sites

The EPA's (1992a) analysis of the impact of environmental hazards on minorities and low-income groups identified residence near CERCLA sites and operating hazardous waste facilities as a matter of environmental inequity. Their conclusion was based on studies conducted by the United Church of Christ (CSJ, 1987) and GAO (1983).

Water Contamination Problems

Scientists from the EPA and public health agencies formed a panel to review the impact of contaminants in water on minorities and low income populations (Calderon

Table 9.4a. Percentages of U.S. Children, Age One to Five Years, with Blood Lead Levels ≥ 10 µg/dL (CDC, 1997).

	Percentage
Race/Ethnicity	
Black, non-Hispanic	11.2
Mexican-American	4.0
White, non-Hispanic	2.3
Income	
Low	8.0
Middle	1.9
High	1.0
Urban Status	
Population ≥ 1 million	5.4
Population ≤ 1 million	3.3

et al., 1993). The panel used the Safe Drinking Water Act and the Clean Water Act as background information against which relevant studies and reports were evaluated. The panel reviewed information about microbial content of water on tribal lands, drinking water quality in migrant worker camps, groundwater contamination from hazardous wastes in poor rural counties, drinking water quality and sanitation along the United States/Mexico border, lead in drinking water, case studies on water quality problems on Navajo lands, and the consumption of fish from contaminated bodies of water. The panel found that most information was anecdotal or case studies and did not therefore lend itself to quantitative comparisons or analyses.

However, the panel concluded, "Despite the sparseness and limitations of the data, the existing data suggest that environmental inequities exist. While the existing data do not support any broad nationwide pattern of inequity, there are, however, clear situations where certain populations are exposed to higher levels of contaminants in water." The panel did not separate factors of low income and race/ethnicity in arriving at their conclusion. The panel advocated collection of additional data on water contamination and populations at health risk. Amendments to the Safe Drinking Water Act in 1996 contain the statutory directive to collect this kind of data.

Other Environmental Hazards

In addition to the EPA's (1992a) analysis of air pollution, children's lead exposure, waste sites, and water contamination problems; consideration was also given to minorities' exposure to pesticides and the consumption of fish caught in bodies of water contaminated with toxicants. However, data were generally lacking that might relate these hazards to any inequities experienced by minorities.

The EPA's Environmental Equity Workgroup developed recommendations to the EPA Administrator on environmental equity issues (EPA, 1992a). They published their findings in a two-volume report entitled *Environmental Equity: Reducing Risk for All Communities*. The findings led to the establishment of environmental equity policies and activities at the EPA.

Environmental Hazards and Hispanics' Health

Metzger et al. (1995) used EPA's ranking to extrapolate the effect of environmental hazards on Hispanics' health. Investigators noted that 22.4 million Hispanics are in the U.S. population, and the number will grow to 31 million by the year 2010. Metzger et al. state, "There are numerous indicators that Hispanics face a disproportionate risk of exposure to environmental hazards. Ambient air pollution, worker exposure to chemicals, indoor air pollution, and drinking water are among the top four threats to human health and all are areas in which indicators point to elevated risk for Hispanic population."

Metzger et al. cite EPA data indicating that Hispanic populations experienced higher risk. They note that 80% of Hispanics live in areas that fail to meet at least one EPA air quality standard compared with 65% of African-Americans and 57% of whites. According to the EPA data on air quality nonattainment areas, Hispanics are also more than twice as likely as either African-Americans or whites to live in areas that have elevated levels of particulate matter. Concerning workers' exposure to chemicals, Metzger et al. observe that 71% of all seasonal agricultural workers are Hispanic, compared with 23% who are white and 3% African-American. The use of pesticides in agricultural applications places Hispanics at elevated health risk. Concerning indoor air pollution, Metzger et al. cite the failure to communicate to Hispanics the risk of radon in indoor air. They note that 61% of Hispanics have never heard of radon compared with 21% of whites. Metzger et al. include the presence of lead and biologic contaminants in drinking water supplies as two examples of problems that some Hispanic populations face.

National Law Journal Study of 1992

Staff of the *National Law Journal* examined 1,177 of the 1,206 NPL sites as of March 1992 (Lavelle and Coyle, 1992). They found it took on average 5.6 years from time of waste site discovery until the site was placed on the NPL, but placing uncontrolled hazardous waste sites on the NPL took 20% longer in minority communities than in White communities. Lavelle and Coyle also analyzed outcomes of all environmental lawsuits filed in federal courts over a seven-year period. The average fine imposed for violating federal toxic waste laws in White residential areas, $335,566, was more than six times the average fine imposed in minority residential areas. The disparity occurred by race alone, not income; the average penalty in areas with the lowest median incomes was only 3% greater than the average penalty in areas with the highest median incomes.

DEMOGRAPHICS INVESTIGATIONS

The preceding section summarized several key reports that gave impetus to assessing environmental inequity concerns. However, these reports often had important methodologic limitations such as the use of ZIP code areas for geographic analyses. Because the U.S. Postal Service developed ZIP codes to facilitate mail delivery, they are subject to change as the Postal Service refines mail delivery patterns. ZIP codes therefore represent variable geographic areas that can lead to uncertainties in demographic analyses. To avoid such methodologic shortcomings, several researchers have conducted more in-depth demographic studies on associations among race, ethnicity, and socioeconomic variables as they relate to siting of hazardous waste facilities and location of CERCLA sites. The key studies are summarized in this section.

Hird Study of 1993

Hird (1993) looked at three broad equity implications of the EPA CERCLA program for environmental policy analysis. He examined three elements of environmental equity: geographic distribution of NPL sites, who pays for site cleanups, and the pace of cleanups. Only the first and third elements of his study will be summarized.[5]

To examine the distributional equity of NPL sites, data were collected on the socioeconomic characteristics of each county in the United States (n = 3,139) and the number of current or proposed NPL sites (n = 788) in each county as of January 1, 1989. This permitted a determination of whether the number of NPL sites in each county was correlated with the socioeconomic characteristics of the surrounding area. Hird argues "The county is both large enough to include the effects of hazardous waste sites, and small enough to record significant socioeconomic variation." His county-level socioeconomic data were obtained from the U.S. Census Bureau. The number of NPL sites per county was the dependent variable in a multivariate Tobit statistical analysis. Independent variables in the Tobit analysis included quantity of hazardous waste generated in each state, percentage of each county's economy attributable to manufacturing, percentage of college educated residents, percentage of housing units occupied by owners, the median housing value, and percentages of county residents that were unemployed, nonwhite, and below the mean poverty level.

Results showed the mean number of NPL sites per county was 0.37 (sd = 1.28, n = 3,139). Manufacturing presence was strongly associated with more county NPL sites. Hird noted, "...the results indicate that more economically advantaged counties (in terms of both wealth and the absence of poverty) are likely to have more Superfund sites." For all 3,139 counties, Hird found no statistically significant association nationally between poor counties and the number of NPL sites they contain. However, counties with high concentrations of nonwhites had *more* NPL sites than did others (holding other socioeconomic factors constant), an outcome that Hird characterized as "...corroborating the United Church of Christ [CSJ, 1987] findings for all hazardous waste sites."

Because multivariate Tobit analysis may have obscured simple relationships between the presence of CERCLA sites and measures of ethnicity and socioeconomic factors, simple bivariate Tobit estimates were calculated. The dependent variable was the number of county NPL sites and the independent variable was either poverty or unemployment rates, median housing values, or percentage of nonwhites in the county. The results of the bivariate analysis showed high numbers of NPL sites strongly related to higher housing values, lower percentages of nonwhites, and the lack of poverty and unemployment.

A different picture emerged about distribution of NPL sites when subsets of all counties were evaluated on the basis of their exceeding high rates of poverty (n = 1,292 counties), unemployment (n = 1,274), ethnicity (n = 1,195), or median housing values (n = 1,254). The average number of NPL sites per county was 0.11 in counties highly represented by persons of low income. For counties with high percentages of unem-

---

[5] This does not imply that who pays for site remediation is unimportant. Indeed, much of the controversy attending CERCLA is about the "polluter pays" principle that undergirds the statute (see Hird, 1993). However, who pays for site cleanups has not been part of the debate on environmental equity.

ployed, the average number of NPL sites per county was 0.23; for counties with high percentages of nonwhites, the figure was 0.33 NPL sites per county. All three averages are therefore *below* the national average of 0.37 NPL sites per county. For the subset of counties with high median housing value, NPL sites per county was 0.74, which were higher than the national county average. Hird concluded, "Therefore, these results indicate that NPL sites are located predominately in affluent areas, and generally irrespective of race."

Regarding equity in cleanup of NPL sites, three measures of site remediation speed were used. Hird examined data from Remedial Investigation and Feasibility studies, Records of Decision for NPL sites, and actual remedial actions. Because these three events occur in temporal sequence, some indication of the remediation speed can be evaluated. The most important indicator of a site's cleanup stage was found to be the Hazard Ranking Score, that is, the higher the hazard scores the faster the cleanups. No association was found between pace of site cleanup and the county's socioeconomic characteristics (which included percentage of nonwhite population).

Anderton et al. Study of 1994

Investigators at the Social and Demographic Research Institute, University of Massachusetts, conducted a comprehensive study of the racial and cultural demographics and income levels of persons living near commercial, controlled hazardous waste facilities (Anderton et al., 1994a, 1994b). These are sites permitted to operate under the Resource Conservation and Recovery Act (Johnson and Weimer, 1995). The investigators focused on commercial facilities that treated, stored, and disposed of hazardous wastes (TSDFs). Note the important difference between TSDFs, which are controlled hazardous waste facilities, and CERCLA NPL sites, which are uncontrolled hazardous waste sites.

The Anderton et al. study is noteworthy because investigators examined the effect of using different geographic units on the outcome of demographics analyses. The investigators chose the census tract[6] as their primary geographic unit to avoid aggregation errors inherent in larger geographic units, such as ZIP code areas.

Commercial TSDFs were identified within census tracts for facilities that had opened for business before 1990 and were still operating in 1992. The investigators defined a TSDF as being privately owned and operated and receiving waste from firms of different ownership; TSDFs were excluded if they were the primary producers of waste. Before the 1990 census, tracts were defined only for Standard Metropolitan Statistical Areas (SMSAs).[7] About 15% of TSDFs are located outside SMSAs and hence were not included in the analysis. Using these criteria, 454 facilities were identified for demographic analysis.

The investigators' first analysis examined how census tracts with TSDFs differ from those without TSDFs. Comparisons were made of census tracts containing TSDFs with tracts that had TSDFs but within SMSAs that contained at least one facility inside their

---

[6] Generally a census tract is a small statistical subdivision of a county. Census tracts have identifiable boundaries and average about 4,000 persons (Anderton et al., 1994b).

[7] SMSAs consist of cities with populations of 50,000 or more persons including surrounding counties or urbanized areas but omitting many rural areas and small cities and towns (Anderton et al., 1994b).

borders. This resulted in analysis of 408 tracts with TSDFs and 31,595 without. The mean percentages of African-Americans were 14.5% in census tracts with TSDFs and 15.2% in tracts without TSDFs; the difference was not statistically significant. The mean percentages of Hispanics were 9.4% in tracts with TSDFs and 7.7% for tracts without facilities, which were not statistically significant. Similarly, no statistically significant difference was found in the median percentage of Blacks residing in tracts with and without TSDFs, which led Anderton et al. (1994a) to observe, "This single finding is sufficient to raise substantial questions about the previously cited research conducted at a zip code level, and about its substantial influence on national policy."

The investigators found a substantially higher mean percentage of persons employed in precision manufacturing located in TSDF tracts (38.6%) than in surrounding areas (30.6%). This suggested that TSDF facilities are located in industrial areas for reasons unrelated to issues of race and ethnicity.

To determine whether environmental inequities differed for large cities, the census tract comparison was repeated for the 25 largest SMSAs. For these areas, TSDF tracts were found to have significantly lower percentages of Blacks, but larger percentages of Hispanics, compared with census tracts without TSDFs. For the 25 SMSAs, TSDF tracts had significantly higher levels of industrial employment, with less expensive and newer houses.

Anderton et al. next constructed larger areal units of analysis, consisting of all tracts with at least 50% of their areas falling within 2.5-mile radii of the center of tracts in which TSDFs were located. The percentage of Blacks in these larger areas (25.7%) was significantly higher than in other tracts (14.5%). For Hispanics, the comparable numbers were 11% versus 7% for other tracts. Furthermore, in these larger areas, industrial development remained significantly higher than in other tracts. A multivariate analysis using "Being a TSDF Tract" as the dependent measure showed, "...the most significant effects in each case are not those of percentage black or percentage Hispanic, but of unemployment and industrial employment within the area. For census tracts, the effects of percentage black and percentage Hispanic are not significant. However, in much larger areas [i.e., 2.5-mile radius areas], both variables appear to be associated with the presence of TSDFs."

In summary, Anderton et al. (1994a), using census tract-level data, found no nationally consistent and statistically significant differences between the racial or ethnic composition of tracts that contain commercial TSDFs and those that do not. The investigators noted that TSDFs were more likely to be found in tracts with Hispanic groups. In a companion paper (1994b), they concluded "We believe our findings show that TSDFs are more likely to be attracted to industrial tracts and those tracts do not generally have a greater number of minority residents."

Zimmerman Study of 1993

Zimmerman (1993) used a unit of geographic analysis different from that used by Anderton et al. (1994a) to examine equity issues of relevance to NPL sites. She focused on social and economic characteristics at the geographic level of communities, which she defined as U.S. Census "Places," or, where places do not exist, as "Minor Civil Divisions" (MCDs). She notes that these communities represent political subdivisions and are the smallest formal level of political decision-making.

In addition to assessing the demographics of populations living near NPL sites, Zimmerman evaluated whether the time taken to develop a Record of Decision (ROD)[8] for a site was associated, as a matter of environmental inequity, with minority communities. Zimmerman initially obtained demographics data for the 1,090 sites on the NPL at the time of her analysis. Her list excluded sites in extremely rural areas whose community populations in 1980 were fewer than 2,500. This resulted in excluding 260 sites, which she asserted, had minimal effect on her overall analysis. Characteristics of NPL communities were compared with the nation and the four census regions (Northeast, Midwest, South, West), based on 1990 census data. Two methods of portraying average percentages for race, ethnicity, and poverty were used. One method was an unweighted averaging of means, counting each community equally regardless of its population. The second method weighted communities according to each community's population (total population as well as minority population). Zimmerman's findings differed according to which method she used.

Using the unweighted averaging method, Blacks represented an average of 9.1% of the population in 1990 for the approximately 800 NPL communities evaluated. This was lower than the national average of 12%. The percentages of Blacks in NPL communities were lower in three of the four U.S. Census regions. In the South census region, percentage of Blacks in NPL communities was 23.7% compared with 18.5% Blacks living in the census region. The percentage of Hispanics in NPL communities was 6.6% compared with 9% nationally; no notable regional differences were found. The mean percentage of persons below the poverty level was 10.6% in NPL communities compared with 13.5% for the nation.

When Zimmerman used the alternate approach of weighting minority populations by total population, the mean percentage of Blacks in NPL communities was 18.7% and for Hispanics was 13.7%. This was based on 622 census places and MCDs that contained 825 NPL sites. These percentages are greater than national percentages for Blacks (12.1%) and Hispanics (9%). She noted that differences in race and ethnicity between the two kinds of population analysis (i.e., nonweighted vs. weighted) reflects the effect of a relatively few large communities with NPL sites that have large Black populations. She concluded, "Thus, racial and ethnic disproportionalities with respect to inactive hazardous waste site location seem to be concentrated in a relatively few areas."

To explore the relationship between Record of Decision (ROD) status for NPL sites and socioeconomic characteristics, a Probit statistical analysis was conducted of all 1,090 NPL sites. In general, the set of independent variables (including race and ethnicity) taken as a group did not contribute much to the variance in the independent variable—existence of a ROD. However, when the analysis was confined to a subset of NPL sites in poor communities with relatively high Black populations, about 20% of sites were found to be without RODs. On this point, Zimmerman concluded, "Disproportionalities with respect to cleanups do exist, but appear to be more a function of the nature of the process of designation of NPL sites in the early 1980s rather than a result of actions connected with cleanup plans *per se*."

---

[8] An EPA Record of Decision discusses the various cleanup techniques that were considered for a site and explains why a particular course of action was selected (EPA, 1992c).

GAO Study of 1995

The General Accounting Office (GAO) conducted a multipurpose study in 1995 of persons living near nonhazardous municipal landfills (GAO, 1995c). The study was requested by Senator John Glenn of Ohio and Congressman John Lewis of Georgia. The primary objective of the study was to evaluate demographics and income levels of persons living near the examined facilities. Another objective was to evaluate ten published demographics studies of persons living near hazardous waste facilities. Other objectives were to examine the EPA's efforts to address environmental justice in their regulations, and to provide information on the extent of data that measure human health effects of waste facilities on minorities and persons of low income. Only findings from the study's first two objectives are summarized here.

To address demographics and income, the GAO identified a potential universe of 4,330 landfills in the United States. This universe was subdivided, using ZIP codes of landfills, into categories of 1,498 metropolitan and 2,832 nonmetropolitan landfills. The GAO used a questionnaire to survey landfill operators with equal probability in each landfill category. The survey elicited information on the geographic location and other characteristics of respondents' landfills. The final sample consisted of 190 metropolitan and 105 nonmetropolitan landfills. The demographics and income levels of persons living within one and three miles of these 295 landfills were evaluated.

Using a Geographic Information Systems (GIS) technique, the latitude and longitude for each site permitted defining two areas that separated landfills from the rest of the county. These areas were within one and three miles from the boundary of the landfill. To determine the demographics of persons within these two boundaries, the GAO used the smallest level of aggregation possible, census block groups,[9] as their units of geographic analysis. The GAO did not use census blocks[10] as their index because Bureau of Census data do not include information on residential income at the census block level.

The number of minorities and nonminorities living in complete and partial block groups was summed, using 1990 U.S. Census data, to determine the total number of persons living in the one- or three-mile areas. These numbers were subtracted from the number of persons living in each county that contained a landfill. The GAO excluded block groups within the one- and three-mile areas that fell outside the county in which the landfill was located. For each one- and three-mile area and the corresponding rest of the county, the GAO developed demographic information in five categories: race/ethnicity, poverty status, median household income, poverty status by race/ethnicity, and median household income by race/ethnicity.

---

[9] Census block groups can be geographic block groups or tabulation block groups. The former are clusters of blocks having the same first digit of their three-digit identifying numbers within census tracts or block numbering areas. Tabulation block groups and geographic block groups may be split to present data for every unique combination of county subdivision, place, American Indian and Alaska Native area, urbanized area, voting district, urban/rural and congressional district shown in the data product (BOC, 1992).

[10] Census blocks are small areas bounded on all sides by visible features such as streets, roads, streams, and railroad tracks, and by invisible boundaries such as city, town, township, and county limits, property lines, and short, imaginary extensions of streets and roads (BOC, 1992).

The GAO found that minorities and persons of low income were not generally over represented near nonhazardous municipal landfills. For 73% of metropolitan landfills and 63% of nonmetropolitan landfills, percentages of minorities living within one mile of landfills were lower than percentages of minorities living in the rest of the counties. The GAO estimated that people living within one mile of about half the landfills analyzed had median household incomes higher than the incomes of residents in the rest of the county. The same result occurred when the three-mile areas were used for analysis.

The GAO's second objective was to evaluate 10 published reports on demographics of communities located near waste sites (GAO, 1995c). The 10 studies are included in this chapter. The GAO noted that three of the 10 studies found minorities more likely than nonminorities to live near hazardous waste sites, four showed either no significant association between the location of a waste site and minority populations or that minorities were *less* likely to live nearby, and three yielded multiple conclusions on whether a disproportionate percentage of minorities lived near the waste sites and facilities.

Seven of the 10 studies also assessed economic factors. The GAO concluded that three studies had found incomes of persons living near hazardous waste facilities were lower than incomes of persons living distant; two reported no significant differences in incomes between those near and those distant from hazardous waste facilities; and two reported multiple results, depending on the analytical method used.

The GAO noted that actual data on exposure to hazardous substances of minorities and persons of low income were generally lacking. Furthermore, they cautioned that comparing demographics and income data across research studies was difficult, because investigators used U.S. Census databases and geographic units of analysis that differed across studies. Overall, the GAO found only marginal support for the argument that minorities and persons of low income are disproportionately located near nonhazardous waste facilities; that is, TSDFs.

Been Study of 1995

A set of particularly noteworthy papers on environmental equity was published by Vicki Been, Professor of Law, New York University School of Law (Been, 1993, 1994, 1995). The papers are noteworthy because of the clarity of writing, clear and logical arguments, and close attention to data analysis. In particular, Been's work on environmental justice issues should be compared to the work of Anderton et al. (1994a,b), because both investigators used similar databases and methods, but with somewhat different outcomes.

Been compared various characteristics of census tracts with TSDFs versus non-TSDF census tracts. In distinction to Anderton et al. (1994a,b), Been identified census tracts hosting TSDFs by examining TSDF listings in the 1994 edition of *Environmental Services Directory* (ESD), which she supplemented with an EPA database, the *Resource Conservation and Recovery Information System*. She used telephone contacts to verify the location and kind of operation of individual TSDFs. From these efforts, Been identified 608 TSDFs for analysis, which she asserts is more accurate than that used by other investigators. Been notes that Anderton et al. (1994a,b) limited their comparison to non-TSDF census tracts in Metropolitan Statistical Areas or rural counties that had at least one facility. Been made no similar restrictions and compared all TSDF tracts with all approximately 42,000 populated non-TSDF tracts within the continental United States, using 1990 census data in her demographics analyses.

Been conducted both univariate and multivariate analyses of characteristics between TSDF and non-TSDF census tracts. Landfills, incinerators, and kilns were separated as a group from other kinds of TSDFs to address whether different kinds of TSDFs were associated with racial, ethnicity, or socioeconomic characteristics. The breakout revealed no statistically significant differences in the mean percentages of African-Americans and lower-income persons living near this particular grouping of TSDF facilities.

To determine whether a smaller geographic comparison would change the nature of differences between TSDF and non-TSDF tracts, Been calculated ratio of the demographics of TSDF sites to demographics of all non-TSDF sites within a state and within a Metropolitan Statistical Area (MSA). The mean of the ratios was tested for significance from unity, the ratio that would occur if the TSDF tracts' characteristics were identical to the non-TSDF tracts' characteristics. Using this approach, and using national demographics data, she found that percentages of African-Americans in TSDF tracts did not differ significantly from percentages in non-TSDF tracts. The percentage of Hispanics, however, was significantly greater for TSDF tracts than for non-TSDF tracts. These results for African-Americans and Hispanics maintained when comparisons were made within states. However, when comparisons were made within MSAs, African-American differences remained statistically insignificant and differences in percentages of Hispanics narrowed. Been noted that differences between median housing values in TSDF and non-TSDF tracts narrows considerably when only the host MSA is studied.

Been extended her own work and that of others by examining not only the means of demographics variables (e.g., racial percentages), but also whether the distribution of TSDF facilities matched the distribution of populations around the mean. Been (1995) assumed that a "fair" distribution of TSDF facilities would be proportionate to the distribution of the population. Using this assumption, she calculated the number of facilities that would be located in particular kinds of neighborhoods if distribution of TSDF facilities were proportionate. The results are fascinating. According to Been (1995),[11] "In terms of raw numbers, if the distribution of facilities followed the distribution of the population, there would be 24 more facilities sited in the neighborhoods with no or very few African-Americans. In neighborhoods where African-Americans made up more than 10% but less than 70% of the population, there would be 34 fewer facilities. Neighborhoods with African American populations of more than 70% would have ten more facilities. Similarly, neighborhoods with Hispanic populations of more than 20% are bearing more facilities than they should if facilities were distributed in the same way in the population." Been also found that neighborhoods with median family incomes of $10,001 to $40,000 bear 62 more facilities than would be proportionate.

Been (1995) concludes[12] "...a more sophisticated comparison of the distribution of facilities to the distribution of neighborhoods with particular demographic characteristics reveals that certain kinds of neighborhoods—those with median family incomes between $10,001 and $40,000, those with African American populations between 10% and 70%, those with Hispanic populations of more than 20%, and those with lower education attainment—are being asked to bear a disproportionate share of the nation's facilities. Analysis of the joint distribution of income and percentage of African Ameri-

---

[11] Excerpted with permission from *J. Land Use & Environmental Law* 11:1–36 (1995). Copyright *J. Land Use & Environmental Law*, College of Law, Tallahassee, Florida.

[12] Excerpted with permission. See previous footnote.

cans in the population suggests that income explains most of the disparity. Multivariate analysis, however, suggests that race is a better predictor of facilities than income. In total, the analysis reveals that environmental injustice is not a simplistic PIBBY—'put it in Black's backyards.' It suggests, instead, a much more ambiguous and complicated entanglement of class, race, educational attainment, occupational patterns, relationships between the metropolitan areas and rural or nonmetropolitan cities, and possibly market dynamics."

Heitgerd et al. Study of 1995

Researchers at the ATSDR used a GIS approach to assess the demographics of populations living near NPL sites (Heitgerd et al., 1995). GIS is a powerful tool that permits the analysis of datasets that have been overlaid on a geographic database. For example, census data can be overlaid on a geographic base to determine housing patterns according to race, income, and geographic location.

Racial and Hispanic origin subpopulations living within one mile of NPL sites were compared with subpopulations in the same county but living outside the one mile border. The investigators extracted census block boundaries from the Bureau of Census' 1990 Topologically Integrated Geographic Encoding and Referencing/Line files and linked them with information on total population, race, and Hispanic-origin data in 1990 census block data. The EPA-defined site boundaries were used to specify the boundaries of 1,200 NPL sites.

Heitgerd et al. found that 670 counties had parts of their area located within one mile of the 1,200 NPL sites they examined. This represents 22% of all counties in the contiguous states of the United States. The areas within one-mile borders of the 1,200 NPL sites comprised 184,191 census blocks, of which 10.6% were within one mile of two or more NPL sites. Each site's demographics were derived for each county by summing over all census blocks within the one-mile range. The investigators assert this shifts the focus of the demographics analysis from NPL sites *per se* to the counties within one mile of NPL sites while retaining block-level data. The comparison population was spatially defined as persons living in the 670 impacted counties but at distances greater than one mile from NPL sites. Population data for the comparison area were obtained by subtracting the site area data from county totals.

A three-factor analysis of variance (ANOVA) model served as the investigators' statistical method. The factors used in the ANOVA analysis were NPL (two levels-within or outside one-mile buffers of NPL sites), State (48 contiguous states), and County (670 counties).

The investigators found approximately 11 million persons live within one-mile boundaries of the 1,200 NPL sites they assessed. As was expected, considerable variation was found in the population figures according to state and regional factors. For example, California ranked highest in the average population per NPL site and average population density, while ranking eighth in average site area per square mile. Furthermore, fewer NPL sites are in the Great Plains states, and those sites accounted for relatively fewer persons compared with other regions.

A separate ATSDR analysis of 972 NPL sites found that 949,000 children six years or younger live within one mile of sites' borders, which represented 11% of the population. This is an average of 980 children six years or younger per NPL site. Given a current total of about 1,300 NPL sites, one can calculate that approximately 1.3 million young children live within one mile borders of NPL sites.

Concerning race and ethnicity, ANOVA analysis revealed a statistically significant difference ($p \leq 0.001$) in the mean percentage for each racial group and persons of Hispanic origin in populations living within one-mile boundaries of sites compared with the remainder of the counties. The analysis controlled for state and county of NPL site. The largest percentage difference was found for African-Americans. They represented 8.3% of the comparison area but 9.4% of the population living within one mile of NPL sites.

To put these percentages in perspective, the ATSDR considered the 33 counties identified as the top 5% of counties adjacent to NPL sites when ranked by the percentages of African-Americans living within one mile of sites (Burg, 1996). There were approximately 221,000 persons of whom 145,000 (66%) were African-Americans. When the racial composition of the remaining areas (greater than one mile from site) of the 33 counties was examined, there were about 7,365,000 persons of whom 2,212,000 (30%) were African-Americans. If the same county rate outside the one-mile limit (i.e., 30% African-Americans) were applied to areas within the one-mile borders of NPL sites, one would expect to find approximately 66,000 African-Americans rather than the 145,000 actually identified. This reveals an additional approximately 79,000 African-Americans who live near NPL sites in the 33 counties considered by Burg (1996).

Heitgerd et al. (1995) concluded, "If it is assumed that the NPL sites are representative of all uncontrolled hazardous waste facilities, then the results support existing environmental inequity research that suggests the location of hazardous waste facilities is more burdensome for minority communities."

Because the Heitgerd et al. study relies on a GIS approach and uses county-based comparison data, it is an important contribution to the literature on environmental equity. Its limitations are the lack of control for sociodemographic variables (e.g., are the observed disparities the result of economic conditions?) and uncertainty about whether the demographic results might be a consequence of the one-mile buffers chosen (e.g., would the results change if some other measure, for example, 0.5 mile, had been used?).

Oakes et al. Study of 1996

Building upon their previous cross-sectional work (Anderton et al., 1994a,b), researchers at the University of Massachusetts conducted the first national longitudinal study of residential characteristics in census tracts that contain TSDFs (Oakes et al., 1996). This study addresses the central issue in environmental justice as it relates to TSDFs, the issue of alleged racism in the siting of waste facilities in minority neighborhoods. Because the Oakes et al. study uses rigorous statistical analysis and current demographics databases it is an important contribution to the scientific literature on environmental equity.

Oakes et al. evaluated community characteristics over a 20-year period before and after new TSDFs were sited. In a follow-up to previous findings (Anderton et al., 1994a,b) that indicated a relationship between TSDFs and the level of industrialization in a community, they compared trends within TSDF communities to other similar industrial communities and to less industrialized communities. Data on 476 commercial TSDFs were compiled from the 1992 edition of *Environmental Services Directory*, using a telephone survey of each facility. Analysis was restricted to census tracts within metropolitan statistical areas and rural counties that each contain at least one TSDF. There were

35,208 census tracts without TSDFs. Data on residential communities came from the 1970, 1980, and 1990 tract-level census files. Census data files contained more than 130 census variables that summarized tract composition. Special efforts were made to reconcile any changes in tract locations or TSDF locations due to changes in census track identification.

Oakes et al. first conducted cross-sectional analysis using 1990 census tract data. They compared racial and economic indicators between tracts that had TSDFs and tracts that did not. Findings showed the average percentages of persons living in TSDF tracts who identified themselves as Black or Hispanic were 17.09 and 10.75, respectively, and for non-TSDF tracts, 16.26% Black and 9.74% Hispanic. These percentages for Blacks and Hispanics were not statistically significant between TSDF tracts and non-TSDF tracks. This result agrees with the researchers' prior report that used 1980 census data (Anderton et al., 1994a,b). Oakes et al. comment, "The largest significant differences that were found between tracts with and without commercial TSDFs were in the average percentage (33.32% in TSDF tracts and 25.28% in non-TSDF tracts) and the median percentage of persons employed in industrial and manufacturing occupations. This again supports the conclusion of Anderton et al. [1994] that TSDF tracts are somewhat more likely to be in industrial working-class neighborhoods."

Oakes et al. then analyzed communities' characteristics across two decades, using 1970, 1980, and 1990 census data. They first assessed demographic and socioeconomic characterizations of communities before TSDFs were sited. They found that TSDFs located in the 1970s and 1980s were, on the average, not systematically sited in areas with unusually high percentages of African-American or Hispanic populations, when compared with other areas with significant industrial development. Moreover, results showed the characteristics of communities, after siting of TSDFs, have trends that parallel those in the population at large.

The researchers performed multivariate methods of analysis to examine whether siting of TSDFs was associated with racial or ethic disparities and other indicators of environmental inequity. No evidence was found to support environmental inequity claims that TSDFs were sited in areas because of racial or ethnic bias. They concluded, "We believe this research, in concert with our earlier findings, suggests that commercial TSDF census-tract communities are best characterized as areas with largely white and disproportionately industrial working-class residential areas, a characterization consistent with what one might historically expect near industrial facilities."

Baden and Coursey Study of 1997

The environmental equity implications of locating waste sites within the city of Chicago were investigated through demographic, social, and economic analysis (Baden and Coursey, 1997). The investigators examined locations of three kinds of waste sites: all CERCLIS sites and TSDFs within the city limits, RCRA hazardous waste generators, and historical hazardous waste sites. Sites were examined with regard to racial, ethnic, and income variables; access to transportation; and waste disposal. Sites were linked with corresponding census tract information. Regression analysis was the primary statistical method used to associate geographic and demographic variables. Different kinds of regression analyses were used to investigate which demographic and physical features predicted the location of sites within limited communities, the location of sites within larger neighborhoods, and the geographic concentration of sites. Census and waste site data were specific to the years 1960 and 1990 for comparison.

Baden and Coursey found that waste sites in 1990 tended to be located in Chicago areas of low-population density near commercial waterways and commercial highways. They found no evidence of environmental racism against African-Americans for either CERCLA sites or TSDFs. There was no indication that African-Americans lived in areas with higher concentrations of hazardous waste than did whites or Hispanics. Evidence showed that the percentage of Hispanics in an area was significant with regard to the location of CERCLA sites and solid waste disposal facilities, perhaps because of recent migration of Hispanics into white ethnic neighborhoods. Baden and Coursey observed, "Surprisingly, areas where RCRA and solid waste disposal sites are located tend to have higher incomes; this is likely the result of the recent trend in construction of high price river-front residences in previously industrial areas." In summary, Baden and Coursey found little to no indication of environmental injustice in the location of hazardous waste sites and facilities within Chicago census tracts.

Tabulation of Studies

Key reports and studies that have shaped much of the environmental equity debate are listed in Table 9.5. Other publications merit examination, but those in Table 9.5 are most often cited in the published literature and conferences.

One reviewer, who examined the literature on race, class, and environmental health including much of the material described in this chapter (Brown, 1995), overlaid his review with editorial comments and recommendations on study designs and future research directions pertaining to environmental equity concerns. Brown concluded, "The overwhelming bulk of evidence supports the 'environmental justice' belief that environmental hazards are inequitably distributed by class, and especially race." He recommends that investigations into class and race issues move away from traditional epidemiologic designs and toward in-depth ethnographic analysis of communities and neighborhoods. Brown's recommendation is predicated on the belief that traditional epidemiologic designs exclude community input as a means of minimizing bias in the conduct of the epidemiologic investigation. He offered no data to support his belief.

EXECUTIVE ORDER ON ENVIRONMENTAL JUSTICE

Any unjust inequities across cultural and racial groups related to exposure to environmental hazards must be prevented as a matter of fairness and social justice. On February 11, 1994, at the urging of environmental justice advocates, President Clinton signed Executive Order 12898 entitled "Federal Actions to Address Environmental Justice in Minority Populations and Low-Income Populations." As with all executive orders, this order applies only to federal agencies. Although state and local governments and private sector entities are not directly subject to executive orders, actions that federal agencies take under executive order can have substantial ripple effects on other levels of government and the private sector.

The Clinton Executive Order directs each federal agency to "...make achieving environmental justice part of its mission by identifying and addressing, as appropriate, disproportionately high and adverse human health or environmental effects of its programs, policies, and activities on minority populations and low-income populations..." (Clinton,

Table 9.5. Key Reports and Studies Bearing on Environmental Equity.

Report	Scope of Study	Geographic Unit	Findings on Race/Ethnicity	Findings on Income
GAO (1983)	4 operating hazardous waste landfills in the southern United States	Census area	Blacks were the majority population in 3 of the 4 census areas containing landfills	For all 4 sites, Black populations had lower mean income than did all races combined in each landfill's census area
Commission on Social Justice (CSJ, 1987)	Commercial TSDFs (n=415) and 18,164 uncontrolled hazardous waste sites on EPA's CERCLIS	ZIP code area	(1) Percentage minority population in ZIP code areas with one commercial TSDF was twice that in ZIP code areas not containing a TSDF; (2) 3 of 5 Black and Hispanics lived in ZIP code areas with CERCLIS sites	Percentage of minorities was a stronger predictor of TSDF activity than was household income
Bullard (1990, 1994)	The book *Dumping in Dixie* describes 5 communities expressing environmental concerns: 2 TSDFs, 1 municipal landfill, 1 lead-contamination area, and 1 chemical-plant emission	No demographic analysis was conducted	Not addressed	Not addressed
Mohai and Bryant (1992)	TSDFs (14 operating, 2 planned) in 3 counties surrounding Detroit, Michigan	Populations within 1.5 miles of facilities were compared with populations farther than 1.5 miles	Percentage of minorities increased from 18% in areas farther than 1.5 miles to 48% within 1 mile of TSDFs; race was more strongly associated with proximity to TSDFs than was income	Percentage of persons below poverty level increased from 10% (farther than 1.5 mile) to 29% (within 1 mile)
Johnson et al. (1992)	Proceedings of first national conference on minorities and environmental contamination	Not addressed	Identified key research needs and data gaps; identified several minorities potentially at risk	Not addressed

Table 9.5. Key Reports and Studies Bearing on Environmental Equity (continued).

Report	Scope of Study	Geographic Unit	Findings on Race/Ethnicity	Findings on Income
EPA (1992a)	Evaluated environmental and health data pertaining to minorities and environmental hazards, including air pollution, lead exposure, waste sites, pesticides in workplaces, and consumption of fish	Not addressed	Elevated exposures of minorities to air pollution, lead (in children), and waste sites are likely; exposure and health data are generally lacking	Not addressed
Lavelle and Coyle (1992)	All U.S. environmental lawsuits filed over 7-year period	Not addressed	Value of fines imposed for violating toxic waste laws in minority communities was 1/6 the value of fines in white residential areas	Average penalties were independent of median income levels
Sexton and Anderson (1993)	Special issue of collected papers prepared in advance of the National Conference on Environmental Justice	Not addressed	Papers summarized findings on minorities and specific environmental hazards	Not addressed
Hird (1993)	U.S. counties containing 788 NPL sites, as of January 1989, were compared with all U.S. counties	Counties	On a statistically broad analysis, counties with nonwhite populations had more NPL sites; this difference disappeared using finer analyses	No association was found between poor counties and the number of NPL sites they contained
Zimmerman (1993)	1,090 NPL sites; 830 were evaluated on socioeconomic characteristics	Communities, defined as Census Places and Minor Civil Divisions; national and regional comparisons	Percentages of Blacks and Hispanics were lower in NPL communities than national population percentages; weighing communities by population gave percentages of Blacks and Hispanics greater than national population percentages	Mean percentage of persons below the poverty level was lower in NPL communities than for the nation

Study	Description	Geographic unit	Results (race/ethnicity)	Results (income)
Anderton et al. (1994a, 1994b)	454 Commercial, operating TSDFs	Census tract	No consistent national-level association found between location of facilities and number of minorities living in vicinity of sites	No consistent national-level association found between location of facilities and percentage of low-income persons there
GAO (1995c)	Assessed the race/ethnicity and median income of persons living within 1 and 3 miles of 295 nonhazardous landfills	Census block group within county	Minorities were not generally overrepresented within 1- and 3-mile areas	Persons of low income were generally not over-represented within 1- and 3-mile areas
Been (1995)	Addressed whether different kinds of TSDFs were associated with racial, ethnicity, or socioeconomic characteristics. (1) Compared 608 TSDFs with 42,000 non-TSDF census tracts, (2) Compared TSDF sites to all non-TSDF sites within a state and within the MSA	(1) Census tract hosting TSDFs compared with U.S. census tracts without TSDFs (2) TSDF Census tracts compared with non-TSDF sites within states and within Metropolitan Statistical Areas (MSAs)	(1) Percentage of African-Americans in the TSDF tracts did not differ from that in non-TSDF tracts; percentage of Hispanics was greater in TSDF tracts than in non-TSDF tracts. (2) Same results as in (1) when tracts within states were compared; comparisons within MSAs showed African-Americans not overrepresented; difference narrowed between TSDF and non-TSDF tracts for Hispanics	"...certain kinds of neighborhoods—those with median family incomes between $10,001 and $40,000, those with African American populations between 10% and 70%, those with Hispanic populations of more than 20%, and those with lower education attainment—are being asked to bear a disproportionate share of the nation's facilities."
Heitgerd et al. (1995)	Demographics of populations living within 1-mile buffers around 1,200 NPL sites	Census block	The mean percentage for each racial group and persons of Hispanic origin in the NPL population was higher than for comparison areas	Not addressed
Oakes, et al. (1996)	473 commercial TSDFs; 35,208 census tracts with no TSDFs	Census tract	Using a longitudinal approach, no evidence found of environmental inequity or disparate impact	Persons of low income were not overrepresented in TSDF areas

Table 9.5. Key Reports and Studies Bearing on Environmental Equity (continued).

Report	Scope of Study	Geographic Unit	Findings on Race/Ethnicity	Findings on Income
Baden and Coursey (1997)	All CERCLIS sites within Chicago city limits as well as RCRA TSDF sites, RCRA hazardous waste generators, and historical hazardous waste sites	Census tract	No indication that African-Americans lived in areas with higher concentrations of hazardous waste; percentage of Hispanics, however, was significantly higher in areas with CERCLA sites and solid waste disposal facilities	Areas where RCRA and solid waste disposal sites are located tended to have higher incomes

1994). The several responsibilities prescribed in the executive order for federal agencies are outlined in the following section.

Creation of an interagency working group—The Administrator of the EPA was directed to convene and chair an interagency federal working group on environmental justice. Members of the group include the EPA, the Departments of Defense, Energy, Health and Human Services, Commerce, Housing and Urban Development, Agriculture, Transportation, Labor, Justice, Interior, and various White House offices.

The working group will serve as the federal government's primary environmental justice body with responsibility to: (1) guide federal agencies on criteria for identifying human health or environmental effects that are disproportionately high and adverse on minority and low-income populations; (2) be a clearinghouse for federal agencies as they develop their environmental justice strategies; (3) assist in coordinating research, (4) assist in the collection of data required by the executive order; (5) examine existing data and studies on environmental justice; (6) hold public meetings; and (7) develop interagency model projects on environmental justice. Elaboration on some of these responsibilities follows.

Development of agency strategies—The executive order directs each federal agency to develop an agency-wide environmental justice strategy. Each agency's strategy must identify and address disproportionately high and adverse human health or environmental effects of its programs, policies, and activities on minority populations and low-income populations. Each strategy must also list programs, policies, planning and public participation processes, enforcement, and/or rulemakings related to human health or the environment that should be revised in light of environmental justice concerns.

Federal agency responsibilities for federal programs—Each federal agency is directed by the executive order to conduct its programs, policies, and activities that substantially affect human health or the environment in a manner that ensures against the effect of excluding persons from participation on the basis of race, color, or national origin.

Research, data collection, and analysis—The executive order mandates each federal agency, whenever practicable and appropriate, to collect, maintain, and analyze information assessing and comparing environmental and human health risks borne by populations identified by race, national origin, or income. Furthermore, agencies are directed to collect such population data for areas around federal facilities and similar areas expected to have a substantial impact on the environment, human health, or economy. Federal agencies must use this population-based data to determine whether their programs, policies, and activities have disproportionately high and adverse human health or environmental effects on minority populations and low-income populations.

Subsistence consumption of fish and wildlife—The executive order requires federal agencies, whenever practicable and appropriate, to collect, maintain, and analyze information on the consumption patters of populations who principally rely on fish and/or wildlife for subsistence. Agencies are directed to communicate to the public the risks of these consumption patterns. This directive is intended to address the human health issues of eating fish and wildlife that contain hazardous substances in their tissues.

Impact of the Executive Order

It is too soon to assess the impact of the executive order on federal programs, policies, and activities. As of December 1996, the Federal Working Group on Environmental Justice had not released an integrated program of work in support of the executive order. Presumably the federal program will represent a synthesis of each federal department's environmental justice strategies. However, an indication of one department's strategies is provided by the Department of Health and Human Services.

INTERNATIONAL PERSPECTIVE

As described in this chapter, environmental equity has developed as an agenda within the American civil rights movement. On reflection, linkage between social justice and environmental equity seems a natural relationship within the U.S. political arena because both movements are rooted in issues of fairness and social change. But is environmental equity confined to the American landscape? Do similar concerns exist in other countries about the impact of hazardous waste on the health and well-being of minority groups and persons of low income? To seek answers to these questions, an analysis was conducted of the single, most comprehensive international statement on environmental protection, which is known as Agenda 21.

Agenda 21 was developed at the United Nations Conference on Environment and Development, which was held in Rio de Janeiro in 1992. Agenda 21 is a document of 40 chapters, consisting of about 600 pages. A review of relevant chapters that might address environmental equity as a matter of national or international policy reveals no statements specific to this issue. In particular, Agenda 21's Chapter 20, entitled "Environmentally Sound Management of Hazardous Wastes Including Prevention of Illegal International Traffic in Hazardous Wastes," makes no mention of the importance of ensuring that hazardous waste generation, transportation, storage, and disposal do not cause unjust environmental inequities. A review of Agenda 21's other chapters reveals a concern for the rights of indigenous peoples and acknowledges their historical commitment to protecting the environment, although this support for respecting the rights of indigenous peoples was not couched in the language of environmental equity.

Why environmental equity did not elicit concern in the major international statement on the environment is curious. However, assuming that environmental inequities can be prevented through the actions of informed persons who have access to political processes, Agenda 21 provides some tangential support for environmental equity concerns. Specifically, the Preamble to Section III of Agenda 21 states (UNEP, 1992):

> One of the fundamental prerequisites for the achievement of sustainable development is broad public participation in decision-making. Furthermore, in the more specific context of environment and development, the need for new forms of participation has emerged. **This includes the need of individuals, groups and organizations to participate in environmental impact assessment procedures and to know about and participate in decisions, particularly those which potentially affect the communities in which they live and work** (emphasis added). Individuals, groups, and organizations should have access to information relevant to environment and development held by national authorities, including information on products and activities that have or are likely to have a significant impact on the environment, and information on environmental protection measures.

Adherence to the philosophy highlighted in this statement from the United Nations Environment Program would help prevent unjust environmental inequities.

SYNTHESIS

This chapter summarizes key events that have forged what is called environmental equity or environmental justice. The two terms are not synonymous. Environmental equity is used by those who seem motivated to measure the outcome of their policies and actions. Environmental justice is preferred by those who want to instill cultural sensitivities and social justice concerns into their policies and actions. It is too soon to predict if either term will ultimately prevail, but the final term must be based on evidenced need and garnered public support.

The conferences, studies, and reports cited in this chapter resulted from serious questions about inequities along racial and low-income lines in placing TSDFs and in remediating CERCLA sites. Bullard (1990, 1994), who conducted comprehensive sociologic evaluations of several communities near waste sites and waste disposal facilities, concluded, "...there is mounting empirical evidence that people-of-color and low-income communities suffer disproportionately from facility siting decisions involving municipal landfills, incinerators, and hazardous-waste disposal facilities." Studies by the GAO (1983) and the United Church of Christ (CSJ, 1987) raised similar concerns about racially or culturally discriminatory actions that allegedly led to locating waste facilities in minority and low-income communities.

More than 15 years have passed since the GAO study of 1983. What do the data now reveal about environmental inequities and hazardous waste sites? First, data are sufficient to show that some environmental hazards are not shared equally across racial and ethnic groups of Americans. For example, data are compelling that African-American and Hispanic children are exposed to environmental lead sources in percentages that exceed their White counterparts. The presence of lead in the paint of older housing is a primary reason, and older housing is associated with low household income. As the problem of lead exposure for young children illustrates, race and ethnicity are intertwined with socioeconomic factors for some environmental hazards. From a public health perspective, these intertwined relationships should be understood and factored into public health interventions that eliminate or reduce health risks.

Second, environmental equity issues about the location of hazardous waste sites have led to several demographic studies conducted since the late 1980s. The siting of hazardous waste treatment, storage, and disposal facilities has stimulated a debate about the equity of site locations. Research on the demographics and socioeconomic characteristics of populations near hazardous waste sites has led to a better picture of what kinds of hazardous waste facilities and sites disproportionately impact minority groups and persons of low income.

The work of several investigators indicates that the method of selecting the geographic unit of analysis and the comparison data will determine how environmental equity questions are answered. Investigations that used large geographic units (e.g., ZIP code areas) generally show that African-American and Hispanic populations live near hazardous waste facilities (TSDFs and uncontrolled hazardous waste sites) in percentages that exceed their percentages in the American population.

As the geographic unit of analysis is reduced in size from ZIP code areas to census tracts and census blocks, the question of minorities and low-income populations living near uncontrolled hazardous waste sites and TSDFs becomes more clear. Minorities appear to live disproportionately near uncontrolled hazardous waste sites (Heitgerd et al., 1995; Zimmerman, 1993; Hird, 1993). The finding of inequity in the percentages of Blacks, and to a lesser extent, Hispanics, living near NPL sites may, however, be primarily because of a subset of NPL sites with large minority populations (Zimmerman, 1993).

However, in distinction to uncontrolled hazardous waste sites, using geographic units smaller than ZIP code areas— primarily census tracts—investigators found no disproportionate association of minorities with living near TSDFs versus living elsewhere (GAO, 1995c; Anderton et al., 1994a,b).

The association between household income and persons' residential proximity to hazardous waste sites does not appear to be particularly significant. That is, household income is not a good predictor of a population's proximity to uncontrolled hazardous waste sites and TSDFs. The most consistent socioeconomic characteristic of communities near hazardous waste facilities and waste sites is the presence of manufacturing facilities, that is, hazardous waste facilities are located in industrial areas. What this finding portends for questions of environmental equity is unclear.

Only one study summarized in this chapter directly addressed the matter of racism and discriminatory actions putatively related to how TSDF facilities are sited. Oakes et al. analyzed communities' characteristics across two decades, using 1970, 1980, and 1990 census data. When they assessed the composition of communities before TSDFs were sited, they found, on average, that TSDFs sited in the 1970s and 1980s were not systematically in areas with unusually high percentages of black or Hispanic populations, compared with other areas that had significant industrial development. Moreover, results showed that characteristics of communities, after siting of TSDFs, have trends that mirror those in the population at large.

Some persons have expressed skepticism about the future of environmental justice. For example, Foreman (1996) described what he considers to be a crucial problem facing the environmental justice movement. He characterized the movement as being a "big tent" that embraces African-American, Latino, Asian, and Native American grassroots organizations and their allies. He contends that maintaining harmony within the environmental justice movement has led to avoiding difficult but necessary decisions. He states, "The movement presumes that any person of color voicing any environmental-related anxiety or aspiration represents a genuine environmental justice problem. Indeed, a broader redistributative and cultural agenda, as well as a profound discomfort with industrial capitalism generally, lurks just behind the concerns over unequal pollution impacts." This strong assertion is given without supporting factual data. Foreman casts environmental justice as a movement with fragile political support and an uncertain future.

Accepting the premise that the search for environmental justice, framed as a political and societal movement, leads inevitably to confrontation, conflict, and litigation raises the question whether a better approach exists. A lesson from the CERCLA statute is instructive. The statute contains the "polluter-pays" principle, which necessarily leads to assessing blame; that is, which parties are responsible for paying remedial costs at hazardous waste sites? Assessing blame has resulted in confrontation, conflict, and litigation. Will environmental justice, with its tendency to assign blame for allegedly targeting minority communities as locales for hazardous waste sites, suffer the same fate as CERCLA? The question cannot be answered yet, but some would argue that a philosophical focus on equity rather than justice would achieve better results.

Given the level of concern that unjust environmental inequities be prevented, the policies and actions already set in motion will likely result in fewer minorities and persons of low income being exposed to environmental hazards. This is a very desirable outcome in terms of both public health and social justice. Government efforts to collect data on the impacts of environmental hazards on minorities and to ensure that federal policies do not create environmental inequities will be important. Furthermore, strengthening the training and education programs that inform communities about their environmental status will be required. A commitment to enhance the number and diversity of minority health and environmental professionals must be allied with such efforts.

Increased diversity in the U.S. population is another important reason to strive for environmental equity. According to census projections, the percentage of nonwhite population will continue to increase. In the first decades of the twenty-first century, growth in the United States will be concentrated in nonwhite populations, primarily in the South and West and in suburbs of the 40 largest metropolitan areas (Hodgkinson, 1992). As diversity of the American population continues to increase, minority groups will place greater attention on assuring that environmental inequities are not occurring.

The importance of fairness will be the centerpiece of whatever gets done. The American public gives strong support to actions predicated on what is fair. Changes in voting rights legislation and fairness in job opportunities are examples of fairness translated into statutory action. Having minorities and persons of low income experience a disproportionate burden of environmental health hazards is unfair. Environmental equity can redress many of these inequities through improved data collection, culturally sensitive regulations, and raising consciousness.

Chapter Ten

Comparative Risk Assessment

Is there general agreement on the nation's environmental hazard priorities? Moreover, what criteria should be used to prioritize environmental hazards? How do human health risks of hazardous waste compare with those of other environmental hazards? The EPA and some states, among others, have used comparative risk assessment to prioritize individual environmental hazards. Legislators have expressed interest in using risk comparisons to help shape legislative and budget decisions. This chapter summarizes approaches to comparative risk assessment and synthesizes a set of national environmental hazard priorities, using data from 15 states.

How individuals and social structures, such as legislatures, compare environmental hazards has become a subject of great interest. As individuals, we make personal decisions that, knowingly or not, constitute a comparison of health risks. Some persons unwisely choose to smoke tobacco, perhaps unaware that nicotine in tobacco smoke is addictive. How persons select their living arrangements can result from comparing health risks. Some locales, such as in urban areas, present higher health risks because of higher levels of air pollutants there than in rural areas. However, health risks due to urban environmental pollution may be outweighed by risks that come with commuting to work from less environmentally polluted suburban or rural areas. Sometimes these personal choices are deliberate and based on factual information; other times, personal-choice decisions stem from fears or perceptions not based on scientific data or fact. As discussed in Chapter 8, the perception of environmental hazards differs markedly between the lay public and technical experts.

What priorities exist for the nation's environmental hazards? Is there a consensus method for ranking them? These two questions have become increasingly important to legislators, government officials, and public service groups. Legislators, in particular, assert their need to match legislative actions with environmental priorities. One approach, called comparative risk assessment, compares risks across various environmental hazards. Many consider this the best approach to establishing priorities for environmental and public health programs.

How did comparative risk assessment become a prominent method of setting priorities for environmental hazards? In 1987, the EPA released its *Unfinished Business* report, which compared and prioritized 31 national environmental hazards for which the EPA had regulatory jurisdiction (EPA, 1987). Since then, several states, some counties and cities, and at least one tribal organization, have conducted similar comparative risk assessments of environmental hazards.

COMPARING ENVIRONMENTAL HAZARDS

Like individuals, social structures such as legislatures must make decisions about environmental hazards. This occurs when legislators craft environmental and public health legislation and appropriate budgets to government programs. Private industry performs similar comparisons of environmental hazards. Companies must budget according to their own environmental priorities and those required to meet government regulations and community concerns.

Because environmental protection and remediation programs are costly, legislators must seek better methods on which to predicate legislative actions. Comparative risk assessment has most often been suggested as the lamp to light the way to improved legislative decisions. This stems from the belief that because risk assessment is a quantitative and systematic approach to characterizing risks, using it to compare individual environmental hazards would lead to more precise ways of prioritizing hazards.

Two examples illustrate how some government officials view comparative risk assessment of environmental hazards. Governor John Engler of Michigan said, "Too often in the past, Michigan's environmental priorities have been set by the crisis of the moment, budget uncertainties, media attention, or conflicting data. I am convinced that it is time to carefully review and evaluate our priorities and base those priorities on careful thought and scientific information. We must do this in order to efficiently apply our limited resources to addressing the most serious environmental risks that our state faces" (MDNR, 1992). In a similar vein, Jan Eastman, Secretary, Agency of Natural Resources, State of Vermont, stated, "The Agency will seek to reduce risks to Vermont and Vermonters by exploring approaches to environmental management, including pollution prevention, toxics reduction, market incentives, and the continued use of public information and education. These approaches may help the Agency make the best possible use of its increasingly scarce resources. The Advisory Committee's ranking of the relative severity of Vermont's environmental problems helps to provide a useful foundation for action" (VANR, 1991). In both examples, linkage between government resources and ranking of environmental hazards is evident.

A perception that "low priority" risks have received too much attention and too many resources underlies the desire for better legislative decisions on environmental hazards. Proponents of this thesis often cite the cost of remediating uncontrolled hazardous waste sites as an example of the imbalance between environmental benefits and costs. They assert that the multibillion dollar costs of remediating uncontrolled hazardous waste sites outweigh the beneficial effects to human health, ecologic systems, and environmental quality. To what extent is this assertion legitimate and where do the risks of hazardous waste sites rank in comparison to those of other environmental hazards? For example, do hazardous waste sites present greater risks to human health than do indoor air pollutants?

Lee Thomas, EPA Administrator, summarized his views on risk comparison by observing, "Although EPA's mission enjoys broad public support, our agency nonetheless must operate on finite resources. Therefore, we must choose our priorities carefully so that we apply those resources as effectively as possible. While we have made much progress to date, the cost of further environmental improvements in many areas will be high. For example, removing additional increments of toxics from industrial effluents or cleaning up contaminated ground water to background levels can be enormously expensive. The unit cost of moving ever closer to the point of zero discharge, zero contamination, and zero risk increases exponentially. Yet this agency must proceed to carry out its mandates and to set its priorities" (Thomas, 1987).

What then is comparative risk assessment? One source states, "Comparative risk [assessment] is an environmental policy and planning process which attempts to bridge the gap between technical environmental risk analysis and public values and perceptions about environmental risk. Simply stated, **the comparative risk process attempts to identify those aspects of the environment which both technical and public groups feel are of top priority. These issues, in turn, form the basis for public environmental expenditure allocations and environmental planning and risk reduction strategies**" (emphasis added) (Ohio University, 1995). This encompassing statement holds out the promise of an orderly, systematic approach to ranking and controlling environmental hazards. As subsequent sections describe, this promise is being partially realized by states and others.

Methods and findings are reviewed in this chapter from comparative risk assessments that the EPA, several states, some local government agencies, and one group of Native American tribes conducted. This book emphasizes the human health risks posed by hazardous waste compared with those of other environmental hazards. Because the EPA's *Unfinished Business* project played such a key role in setting the agenda and process for comparative risk assessment, the next section describes their work. This is followed by a description of how state and local governments conduct comparative risk assessments and the findings from their assessments, including an analysis of states' highest ranked environmental hazards.

EPA'S *UNFINISHED BUSINESS* ANALYSIS

In 1987, the EPA released its *Unfinished Business* report (EPA, 1987), which was prepared by 75 career EPA managers who ranked 31 environmental hazards for which the EPA had regulatory jurisdiction. Because of lack of jurisdiction, the EPA did not rank some key environmental hazards that state and local governments later considered important in their comparative risk assessments, including food safety, lead contamination, and natural hazards.

For the 31 environmental hazards, the EPA assessed four different kinds of risk: cancer, noncancer health effects, ecologic effects, and welfare effects (e.g., materials damage, aesthetic degradation). Quantitative cancer risk estimates were developed where possible; other risks were qualitatively estimated through a process involving professional judgment and consensus. An overall risk assessment priority for each environmental hazard resulted when rankings from the four risk criteria were combined. Table 10.1. shows the 10 top-ranked (not in rank order, according to the EPA) environmental hazards in terms of overall risk. The EPA's analysis produced the following five summary conclusions (EPA, 1987):

Table 10.1. Top-Ranked Environmental Hazards in the EPA's Comparative Risk Assessment Study (not in rank order) (EPA, 1987).

- Criteria air pollutants
- Hazardous/toxic air pollutants
- Other air pollutants
- Radon—indoor
- Indoor air pollution other than radon
- Radiation from sources other than indoor radon
- Substances suspected of depleting the stratospheric ozone layer
- CO_2 and global warming
- Direct, point-source discharges to surface waters
- Indirect, point-source discharges to surface waters

- No environmental hazards ranked relatively high or relatively low in all four types of risk. Whether a hazard appears large or not depended critically on the type of adverse effect with which one is concerned.
- Hazards that ranked relatively high in three of four risk types, or at least medium in all four included: criteria air pollutants, stratospheric ozone depletion, pesticide residues on food, and other pesticide risks (runoff and air deposition of pesticides).
- Hazards that ranked relatively high in cancer and noncancer health risks but low in ecologic and welfare risks included: hazardous air pollutants, indoor radon, indoor air pollution other than radon, pesticide application, exposure to consumer products, and workers' exposure to chemicals.
- Hazards that ranked relatively high in ecologic and welfare risks, but low in both health risks, included: global warming, point and nonpoint sources of surface water pollution, physical alteration of aquatic habitats (including estuaries and wetlands), and mining wastes.
- Groundwater-related environmental hazards consistently ranked medium or low.

The EPA's comparative risk assessment ranked inactive hazardous waste sites (i.e., CERCLA sites) 17th overall among 31 environmental hazards. They concluded that CERCLA sites ranked 8th out of 29 environmental hazards[1] on cancer risk, low risk on noncancer health effects, medium risk on ecologic effects, and medium risk for welfare effects. The EPA's analysis was based on environmental contamination data from 35 CERCLA sites. Six contaminants (arsenic, benzene, 1,2-dichloroethane, TCE, tetrachloroethylene, and vinyl chloride) were selected for risk assessment. Their releases from CERCLA sites were estimated to cause approximately 1,000 cases of cancer annually. An interesting affirmation of the wisdom of the EPA's selection is the fact that all six substances are priority hazardous substances that have been found in completed exposure pathways (Table 2.3).

The final risk ranking in the EPA's comparison of individual environmental hazards synthesized all four criteria into one list of ranked environmental hazards. In view of the public concern about effects of environmental hazards on human health, the human health rankings from the EPA's *Unfinished Business* report are interesting. The EPA ranked each environmental hazard separately on carcinogenic risk and noncarcinogenic risk. The author assigned points to EPA's risk rankings on a scale of one to ten. For example,

---

[1] The 31 hazards were not all ranked on all four risk dimensions.

the EPA's rank of high cancer risk for consumer exposure to chemicals rated an assignment of ten points; seven points were assigned for high noncancer risk with medium EPA confidence. The summary health ranking shown in Table 10.2 gives equal weight to cancer and noncancer risks.

The highest-ranked human health risk in Table 10.2 is workers' exposure to chemicals, followed by a group that includes indoor air pollutants (minus radon), hazardous air pollutants, consumer products, pesticide residues on food, drinking water quality, and application of pesticides. The common theme in this set of hazards appears to be large numbers of persons at potential risk of exposure. When both noncarcinogenic effects and carcinogenic risks are considered together in the EPA analysis, inactive hazardous waste sites were ranked 16th among 24 environmental hazards.

It is interesting to observe that the EPA ranked global warming high on ecologic and social welfare risks but not on human health risk. Some climatologists and health officials have a different view. The World Health Organization (WHO) speculates that under the most dire scenario, a 2°–3°C rise in the earth's average temperature would cause serious human health effects, including: (McMichael et al., 1993)

- Direct thermal effects—heat illnesses
- Weather-related disasters
- Insect migration and infestation
- Decreased yields of grain crops

By contrast, the EPA did not consider global warming a cancer risk and did not rank it on noncancer risk. This supports Fischhoff's assertion—described in a subsequent section—that priorities will necessarily reflect the bias and values of the persons and organizations conducting the analysis.

In 1988, the EPA asked its Science Advisory Board (SAB) to review and comment on the *Unfinished Business* report. The SAB endorsed both the EPA's comparative risk assessment methodology and the findings in the report. They encouraged the EPA to target its environmental protection programs on the basis of risk reduction (Minard, 1996). Furthermore, the SAB opined that no amount of science could or should completely exclude professionals' judgment in the comparative risk assessment process.

Although the EPA's ranking of environmental hazards, including hazardous waste sites (active and inactive), was reasonable for the data and conditions of circa 1986, it should be updated because more complete toxicologic, epidemiologic, and environmental quality data are now available for several of the hazards the EPA ranked in 1987.

COMPARATIVE RISK ASSESSMENT METHODOLOGY

As the EPA's *Unfinished Business* study illustrates, comparative risk assessment is a logical extension of hazard-specific risk assessment. For many years, risk analysts have used actuarial (i.e., life expectancy) data to compare lifestyle behaviors. From their analyses have come comparative risk estimates for several lifestyle behaviors, including tobacco smoking, failure to wear automobile seatbelts, and diets deficient in vegetables and grains. These comparisons are very important because they are based on actuarial data, and the risks can be communicated as personal public health messages in health promotion programs (Chapter 8).

Table 10.2. Health-Based Priorities of Environmental Hazards in *Unfinished Business* (adapted from EPA, 1987).

Environmental Hazard	CR	NCR	Sum	Environmental Hazard	CR	NCR	Sum
Workers' exposure to chemicals	10	10	20	UV radiation/ozone depletion	7	4	11
Indoor air pollutants (minus radon)	10	7	17	Industrial nonhazardous waste sites	7	4	11
Hazardous air pollutants	10	7	17	Municipal nonhazardous waste sites	5	6	11
Consumer products	10	7	17	Inactive hazardous waste sites	7	2	9
Pesticide residues on food	10	7	17	Active hazardous waste sites	7	2	9
Drinking water	7	10	17	Indirect discharges	3	6	9
Application of pesticides	7	10	17	Contaminated sludge	5	2	7
Criteria air pollutants	5	10	15	Discharges to estuaries and oceans	1	6	7
Radon–indoor air	10	4	14	Mining waste	5	1	6
Radiation–not radon	7	6	13	Nonpoint sources	5	—	5
Other pesticide risks	7	6	13	Releases from storage tanks	5	—	5
Accidental releases–toxics	3	10	13	Direct discharges (industrial)	3	2	5

CR = points assigned for cancer risk; NCR = points assigned for noncancer risk; Sum = sum of CR and NCR point scale used by author (BLJ): high risk (10 points), medium–high risk (7 points), medium risk (5 points), medium–low (3 points), low risk (1 point).

Comparing risks of environmental hazards is a more recent undertaking. Approaches for conducting comparative risk assessments of environmental hazards are still evolving, with the role of government being integral to these efforts. Fischhoff observes that government has some natural advantages in facing the tasks that attend comparative risk assessment, such as the possession of more resources than individuals and public interest groups have (Fischhoff, 1995).

Fischhoff cautions that government analysts, and other analysts as well, cannot rank health and environmental risks unambiguously. He says a matter of ethics, not science, determines which risks can even be considered for ranking, how "risk" is to be measured, how the different dimensions of risk should be weighted, and how uncertainty should be treated (Fischhoff, 1995). Indeed, "The rankings of risk, and of risk-reduction strategies, will depend on how these issues are resolved. As a result, there will be differences in whose welfare is protected and whose risk-producing activities are restrained."

The Western Center for Environmental Decision-Making, an organization that helps government agencies and others plan and conduct comparative risk assessment projects, defined four models for comparative risk projects by who initiates the project: (1) a government agency, (2) a concerned citizen, (3) a legislative body, and (4) a public-private partnership (WCEDM, 1996a). The first model is exemplified by the EPA's initiation and conduct of its *Unfinished Business* project. An example of the second model is the Michigan comparative risk project initiated by an academic scientist with input from private consultants who collaborated with the state's Department of Natural Resources. The third model is illustrated by states and cities/counties in which elected bodies initiated the comparative risk project, such as the City of Seattle's comparative risk assessment project. The fourth model is illustrated by projects undertaken by Cleveland, Ohio, and the state of Maine. In these latter cases, a government-private sector partnership conducted the comparative risk assessment. The Center noted that the fourth model produced the greatest amount of public participation but also required the most time to conduct. It is too soon to determine whether any of these four models will become the norm for stimulating the initiation of comparative risk assessment projects.

Fischhoff (1995) developed four strategies for government agencies that conduct comparative risk assessments. At the heart of this set of strategies is how government involves the public when prioritizing risks:

- Strategy 1: Make as few assumptions as possible, reporting the results in something approaching raw form, by assembling but not digesting the data. Any integration and interpretation of the data is left to consumers.
- Strategy 2: Integrate the evidence in several different ways, each reflecting an alternative value system. Consumers can locate themselves in the "space" created by alternative value systems.
- Strategy 3: Elicit values from citizens and then derive analytically appropriate rankings implied by these values. Success depends on individuals' ability to express their values in the abstract form required by analytic models.
- Strategy 4: Have panels of citizens determine rankings with the help of technical staff who provide risk data and perhaps suggest alternative perspectives. The credibility of citizen panels depends on how well they reflect the values of the populations they represent.

These four strategies show that getting stakeholders involved is a key element in conducting comparative risk assessments, and how stakeholders are involved will shape the outcome of the risk comparison because they bring their values to the process. This is, of course, overlaid by the values of government. Thus, any comparison of risks, environmental or otherwise, will necessarily be values-laden because the analysis will always be conducted by individuals who bear intrinsic values.

No standard exists for conducting comparative risk assessments of environmental hazards. The methodology used depends on the needs, motives, and resources of the organization conducting the assessment. One group of risk assessors proposed a framework for federal government agencies that involves creating intergovernment task forces and companion technical workgroups to rank specific risks by using an analytic process described elsewhere (Morgan et al., 1996). The developers assert that using this framework would reduce the likelihood that federal agencies might develop and promote disparate sets of ranked environmental hazards.

The EPA and state and local government agencies have conducted several comparative risk assessments of environmental hazards. A core set of general methodologic principles for comparing environmental risks can be gleaned from this experience and database. Seven steps, which have emerged as components of the comparative risk assessment process, will be evident in subsequent sections of this chapter when assessments conducted by states and others are described.

Essential Steps in Conducting Comparative Risk Assessments

- Develop infrastructure
- Involve stakeholders
- Form workgroups
- Select environmental hazards
- Develop ranking criteria
- Assess and compare risks
- Communicate findings

First, the agency or organization performing the assessment develops a support infrastructure. Because the outcome of a comparative risk assessment is likely to attract considerable public interest, adequate resources (people, budget, authority) must be available to ensure competent and meaningful assessments.

Second, the lead agency determines the extent of stakeholder involvement in the assessment process; particularly, how the public will be involved in actual conduct of the comparative risk assessment. Experience has shown that the public's involvement is essential if general agreement on environmental hazard priorities is to be achieved. As environmental issues become more political, achieving general agreement on environmental priorities is important for societal acceptance and support.

Third, workgroups are formed of public and private sector representatives to address specific environmental hazards. In most instances, a parent workgroup (e.g., steering committee) is created to direct the activities of subordinate workgroups that focus on specific aspects of the comparative risk assessment being conducted.

Fourth, workgroups consider the universe of environmental hazards and select those to be compared. To the extent practicable, workgroup deliberations should be conducted as public meetings. The number and breadth of public meetings is determined by the needs and interests of the conducting agency and its infrastructure groups.

Fifth, workgroups develop criteria for ranking environmental hazards. Three general risk dimensions (or categories of risk) have emerged as the basis of an overall risk comparison: risk to human health, risk to ecologic health, and risk to social welfare (sometimes stated as effects on quality of life). The importance of each risk criterion is usually weighted, and rankings from each risk dimension are combined if an overall ranking of environmental hazards is desired.

Sixth, workgroups use the best available information to quantitatively (preferred) or qualitatively (less preferable) assess risks to human health, ecologic health, and social welfare. To the extent practicable, workgroups should conduct their work in public meetings. For human health risk, quantitative cancer risk assessments can be performed for environmental hazards known to have carcinogenic properties. Analysts use the four steps of risk assessment (Chapter 3), augmented by a fifth step—risk-ranking—to characterize environmental hazards:

1. *Hazard identification*—Which environmental hazards are known or suspected to harm human health? What kind of harm is it? Which hazards should be considered for in-depth analysis?
2. *Dose-response assessment*—What does or could happen to humans exposed to different levels of specific environmental hazards? Does cancer result? What are the noncancer effects?
3. *Exposure assessment*—What are the sources and durations of exposures? How many persons are exposed? What is the range of magnitude of exposure? What is the timing of exposures? Are susceptible populations being exposed (e.g., children)?
4. *Risk characterization*—What are the impacts of current and past exposures? What is the risk to an individual? What is the risk to a population? Are some subpopulations more impacted or susceptible than others? What confidence can be placed in the overall risk analysis? What uncertainties and limitations are in the data, and how will they be communicated to the public?
5. *Risk-ranking*—How serious or consequential is a particular environmental hazard, relative to other environmental hazards or problems?

Seventh, products of the comparative risk assessments are communicated with the public, emphasizing the needs of stakeholders. This effort can include public meetings, briefings of legislators, news media coverage, and meetings with focus groups. Comparative risk assessments are part science and part art. The process of comparing environmental hazards puts topical technical information in the hands of legislators, policy makers, and the interested public in ways that enhance political and personal decisions on risk (Minard, 1996).

Because comparative risk assessment projects are a relatively recent development, only limited effort has been made to collect, collate, and digest the implications of various projects that states, local agencies, and the private sector have conducted. The Western Center for Environmental Decision-Making analyzed comparative risk projects conducted by local governmental agencies (WCEDM, 1996b). They concluded, "In short, the five years of experimentation with local comparative risk has demonstrated the technique's potential to help people address environmental concerns. A major selling point of the approach—the promise of re-orienting government attention from low to high risk concerns—remains largely unrealized. However, positive outcomes such as increased communication with the public, stronger connections within government and between government and the public, and greater community understanding of environmental conditions demonstrate the value of comparative risk." The Center's conclusion was based on eight comparative risk projects that local government agencies conducted. Impacts of state-conducted comparative risk projects have not been similarly analyzed.

STATES'/TERRITORIES' COMPARATIVE RISK ASSESSMENTS

Several states and local governments, in addition to the EPA, have conducted comparative risk assessments of environmental hazards. Through December 1996, at least 15 states and one U.S. territory (Guam) have completed comparative assessments of their major environmental hazards (Table 10.3). Generally speaking, states seem to have adopted Fischhoff's Strategy 4 as their approach to conducting risk comparisons, with state agencies providing infrastructure and technical assistance and state government analysts applying risk assessment methods to a list of major environmental hazards. Focus groups and workgroups provide public input to the risk assessment process and help to achieve general agreement.

Like the U.S. Congress, state legislatures have advocated linking environmental priorities to the allocation of budgets and other legislative actions. The EPA has provided states grants to conduct comparative risk assessments of their environmental hazards. States have assessed risks to human health, ecology, and quality of life (sometimes called social welfare), generally following the risk dimensions the EPA used in its *Unfinished Business* analysis (EPA, 1987).

Because some legislative bodies and government agencies prioritize health and environmental risks to determine budget and legislative actions, searching for any national agreement on human health risks from environmental hazards is important. The EPA *Unfinished Business* list (Table 10.2) is arguably one possible expression of general agreement, given that agency's federal primacy in environmental affairs.

However, another possibility for reaching general agreement on environmental hazards is to synthesize a list of priorities from the outcomes of state-conducted comparative risk assessments. A later section contains the results of such a synthesis.

It is important to emphasize that states generally considered risk in the context of residual risk, or the health risk that remained after risk reduction actions were implemented, such as risks associated with sources of air pollution that remain after maximum achievable control technology was implemented (CRARM, 1997). The following section summarizes 15 states' comparative risk assessments, focusing on the states' human health risk priorities. Wording about specific environmental hazards has been excerpted from the states' reports.

Summaries of States' Rankings

Arizona: In February 1993, Governor Fife Symington established *The Arizona Comparative Environmental Risk Project* by executive order (Arizona, 1995). A Steering Committee created three committees of technical experts on ecosystems, human health, and quality of life. They developed initial lists of environmental hazards in Arizona, using comparative risk projects conducted by other states as starting points. A Public Advisory Committee, consisting of stakeholders involved in the state's environmental issues, joined the Steering Committee to refine the initial list of hazards into a resulting list of 14 environmental hazards.

Table 10.3. States and Territory Comparative Risk Assessments Through December 1996.

• Arizona	• Louisiana
• Arkansas	• Maine
• California	• Michigan
• Colorado	• Ohio
• Florida	• Texas
• Guam	• Utah
• Hawaii	• Vermont
• Kentucky	• Washington

The environmental hazards in Arizona ranked highest as risks to human health were: High risk—(1) allergens and Valley Fever, (2) environmental tobacco smoke, (3) fine particulate matter, (4) food safety, (5) ionizing radiation, (6) lead poisoning, and (7) radon. Medium risk—(1) drinking water, (2) hazardous air pollutants, (3) medical exposure to ionizing radiation, (4) natural hazards, and (5) occupational exposure to toxic materials.

Federal (i.e., CERCLA) and state Superfund sites were ranked low as risks to human health, as were accidental releases of toxic substances, leaking underground storage tanks, and RCRA sites.

Arkansas: The Arkansas Department of Health conducted the state's comparative risk project (ADOH, 1995). The state's assessment considered only human health risks. Ecologic and quality-of-life risks were not addressed. Staff from the Arkansas Department of Health constituted the Human Health and Environment Committee (HHEC). The HHEC developed a list of environmental hazards, developed risk assessment methods, conducted risk assessments of individual environmental hazards, and issued an initial ranking of hazards. The resulting draft report of "environmental health" priorities was reviewed by public groups and then revised. (Environmental health was not defined in their report, but in this book it is used to denote the area of public health that is concerned with the effects on human health of socioeconomic conditions and physical factors in the environment.)

The environmental hazards in Arkansas ranked highest as risks to human health were: High risk—(1) residential and consumer product source releases to indoor air. Medium-to-High risk—(1) inactive hazardous waste sites. Medium risk—(1) nonpoint-source releases to water, (2) stationary and commercial area source releases to air, and (3) mobile source releases in air.

Inactive hazardous waste sites were ranked as a medium-to-high risk to human health. The Arkansas project found, "Significant added cancer risks to small populations around some site types. Several noncancer symptoms (respiratory irritation, nausea) associated with proximity to sites. Unknown impact on more severe noncancer endpoints such as reproductive or immune system toxicity" (ADOH, 1995). Hazardous waste treatment, storage and disposal facilities, and storage tank releases were deemed to be low risks to human health.

California: The California Environmental Protection Agency sponsored *The California Comparative Risk Project* (CEPA, 1994). The project was conducted over a two-year span with the assistance of nearly 300 volunteers. Included were representatives from agriculture, community groups, county and state government, environmental organizations, industry, and universities. The Statewide Community Advisory Committee provided input to the project.

Residual risks to human health, identified according to sources of environmental release, were: High risk—(1) releases to air from mobile sources, (2) releases to groundwater from natural sources, (3) releases to air from residential and consumer products, and (4) releases to air from stationary and commercial area sources. Medium risk—(1) releases to groundwater from anthropogenic sources, (2) releases from inactive hazardous waste sites, and (3) releases to surface water from nonpoint-sources.

Inactive hazardous waste sites were ranked as a medium priority risk to human health. The project found "The human health risks from land releases are generally low to medium, because exposures are not high. The highest risks are associated with exposures to uncontrolled inactive hazardous waste sites. The volume of hazardous waste generated annually (1.9 million tons) carries the potential for high human health risks to large populations, if regulatory oversight programs are not in place" (CEPA, 1994). Treatment, storage, and waste-disposal facilities were ranked as a low risk to human health. The project was unable to rank human health risks attributable to active hazardous waste generators.

Colorado: The *Colorado Environment 2000 Project* was conducted to identify the state's most important environmental hazards and to develop ways to reduce the impacts of environmental hazards. The governor's office, the Colorado Department of Health, and the Colorado Department of Natural Resources coordinated the project. The *Environmental Status Report* (CDOH, 1989) from the project provided background information about the state's environmental quality. Volunteers from government, business, and citizen organizations constituted the work groups.

Technical work groups examined risks associated with pollutants in their subject areas. The Air Technical Work Group ranked air pollution and indoor radon as the highest current health risks in Colorado. The Land and Multi-Media Technical Work Group ranked environmental lead, active and inactive mining and milling sites, and pesticides as the hazards presenting the greatest current risks to human health. The Water Technical Work Group ranked groundwater contamination as the highest human health risk. No list was developed of environmental hazards ranked according to human health risk.

Concerning inactive hazardous and radioactive waste sites, the Colorado project found, "Though the risks associated with these abandoned sites are usually localized, the types, quantities, and risks are unknown without extensive and costly site investigation. ...Health effects from these sites are generally localized, and only a small number of persons may be affected. Of concern are the uncertainties about the number and location of unidentified inactive hazardous and radioactive waste sites and therefore the number of people who may be exposed" (CDOH, 1989).

Florida: The Florida Department of Environmental Protection coordinated a project that assessed and compared the residual risks of environmental hazards to human health, ecosystems, and quality of life (FDEP, 1995). The project had three phases. In the technical-assessment phase, environmental hazards were identified, information was collected for each problem, the risks posed by each hazard were estimated, and an initial ranking of hazards was developed. Three technical advisory committees composed of experts from universities and state agencies performed these efforts. In the public-review phase, technical experts and lay persons reviewed the initial rankings of environmental hazards and background information and prepared a final, integrated ranking. In the final phase of the project, strategies were developed to reduce risks from environmental hazards.

The environmental hazards in Florida ranked highest as residual risks to human health were: High risk—(1) indoor air quality. Medium risk—(1) ambient air quality, (2) food quality (3) groundwater quality, (4) transportation/storage of hazardous materials and wastes, and (5) soil quality.

Inactive hazardous waste sites were not ranked as a separate environmental hazard but instead were subsumed in the categories of groundwater quality and soil quality (FDEP, 1995). Both these categories were ranked as medium health risks. The Florida report notes, "There are currently 21 active state-listed hazardous waste sites and 54 active federal National Priorities List (NPL) sites in Florida. Many of these sites have groundwater contamination. In addition, there are at least 1,500 other sites in Florida that have confirmed groundwater contamination" (FDEP, 1995). Florida also ranked the transportation/storage of hazardous materials and wastes as medium risk to human health. A state database on hazardous materials spills listed 2,105 releases from September 1987 to June 1995 involving highway transport, 200 releases from waterway vessels, and 82 incidents involving railroad transport (FDEP, 1995). The number of persons impacted by these incidents was not reported.

Guam: One U.S. territory has conducted a comparative risk analysis of its environmental hazards. The University of Hawaii (1993), in conjunction with the Guam Environmental Protection Agency, analyzed 17 environmental hazards for their impact on cancer, noncancer health effects, and risk to Guam's ecosystems. The health risks were ranked on the number of cancers and other illnesses estimated to result from exposure to each environmental hazard.

When sufficient data were available, risk assessments of environmental hazards were performed using the EPA's CERCLA guidelines. "The analysis of human health risks was very simple, and agreement was easily reached among health experts on Guam" (University of Hawaii, 1993). The risk ranking for human health was done separately for cancer and noncancer health risks. Overall, the analysis showed few cancer and noncancer health risks resulting from pollution in Guam. Highest estimated cancer risks were from (1) indoor air pollution from radon and environmental tobacco smoke, (2) radioactive materials, waste and radiation, and (3) toxic substances in workplaces. For noncancer health risks, the highest ranked environmental hazards were (1) food safety, (2) pests (bites), and (2) toxics in the workplace. Medium ranked noncancer health risks were from (1) polluted marine waters, and (2) toxic substances in the household.

The group "hazardous materials and waste" was ranked as a low risk for both cancer and noncancer health effects.

Hawaii: At the request of the Hawaii Department of Health, the Hawaii Association of Environmental Professionals conducted the state's environmental risk ranking study (HDOH, 1992). The study assessed environmental hazards along three risk dimensions: public health, ecosystems, and economic welfare. An initial list of environmental hazards was circulated for public reaction, and 70 environmental hazards were the subject of risk assessment.

The environmental hazards in Hawaii ranked highest as residual risks to human health were: High risk—(1) indoor air pollutants (general population), and for special or smaller populations; (2) lead contamination; and (3) pesticide residues and metals in some dietary fish. Medium risk—(1) benzene exposure (general population) and for special or smaller populations; (2) skin cancer from ultraviolet radiation; (3) carbon monoxide in confined spaces; and (4) lead for users of roof catchments.

Toxic and hazardous materials and wastes and accidental releases of "toxics" were ranked low as hazards to human health. The final report states, "Hazardous wastes are now managed effectively within a regulated plan for collection, transportation, treatment, and for disposal on the mainland. Past improper management has created some hazardous waste sites that are now being evaluated for clean-up action, but these sites are not accessible by, and do not pose a threat to, the general public."

Kentucky: The Kentucky Natural Resources and Environmental Protection Cabinet coordinated the state's first comparative risk assessment of environmental hazards (KNREPC, 1995). The assessment involved three committees, Steering, Public Advisory, and Technical. The Public Advisory Committee, which included farmers, business leaders, academicians, and environmentalists, produced an initial list of 110 items that committee members believed could harm human health, ecologic health, or quality of life in Kentucky. The Technical Committee refined the list and then evaluated and ranked risks, after a Subcommittee on Human Health assessed the health risks of individual environmental hazards.

The following residual risks to human health were identified: High risk—(1) air quality (air toxics from stationary sources, ozone, VOCs from mobile sources, and VOCs from area sources), (2) indoor air (active tobacco smoke, passive tobacco smoke, radon, and lead), and (3) waste (household hazardous waste). Medium risk—(1) air quality (stationary sources: particulate matter, ozone, VOCs; mobile sources: nitrogen dioxide, carbon monoxide, particulate matter, and air toxics; area sources: air toxics and particulates), (2) indoor air (formaldehyde, incomplete combustion products, and microbiology agents), and (3) water (microbiologic contamination of drinking and surface water).

Hazardous waste in households was assessed to be high risk to human health. Toxic and hazardous waste, low-level radioactive waste, underground storage tanks, and open and illegal dumps were all ranked as low risks to human health.

Louisiana: *The Louisiana Environmental Action Plan to 2000* project ranked the state's environmental hazards. Involved were representatives from 12 state agencies, three federal government agencies, and more than 30 interest groups that included industry, environmentalists, fish and wildlife interests, labor, and tenant advocates. This group, called the Public Advisory and Steering Committee (PASC), negotiated and ratified the final project report. The ranking took part in three steps. A Technical Committee, composed of state government representatives, reviewed environmental data and assessed the risks to human health, Louisiana's ecosystems, and the quality of life. The PASC reviewed the Technical Committee's findings and, combining the three kinds of risk, compiled a provisional ranking,. The PASC considered the views that state residents expressed in a statewide environmental summit and at 11 town meetings and then compiled the final ranking of 33 environmental hazards (DEQ, 1991a).

The project's Health Effects Workgroup combined cancer and noncarcinogenic residual health risks into a single risk score for each of 26 environmental hazards. The environmental hazards in Louisiana ranked highest as risks to human health risks were: Very High—(1) indoor air. High—air toxics. Medium–High—(1) industrial wastewater discharges, (2) drinking water supplies, (3) abandoned/inactive sites, (4) accidental releases, (5) pesticides, and (6) ozone, nitrogen oxides, and carbon monoxide. Medium—(1) municipal wastewater discharges, (2) nonpoint sources, (3) groundwater contamination, (4) active hazardous waste (RCRA) sites, (5) radiation, (6) ozone depletion, (7) deep well injection, (8) recreation finfish contamination, and (9) consumer exposure. Low—(1) storage facilities, (2) municipal/industrial solid waste sites, (3) sulfur oxides, (4) lead, (5) particulate matter, and (6) naturally occurring radioactive material (DEQ, 1991b).

Abandoned/inactive hazardous waste sites were ranked medium–high on human health risk, as were accidental releases of toxic substances. Active hazardous waste sites were ranked as a medium health risk, and municipal/industrial solid waste sites were ranked as low risk to human health.

Maine: *The Maine Environmental Priorities Project* began following a recommendation in 1993 to the governor from a group representing government, business, environmental groups, and academic institutions (MEPP, 1996). The project was governed by a Steering Committee of 37 members from diverse private- and public-sector organizations. This Steering Committee developed a comprehensive list of environmental hazards to be assessed. Three working groups of technical experts provided input to the list. Views of the general public were obtained from focus groups, public dialogue, and a survey of more than 900 households. The technical working groups assessed risks to human health, ecology, and quality of life for the listed environmental hazards. The Steering Committee derived the final list of priority environmental hazards.

The environmental hazards in Maine ranked highest as human health risks were: High risk—(1) discharges to surface water, (2) toxic air pollution, (3) radon, (4) indoor air pollution, (5) accidental release of toxics, (6) releases from storage tanks, and (7) inactive hazardous waste sites. Medium–High risk—(1) criteria air pollutants, and (2) drinking water at the tap (MEPP, 1995).

Uncontrolled hazardous waste sites represented a "high level of concern" for persons living near them. Accidental releases of hazardous substances were ranked as a high concern for health risk. RCRA facilities were ranked as a low health concern.

Michigan: Michigan's Department of Natural Resources administered the state's Relative Risk Analysis Project (MDNR, 1992). The project focused on residual risks, that is, risks remaining after considering the effects of current risk management actions. The project's Steering Committee oversaw the work of three other committees, Citizen, Agency, and Scientist. The Steering and the Citizen committees identified 23 environmental hazards. In 1992, the Scientist Committee prepared technical papers for each of the 23 hazards. Generation and disposal of hazardous and low-level radioactive wastes later became the 24th ranked hazard. The committees combined quality of life, human health, and ecologic risks into one ranking. Human health risks were not reported separately.

The highest ranked environmental issues were: High–High risk—(1) absence of land-use planning that considers resources and the integrity of ecosystems, (2) degradation of urban environments, (3) energy production and consumption, (4) global climate change, (5) lack of environmental awareness, and (6) stratospheric ozone depletion. High risk—(1) alteration of surface water and groundwater hydrology, (2) atmospheric transport and deposition of air toxics, (3) biodiversity and habitat modification, (4) indoor pollutants, (5) nonpoint-source discharges to surface water and groundwater, and (6) trace metals in the ecosystem.

Concerning overall risk of hazardous waste, the following environmental hazards were ranked Medium–High: (1) contaminated sites, generation and disposal of hazardous waste, and generation and disposal of high-level radioactive waste. The Michigan assessment concluded for contaminated sites, "This problem should be considered on a case-by-case basis. The risk depends upon how persons are exposed, such as through groundwater, surface water, soil, etc. Individual sites can and do pose local risks, depending on the fate and transport of and exposure to contaminants. Considerable state and private efforts are key to reducing local exposure; continuation of current programs is critical" (MDNR, 1992.).

Ohio: The Ohio Environmental Protection Agency was responsible for *The Ohio Comparative Risk Project*, which began in 1994 (OEPA, 1995). Three technical groups assessed human health, ecologic, and quality-of-life residual risks. A public comment period identified more than 700 environmental hazards or issues of concern. The Public Advisory Group and the three technical groups reduced the list to 11 general environmental categories. Human health risks consisted of cancer and noncancer health endpoints, each of which could be acute or chronic. The public was extensively involved through public meetings and membership on workgroups.

The environmental hazards in Ohio ranked highest as risks to human health were: High risk—(1) abandoned industrial sites, (2) drinking water at the tap, (3) exposure from lack of consumer awareness, (4) inadequate infrastructure, (5) indoor air quality, (6) industrial/commercial wastewater discharges, (7) mobile source emissions, (8) municipal waste disposal facilities, (9) ozone-depleting substances, and (10) unregulated/abandoned hazardous waste facilities. Medium risk—(1) abandoned water wells, (2) natural food toxins, (3) oil and gas exploration, (4) pesticide residues on foods, (5) tire management, and (6) underground storage tanks.

Abandoned industrial sites and unregulated/abandoned hazardous waste facilities were ranked as high risk to human health. The Ohio report notes, "There are 1,180 abandoned or uncontrolled hazardous waste sites in Ohio. Health risk from exposure to these sites occurs mainly as a result of drinking contaminated ground water; some substances associated with these sites are known carcinogens" (OEPA, 1995).

Texas: The Texas Natural Resource Conservation Commission coordinated the state's environmental comparative risk project, *The State of Texas Environmental Priorities Project*. Environmental hazards were ranked for impacts on human health, ecologic systems, and socioeconomic welfare (TNRCC, 1995). Technical workgroups composed of persons from universities, private research institutions, and government agencies reported to the Public Advisory Committee. They analyzed 27 environmental hazards. The final set of ranked environmental priorities was peer reviewed, but how involved the public was in determining the final set of priorities is unclear in their report.

Human health risks included cancer and noncancer endpoints. The environmental hazards in Texas ranked highest as risks to human health were: High risk—(1) indoor air pollution, (2) particulate matter, (3) ground-level ozone, and (4) lead contamination. Medium risk—(1) groundwater quality/food safety (tied in priority), (2) air toxics, (3) surface water quality/water availability/waste management, (4) global climate change/stratospheric ozone depletion, and (5) drinking water quality.

Waste management issues were ranked as a medium health risk. However, several environmental hazards were included in the waste management category. This included hazardous waste, toxic waste, industrial solid waste, municipal solid waste, oil and gas waste, and medical waste. RCRA facilities and CERCLA waste sites were also included in the waste management category. Regarding uncontrolled hazardous waste sites, the state reported, "Superfund and other abandoned sites exist in many counties in Texas. The average population within one mile of State and Federal sites was estimated to be approximately 45,000 persons. It is not possible to quantify the risk posed by these sites *and* extrapolate the risk to people in the surrounding areas that may be affected because exposure to site contaminants is unknown" (TNRCC, 1995).

Utah: In 1992, the Utah Department of Environmental Quality undertook the state's comparative risk assessment of environmental hazards (UDEQ, 1995). The department's five statutory boards identified 16 environmental concerns. Seven statewide meetings were held with elected officials, business leaders, local health officers, environmentalists, and members of the general public. These meetings honed the list of Utah's environmental hazards and their relative rankings, resulting in 21 environmental hazards that were evaluated for risks to human health, ecology, and quality of life.

Human health risks were based on cancer and noncancer endpoints. The environmental hazards in Utah ranked highest as risks to human health were: High risk—(1) outdoor air pollution, (2) groundwater quality, and (3) drinking water at the tap. Medium–High risk—(1) indoor air quality, (2) hazardous air pollutants, surface water quality, and (3) uncontrolled waste sites, Medium risk—(1) leaking storage tanks, (2) spills/releases of hazardous materials, (3) releases of unique chemical/biologic materials, (4) solid waste disposal activities, (5) hazardous waste generation and disposal, and (6) radioactive waste disposal.

Uncontrolled hazardous waste sites were ranked as medium–high human health risk, whereas Superfund waste sites were ranked medium–low. Utah had 14 NPL sites at the time of their study.

Vermont: The Vermont Agency of Natural Resources coordinated a review of the state's environmental hazards and priorities (VANR, 1991). The two-year project addressed risks to ecosystems, human health, and quality of life. An advisory committee selected the 20 most serious environmental hazards and conducted extensive public outreach. Three technical workgroups composed of Vermont state employees prepared risk estimates.

Residual risks to human health were ranked on the basis of cancer and noncancer health effects. The environmental hazards in Vermont ranked highest as risks to human health were: Higher risk—(1) indoor air pollution, (2) toxics in the household, and (3) toxics in the workplace. Medium risk—(1) ozone layer depletion, (2) air pollution, (3) drinking water at the tap, and (4) food safety.

Hazardous waste and solid waste ranked relatively low as health risks to Vermonters. The state's report comments, "The Advisory Committee concluded that health risks from hazardous and radioactive waste are relatively low in Vermont, primarily because the number of persons is small. Risks to some individuals may be fairly high, however. There is considerable uncertainty in these estimates."

Washington: The Washington state comparative risk project was initiated in 1988 as part of a long-range planning effort called Washington Environment 2010. The state's Department of Ecology oversaw the project (WDOE, 1989a). The Steering Committee, which consisted of the directors of 13 state agencies and representatives from two federal agencies, had overall responsibility for the project. The project's Technical Advisory Committee comprised a team of more than 25 technical experts from the state's environmental protection and natural resource management agencies. They assessed 23 environmental hazards, focusing on human health, ecologic, and economic risks. The Public Advisory Committee, composed of approximately 30 individuals—farmers, educators, legislators, environmentalists, business people and others—compared and categorized the 23 environmental hazards. The final report contains only the overall ranking of the 23 hazards; rankings according to human health risks are not readily evident (WDOE, 1989a). However, a background document provides human health rankings and, for the purposes of this chapter, was used to characterize the state's human health priorities for environmental hazards (WDOE, 1989b).

The environmental hazards in Washington ranked highest as risks to human health were: High risk—(1) ambient air pollution (in which the state includes criteria air pollutants and hazardous air pollutants), and (2) indoor air pollution. Medium–High risk—(1) inactive hazardous waste sites, (2) point-source water pollution, (3) drinking water contamination, and (4) nonpoint-source water pollution. Medium Risk—(1) pesticides, (2) indoor radon, (3) nonhazardous waste sites, (4) active hazardous waste sites, (5) accidental releases, and (6) materials storage.

Inactive hazardous waste sites were ranked as medium–high human health risk, and nonhazardous waste sites, active hazardous waste sites, accidental releases of substances, and materials storage (which included storage tanks) were each ranked as medium human health risks. For inactive hazardous waste sites, the Washington Environment 2010 background report notes, "High cancer and noncancer risk to the maximum exposed individual, medium to low cancer and noncancer risks to population" (WDOE, 1989b).

States' Environmental Hazard Priorities

A review of states' environmental hazard priorities will provide a partial answer to an earlier question of whether general agreement exists on the nation's environmental hazard priorities. The question can be only partially answered because data are available from only 15 states. Hazard rankings from the states listed in Table 10.3 were reviewed and compiled into a consolidated ranking of environmental hazards. **Because this book is concerned with human health (and hazardous waste), only the human health priorities were compared across states and synthesized into one list.** After excerpting human health priorities from each state's comparative risk assessment report and placing them in a spreadsheet, the author (BLJ) assigned points according to high risk (five points), medium risk (three points), and low risk (one point).[2] Professional judgment was used when needed to characterize a state's priorities. For instance, because some states lumped "pesticides" into a single category, a careful reading was conducted of these states' final reports and background materials to distinguish among different areas of pesticide risks.

A consolidated list of 15 states' environmental hazard priorities is presented in Table 10.4. Because the risks assigned by states are primarily "residual risks," that is, risks remaining after extant programs have achieved their current level of risk reduction, some serious environmental hazards that were thought to be under control were ranked lower in human health risk. With the stipulation that these are human health-based priorities and do not necessarily reflect states' ecologic and quality-of-life priorities, the following observations can be drawn from Table 10.4:

- The standout, highest-ranked environmental hazard to human health is indoor air pollution and is labeled high–high risk in Table 10.4. Indoor air includes environmental tobacco smoke and other hazardous substances but excludes indoor radon. The population at risk is presumed to be very large because most people spend more time indoors than out.

---

[2] The spreadsheet is available from the author.

Table 10.4. Consolidated List of 15 States[a] Environmental Hazard Priorities.

Environmental Hazard	Pts.[b]	Rank[c]	Environmental Hazard	Pts.	Rank
Indoor air (excluding radon)	64	H-H	Application of pesticides	24	M
Criteria air pollutants	47	H	Hazardous waste sites–active	21	M-L
Hazardous air pollutants	47	H	Food safety	20	M-L
Radon–indoor	43	H	Waste sites–municipal	19	M-L
Lead contamination	40	M-H	Indirect, point-source discharges to surface water	19	M-L
Hazardous waste sites–inactive	39	M-H	Accidental releases of toxics	18	M-L
Drinking water at the tap	37	M-H	Releases from storage tanks	18	M-L
Nonpoint-source discharges to surface water	30	M	Radiation (excluding radon)	17	M-L
Other groundwater contamination	28	M	Other air pollutants	17	M-L
Ozone depletion	27	M	Pesticide residues on food	17	M-L
Direct, point-source discharges to surface water	27	M	Workers' exposure to chemicals	11	L
Exposure to consumer products	26	M	CO_2 and global warming	11	L
Other pesticide risks	25	M	Mining wastes	9	L

[a] States: AK, AZ, CA, CO, FL, HI, KY, LA, ME, MI, OH, TX, UT, VT, WA.
[b] Sum of points (pts) from 15 states, using the following values for risks: high=5, medium=3, low=1.
[c] Ranking of risk as high (H), medium (M), or low (L).

- Three high-risk environmental hazards cluster closely together: criteria air pollutants, toxic/hazardous air pollutants, and radon. Air pollutants rank high because of the large number of persons at risk, especially vulnerable populations like children, and radon ranks high because of its carcinogenicity.
- Inactive hazardous waste sites, drinking water contamination, and lead contamination rank as medium–high risk to human health. These risks rank higher than the highest-ranked human health risks in the EPA's *Unfinished Business* report (Table 10.2) (EPA, 1987).
- Food safety and lead contamination are environmental hazards of concern to the states but were not listed as priorities in the EPA *Unfinished Business* comparative risk assessment of 1987 because the EPA had no regulatory jurisdiction over them.
- States rank pesticide risks as medium to medium–low risk to human health, which is lower than the EPA ranked pesticide risks in their *Unfinished Business* report (EPA, 1987).
- Occupational (i.e., workplace) risks were ranked much lower by states than by the EPA (Table 10.2). This may be because some states might not consider workplace hazards as part of the "environment," viewing them as separate entities regulated under occupational safety and health statutes.
- States rank climate change as low risk to human health as did the EPA in its *Unfinished Business* report. This differs from the findings of the World Health Organization discussed earlier in this chapter.

It is not known to what extent states have used the findings and recommendations from their comparative risk assessments of environmental hazards. The states have obviously expended considerable effort to build general agreement on their environmental priorities, but to what extent this may have influenced legislators and state government agencies on legislative actions and program decisions awaits analysis.

COUNTIES'/CITIES' COMPARATIVE RISK ASSESSMENTS

In addition to states setting their environmental hazards priorities, several local governments have conducted similar assessments, including county and city health agencies. Health agencies perform myriad actions for their communities, including a range of environmental health responsibilities such as food and sanitation inspections, pest control, and audits of environmental hazards. Local health departments can also become involved in problems of hazardous waste management. Indeed, citizen concerns about uncontrolled hazardous waste sites often come to the attention of local health departments, as was illustrated in Chapters 8 and 9.

Regarding comparative risk assessment, the following section describes a survey conducted by a national association of county health officials to ascertain their members' environmental health priorities. Health risk priorities developed by specific county and county governments are then illustrated. It will be interesting to compare their priorities with those developed by states.

National Association of County and City Health Officials

The National Association of County and City Health Officials (NACCHO), formerly the National Association of County Health Officials (NACHO), provides technical assistance to county and city health departments. They develop and advocate policy and political positions for their membership. As noted above, city and county health departments conduct a host of public health programs, including those in environmental health (NACHO,[3] 1992). NACHO was interested in ascertaining a national picture of environmental health priorities, as expressed by their membership. Their interest was prompted by three trends that affect the management of local environmental health problems: (1) increased public concern about environmental hazards, (2) the growing complexity of environmental health issues, and (3) the frequent separation of public health agencies from environmental agencies. This separation refers to the frequent placement of environmental health responsibilities in departments of environmental quality or protection, rather than in health departments.

In 1990, NACHO conducted a national survey of local health departments to assess needs and resources for environmental health programs (NACHO, 1992). Their survey did not define environmental health and did not follow a comparative risk paradigm; that is, environmental hazards were not ranked on the three risk dimensions that states used. The purpose of their study was to identify: (1) various environmental health issues that challenge local health departments; (2) how these challenges are being met; and (3) the kinds of education, training, and other support local health departments need to adequately assess, communicate, and reduce environmental health risks. Although items 2 and 3 are significant, this chapter will elaborate only item 1 because it relates most directly to priority environmental health problems. The NACHO survey of 1990 still represents the only national data on environmental health priorities specific to local health departments.

NACHO used a questionnaire to survey a stratified random sample of 670 of the 3,169 local health departments operating in the United States at the time of the survey. The sample was stratified according to the size of the population served, which NACHO used as an indicator of resources required by the department. Representative percentages of the total sample within each population range were selected to reflect a national picture.

NACHO received 355 responses, a 53% response rate. Respondents were asked to note the current environmental health issues facing their communities and designate them as long-standing or emerging. The survey identified several common environmental health themes and responsibilities of local health departments. At least 50% of respondents played some role in the following six environmental health issues: contaminated private wells, groundwater and surface water, illegal dumping, radon, and hazardous materials spills/accidental chemical releases (NACHO, 1992). The most frequent environmental health services reported by local health departments are shown in Table 10.5.

Table 10.6 shows an alphabetized list of 28 environmental hazards with the percentage of 355 local health departments reporting them as long-standing or emerging problems.

From the information in Table 10.6, it is possible to compile a prioritized set of top-ranked environmental health issues. Table 10.7 lists the top nine long-standing and emerging health hazards, as identified by local health departments. The environmental health priorities of local health departments are quite different from the EPA's priorities (Table 10.1) (EPA, 1987). However, the priorities of local departments (Table 10.7) cannot be compared with the states' health-based priorities (Table 10.4) because local departments

---

[3] The National Association of County Health Officials changed its name subsequent to the 1992 report. The organization is based in Washington, D.C.

Table 10.5. Environmental Services Most Often Provided by Local Health Departments (NACHO, 1992).

- Food protection (91%)
- Nuisance control (88%)
- Sewage treatment (85%)
- Animal/vector control (85%)
- Private well testing (83%)
- Swimming pool inspection (83%)
- Emergency response (80%)

set "environmental health" priorities that include environmental issues (i.e., recycling, composting) that are not in themselves environmental hazards. That should not be surprising. Local government, which is closest to people and their problems, responds to citizen and community concerns more directly than state or federal governments do. Thus, waste collection, composting, and recycling are more important issues at the local than at the federal government level.

Table 10.7 shows that hazardous waste concerns are important to local health departments. Indeed, waste management problems represent four of the nine reported long-standing environmental health issues. Of the emerging environmental health issues reported, five are related to waste management and response.

Illustrative Examples

Several cities and counties have conducted comparative risk assessments of their environmental hazards. The following examples will illustrate both how environmental hazard priorities are set and the outcomes of the comparisons. The general paradigm used by states to conduct their comparative risk assessments was usually followed by the cities and counties cited here; that is, environmental hazards were ranked for risk to human health, ecologic health, and quality of life.

Seattle, Washington—In 1990, the City of Seattle undertook a comparative risk assessment of its environmental hazards (Seattle, 1991). The assessment was established by the mayor and performed by the Technical Advisory Committee, composed of environmental and other experts, who identified key environmental hazards in the city and associated databases. The Policy Advisory Committee, composed of environmental advocates, business representatives, academics, tribal representatives, and citizen activists, advised the mayor on actions to take in response to the Technical Advisory Committee's findings and environmental priorities.

Relative risk rankings were estimated according to human, ecologic, and quality-of-life risks. An integrated ranking of hazards was developed and reported, with commentary about some individual environmental hazards. Indoor and outdoor air quality concerns were ranked as the greatest risks to human health. Within these two categories, transportation sources of air pollution, wood burning, and environmental tobacco smoke were ranked as high risks. In terms of overall relative risk, use of hazardous materials by business and industry was ranked as high risk; inactive hazardous waste sites were ranked as medium risk.

Table 10.6. Percentage[a] of Local Health Departments Reporting Environmental Health Hazards as Long-Standing or Emerging (adapted from NACHO, 1992).

Hazard	Long-Standing	Emerging	Hazard	Long-Standing	Emerging
Air pollution	54.9%	17.4%	Lead in homes	40.0%	35.7%
Asbestos	45.8	36.0	Lead in water	32.8	43.2
Composting	8.5	45.6	Leaking underground storage tanks	30.9	50.3
Groundwater contamination	73.3	18.4	Mining sites	13.4	5.2
Hazardous materials spills	45.4	40.0	Oil and gas industry pollution	25.2	14.1
Hazardous waste sites	25.2	31.0	Pesticides	48.8	29.4
Hazardous waste transport	29.7	34.8	Private well contamination	75.2	15.2
Household hazardous waste	22.6	55.5	Public water contamination	62.1	19.3
Illegal dumping	83.9	6.5	Radon	16.6	66.6
Illegal dumping of hazardous waste	66.6	15.8	Recycling	11.5	73.9
Industrial pollution	43.3	20.9	Surface water supply safety	66.0	16.4
Lake/ocean pollution	40.6	11.2	Waste collection	63.7	14.4
Landfill siting	49.4	18.7	Waste contamination	55.6	17.5
Lead in the environment	31.0	33.9	Waste incineration	18.2	25.2

[a] Based on 355 responses.

Table 10.7. Local Health Departments' Top Nine Environmental Health Issues (adapted from NACHO, 1992).

Long-Standing Environmental Health Issues and Percent of Respondents		Emerging Environmental Health Issues and Percent of Respondents	
Illegal waste dumping	84	Recycling	74
Private well contamination	75	Radon	67
Groundwater contamination	73	Household hazardous waste	56
Illegal hazardous waste dumping	67	Leaking underground storage tanks	50
Surface water contamination	66	Composting	46
Waste collection	64	Lead in water	43
Public water contamination	62	Hazardous materials spills	40
Waste containment	56	Lead contamination (general)	39
Air pollution	55	Asbestos	36

Columbus, Ohio—In 1994, Columbus, Ohio, undertook *Priorities '95*, a comprehensive project to improve health and environmental protection in the city (OEPA, 1995). A comparative risk assessment of environmental hazards was a component of the project. Three committees were involved with the assessment. The Technical Advisory Committee assembled scientific data and developed informational summaries on various environmental hazards. The Public Advisory Committee was responsible for public outreach and collecting opinions about values associated with the environmental hazards under consideration. The Steering Committee was responsible for key decisions on the project's scope and direction. Environmental hazards were ranked as risks to human health, ecologic health, and quality of life.

Risks to human health were ranked as medium–high, medium–low, and low. The following environmental hazards were ranked medium–high: indoor air contaminants, lead exposure, nitrates, problems with landfill sites, and solid waste reduction facility/dioxin. Medium–low risks were: contaminated sites, drinking water safety, food supply toxins, groundwater contamination, hazardous materials, industrial emissions, municipal solid waste disposal, ozone depletion, transportation pollution, and vector problems.

Athens County, Ohio—In 1995, the Institute for Local Government, Ohio University, Athens, assessed Athens county's environmental priorities (OEPA, 1995). Two committees, the Public Advisory Group and the Technical Advisory Group, examined the county's environmental hazards in preparation for a county-wide forum on environmental priorities. Athens County, which is part of the Ohio Appalachia region, is a rural county with relatively good environmental quality. After reviewing the results of a telephone survey of more than 300 county residents and discussions about specific environmental hazards, county residents ranked the top five environmental hazards at a public forum held in 1995. Groundwater quality was the top-ranked concern, followed in decreasing order by acid mine drainage, environmental justice concerns, absence of county-wide planning, and timber cutting/forest disturbances. Hazardous waste issues were not of concern to the county's residents.

Getting a high degree of county involvement by residents was significant for shaping a list of local environmental priorities. Interestingly, Athens County residents did not interpret the term "environmental justice" racially (Chapter 9) but related it to their vulnerability for having environmentally questionable businesses located in their county (OPEA, 1995).

TRIBAL COMPARATIVE RISK ASSESSMENT

Federally recognized Indian Tribes are sovereign, dependent[4] nations that have developed a trust relationship with the federal government (EPA, 1992b). Most Native American tribes have the right to exercise civil and certain criminal authorities over their reservation lands. States' powers over these lands are limited. Under the U.S. Constitution, the federal government can regulate aspects of Native American life. Court decisions have prescribed a special responsibility of the federal government to act as "trustee" on behalf of and for the benefit of Native American tribes. As any student of American history knows, the relationship of Native Americans with European settlers and their descendants has a difficult and often inglorious history. Indeed, many decisions made by European settlers and their descendants have been at the expense of Native American culture and well-being.

The EPA notes that the unique legal position of Tribal Nations has given them authority to establish environmental standards for their lands (EPA, 1992b). These standards are independent of state standards but must be consistent with any federal minimum standards. The federal responsibility is to contribute to tribes' achievement of standards.

Tribal Nations are well known to have cultural and spiritual respect for the environment and nature. As Chief Joseph of the Nez Perce nation said, "We are the keepers of Mother Earth. If we love her, protect her" (Nez Perce Tribe, 1996). Because of their cultural values, an increasing number of Native American tribes have become involved in environmental protection and health programs. Some of these efforts are a consequence of environmental equity concerns (Chapter 9). Other efforts are based on a return to Native American culture and traditions. Risk assessments of environmental hazards and tribal education about environmental hazards are now included in the environmental activities of several Tribal Nations (e.g., Nez Perce Tribe, 1996; EPA, 1992a). These efforts portend a greater involvement of Native Americans in solving their own environmental problems.

One comparative risk assessment of environmental hazards on tribal lands was conducted from 1991 to 1992 and evaluated environmental hazards faced by the 11 Native American tribes in Wisconsin (EPA, 1992b). Most tribes are native to the state, but two moved to Wisconsin—the Oneida from New York and the Stockbridge from New England—over a century ago. The tribes native to Wisconsin derive mostly from the Chippewa nation.

The EPA staff conducted the comparative risk assessment of environmental hazards among Wisconsin tribes. As the EPA stated, "In the Wisconsin Tribes project, then, we collected and analyzed available data on environmental conditions in the Wisconsin reservations, but supplemented the data extensively with judgment. The judgments derived from work group members' professional knowledge, from results of risk studies in nearby areas thought likely to be similar to the Wisconsin reservations, and from experience gained in previous comparative risk studies elsewhere. ...The process and results of our deliberations in this report were reviewed and endorsed by the eleven Wisconsin Native American Tribes" (EPA, 1992b). The government strategy the EPA used seems to be Strategy 3 of Fischhoff's set of strategies. Because of the historical

---

[4] Under the U.S. Constitution sovereignty is granted to the federal government, states, and recognized Indian Tribes. However, Indian Tribes are constitutionally dependent on the federal government.

relationship of the federal government and Native Americans, greater involvement of the tribes in the process would have been desirable.

Three kinds of risk were considered in conducting the tribal comparative risk assessment: human risk, ecologic risk, and social and welfare damage. The four steps of risk assessment: hazard identification, dose-response assessment, exposure assessment, and risk characterization, (NRC, 1983) were applied to 22 environmental hazards of concern to the 11 Wisconsin tribes. The outcome of the comparative risk assessment is given in Table 10.8.

Human health risks for the Wisconsin tribes are generally consistent with the EPA's *Unfinished Business* priorities, except that food contamination does not appear on the EPA's list (EPA, 1987) and health concerns about water contamination appear higher on the tribes' list of priorities than on the EPA list. However, it is not possible to determine whether the priorities in Table 10.8 are truly a product of tribal concerns or are more a consequence of the EPA analysts using a risk assessment process and a set of environmental hazards familiar to them.

Uncontrolled hazardous waste was not ranked as an environmental health concern for the 11 Wisconsin tribes, but it did rank high as a social and economic concern. This may result from the high cost of remediating abandoned hazardous waste sites and the failure to incorporate more contemporary data on the human health impacts of hazardous waste into the risk assessment (Chapter 5).

SYNTHESIS

This chapter has described comparative risk assessments of environmental hazards. The comparison and ranking of environmental hazards has become an important means for influencing government programs and legislative actions. Some legislators, in particular, view comparative risk assessment as a way to improve budgeting and priorities for environmental programs. No standard method exists on how to conduct a comparison of environmental risks, but states and local government agencies take the direction that includes consensus building through meaningful engagement of the public and stakeholders.

The values of the persons and organizations performing the assessment will influence the products of comparative risk assessments. Therefore, persons who use the products of an organization's comparative risk assessment should know who prepared the risk assessment, under what conditions, and with what values and motivation.

A key to successful comparative risk assessments that achieve public support and general agreement among stakeholders is involvement of the public in addition to technical experts. This is consistent with Fischhoff's Strategy 4, the model states and local agencies that have conducted successful comparative risk assessments seem to prefer.

One analyst's review of state and federal comparative risk projects identified several benefits (Minard, 1996). One of the most important benefits was the evolution of a more sophisticated and cohesive staff in the agencies that conducted the risk assessment. Participants in the comparative risk assessment process became better informed about how their programs related to other environmental programs in their agencies and elsewhere. The principal strength of the comparative risk process, the analyst argues, is the capacity to frame important public policy questions and to engage people in a productive attempt to answer them.

Table 10.8. Priority Environmental Hazards for the 11 Wisconsin Native American Tribes (EPA, 1992b).

Human Health Risk:
- Food contamination
- Nonpoint source pollution of surface water
- Indoor air pollution other than radon
- Indoor radon
- Drinking water contamination
- Groundwater contamination

Ecological Risk:
- Nonpoint source pollution of surface water
- SO_x/NO_x (acid deposition)
- Physical degradation of water and wetlands

Social and Economic Damages:
- Nonpoint source pollution of surface water
- Physical degradation of water and wetland habitats
- Food contamination
- Physical degradation of terrestrial ecosystems
- Unmanaged hazardous waste (past disposal)
- SO_x/NO_x (acid deposition)

Because legislative bodies have expressed support for setting priorities among environmental hazards as a means of influencing legislative actions, some comments were offered about this linkage of legislation and comparative risk assessment.

Comparative risk assessment is a useful structure for organizing information about a set of environmental hazards. Risk assessment methods can be applied to a wide range of environmental hazards, thereby providing a means for informed comparisons. However, several concerns about the use of comparative risk rankings to set priorities and allocate resources have come from sources that completed comparative risk assessments of environmental hazards (e.g., ADOH, 1995). Following are six such concerns:

- Comparative risk assessment will always be influenced by the values and bias of the persons and organizations conducting the assessment. Therefore caution must be exercised when accepting the product of any comparative risk assessment.
- Comparative risk assessment does not provide adequate information on the qualitative aspects of risk such as involuntariness of exposure and equity. For instance, do the risks fall on persons who have little or no voice in assuming risks (e.g., an unborn child)?
- Comparative risk assessment can be of limited value whenever a high degree of uncertainty exists in scientific data and public values. Risk assessments are necessarily best-judgment exercises.
- Budget and resource allocations should not necessarily be directly proportional to the ranking of risks. Although risk management actions and priorities will, of course, build on the outcome of priority risk assessments, the actions attending risk management should have their own criteria for implementation. For example, successful risk management programs may be overlooked simply because they are successful, thereby leading to the possibility of reduced funding in favor of "higher priority" programs. Therefore, environmental hazards should be compared on the basis of

residual risks, or those risks that will remain after risk management controls have been implemented.

- Today's priorities may not be tomorrow's priorities. Scientific databases on environmental hazards can change and public attitudes may shift, thereby altering environmental health priorities. This reality leads one to conclude that comparative risk assessments must be updated periodically if they are to have societal meaning.
- Another reason for using comparative risk assessments cautiously is the generally ineffective oversight of government programs by legislative bodies. Because much of the time of legislative bodies is consumed by the exigencies of new legislation, budget decisions, and political responses to emergent public problems, too little time, interest, and resources are available to review government agencies and their programs. This may make legislative actions based on risk assessment priorities difficult to adjust in subsequent years.

Any set of priorities will necessarily reflect the values and conditions of the performing organization. Federal EPA health-based priorities are different from those articulated by state health departments, and environmental health priorities of local health departments are quite different from the EPA's overall hazard priorities set forth in its *Unfinished Business* analysis. This is not surprising. Conditions and concerns at the federal level differ from those at the local level, and federal and state agencies are not in day-to-day contact with the public as often as local authorities are. For example, although composting is an emerging environmental health concern of local health departments, neither federal nor state agencies identified composting as an environmental priority.

An examination of only the federal- and state-based priorities for environmental hazards shows a general agreement on the top-ranked hazards. They are, in descending order of importance to human health: indoor air pollution (excluding radon), indoor radon exposure, lead contamination, criteria air pollutants, and hazardous air pollutants.

The EPA *Unfinished Business* study of 1987 ranked uncontrolled hazardous waste sites 16th among 24 hazards as a risk to human health. States that have conducted comparative risk assessments since 1987 have generally classified uncontrolled hazardous waste sites as medium-high risk to human health, with a ranking of 6th among 26 environmental hazards. This disparity from the EPA's ranking is due primarily to two factors. First, states involved the public in their comparative risk assessments to a much greater degree than did the EPA, and the public ranks hazardous waste sites as a human health hazard. Second, states had some health effects information, such as that in Chapter 5, that the EPA did not possess in 1987 when they conducted their *Unfinished Business* study.

References

Abadin, H.G., B.F. Hibbs, and H.R. Pohl. 1997. Breast-feeding exposure of infants to cadmium, lead, and mercury: A public health viewpoint. *Toxicol. Indust. Health* 13:495–517.

Abdnor, J. 1985. Congressional Record, 99th Congress, 1st Session. *Cong. Rec.* S 11998 (No. 121):131.

ADOH (Arkansas Department of Health). 1995. *Human Health and Environment Comparative Risk Project*. Little Rock, AR, 4.

Andelman, J.B. and D.W. Underhill. 1987. *Health Effects from Hazardous Waste Sites*. Lewis Publishers, Boca Raton, FL.

Anderton, D., A. Anderson, J. Oakes, and M. Fraser. 1994a. Environmental equity: The demographics of dumping. *Demography* 31:229–248.

Anderton, D., A. Anderson, P. Rossi, J. Oakes, M. Fraser, E. Weber, and E. Calabrese. 1994b. Environmental equity issues in metropolitan areas. *Evaluation Rev.* 18:123.

Andrews, L.P. 1990. *Worker Protection During Hazardous Waste Remediation*. Van Nostrand Reinhold, New York, 6.

Antonio, P., S. Edison, E. Esplain, D. Malone, J. Manygoats, P. Molloy, D. Moore, J. Morris, and G. Rajen. 1992. Assessments of hazards posed by abandoned uranium mines on the Navajo Nation. In: *National Minority Health Conference: Focus on Environmental Contamination*, B. Johnson, R. Williams, and C. Harris, Eds. Princeton Scientific Publishing, Princeton, NJ, pp. 37–51.

Aral, M.M., M.L. Maslia, G.V. Ulirsch, and J.J. Reyes. 1996. Estimating exposure to volatile organic compounds from municipal water-supply systems: Use of a better computational model. *Arch. Environ. Health* 51:300–309.

Arizona. 1995. *Arizona Comparative Environmental Risk Project*. Governor's Office, Phoenix, AZ.

Aschengrau, A., S. Zierler, and A. Cohen. 1989. Quality of community drinking water and the occurrence of spontaneous abortion. *Arch. Environ. Health* 44:283–290.

Aschengrau, A., D. Ozonoff, C. Paulu, P. Coogan, R. Vezina, T. Heeren, and Y. Zhang. 1993. Cancer risk and tetrachloroethylene-contaminated drinking water in Massachusetts. *Arch. Environ. Health* 48:284–291.

Ashley, D.L., M.A. Bonin, F.L. Cardinali, J.M. McGraw, and J.V. Wooten. 1994. Blood concentrations of volatile organic compounds in a nonoccupationally exposed U.S. population and in groups with suspected exposure. *Clin. Chem.* 40:1401–1404.

Assennato, G., P. Cannatelli, P. Emmett, I. Ghezzi, and F. Merlo. 1989. Medical monitoring of dioxin cleanup workers. *Am. Indust. Hyg. Assoc. J.* 50:586–592.

ASTSWMO (Association of State and Territorial Solid Waste Management Officials). 1994. *A Report on State/Territory Non-NPL Hazardous Waste Site Cleanup Efforts for the Period 1980–1992*. U.S. Environmental Protection Agency, Washington, DC.

ATSDR (Agency for Toxic Substances and Disease Registry). 1988. *The Nature and Extent of Lead Poisoning in Children in the United States: A Report to Congress*. Atlanta, GA.

ATSDR. 1989a. *Final Report of the Crystal Chemical Arsenic Exposure Study*. Division of Health Studies, Atlanta.

ATSDR. 1989b. *Public Health Advisory: Forest Glen Mobile Home Park, Niagara Falls, New York.* Office of Health Assessment, Division of Health Assessment and Consultation, Atlanta.

ATSDR. 1992a. Health assessment name change. *Fed. Reg.* 57:9259–9260.

ATSDR. 1992b. *Health Assessment Guidance Manual.* Atlanta, 8-3-8-6.

ATSDR. 1992c. Notice of revised priority list of hazardous substances. *Fed. Reg.* 57:48801.

ATSDR. 1992d. *The Feasibility and Value of Performing Multisite Epidemiologic Studies for Superfund Sites.* Division of Health Studies, Atlanta.

ATSDR. 1992e. Hazardous waste sites: Priority health conditions and research strategies—United States. *Morb. Mort. Weekly Report* 41:72.

ATSDR. 1992f. Notice of the development of community assistance panels. *Fed. Reg.* 57FR 27779–27780, June 22.

ATSDR. 1993a. *Biennial Report to Congress.* Office of Policy and External Affairs, Atlanta.

ATSDR. 1993b. *Toxicological Profile for Lead—Update.* Division of Toxicology, Atlanta.

ATSDR. 1994. Priority list of hazardous substances that will be subject of toxicological profiles. *Fed. Reg.* 59:9486–9487.

ATSDR. 1995a. *Multisite Lead and Cadmium Exposure Study With Biological Markers Incorporated.* Division of Health Studies, Atlanta.

ATSDR. 1995b. *Hazardous Substances Emergency Events Surveillance: Annual Report.* Division of Health Studies, Atlanta.

ATSDR. 1995c. Final criteria for determining the appropriateness of a medical monitoring program under CERCLA. *Fed. Reg.* 60:38840–38844.

ATSDR. 1995d. *Mississippi Delta Project Prospectus.* Office of Urban Affairs, Atlanta.

ATSDR. 1996a. *FY 1995 Agency Profile and Annual Report.* Office of Policy and External Affairs, Atlanta.

ATSDR. 1996b. *Public Health Advisory for 722 Grant Street (A288), Hoboken, Hudson County, New Jersey.* Division of Health Assessment and Consultation, Atlanta.

ATSDR. 1996c. *ATSDR and EPA Issue Public Warning of an Emerging National Pattern of Illegal Use of Methyl Parathion. News Release.* Office of Policy and External Affairs, Atlanta.

ATSDR. 1996d. Update on the status of the Superfund Substance-Specific Applied Research Program. *Fed. Reg.* 61:14420–14427.

ATSDR. 1996e. *Pediatric Environmental Neurobehavioral Test Battery.* Division of Health Studies, Atlanta.

ATSDR. 1996f. Notice of the revised list of hazardous substances that will be the subject of toxicological profiles. *Fed. Reg.* 61:18744–18745.

ATSDR. 1996g. *A Cohort Study of Current and Previous Residents of the Silver Valley: Assessment of Lead Exposure and Health Outcomes.* Division of Health Studies, Atlanta.

ATSDR. 1996h. *Hazardous Substances Emergency Events Surveillance (HSEES): Annual Report.* Division of Health Studies, Atlanta.

ATSDR. 1996i. *Guidelines for Conducting Health Studies.* Division of Health Studies, Atlanta.

ATSDR. 1996j. *Community Health Education Project: Saltville, Virginia.* Division of Health Education and Promotion. Atlanta.

ATSDR. 1997a. Announcement of final data needs for 12 priority hazardous substances and call for voluntary research proposals. *Fed. Reg.* 62:40820–40828.

ATSDR. 1997b. Notice of the revised priority list of hazardous substances that will be the subject of toxicological profiles. *Fed. Reg.* (62 FR 61332), Nov. 11.

ATSDR. 1997c. *Report from the Mixtures Expert Panel Review Meeting.* Division of Toxicology, Atlanta.

ATSDR. 1997d. *Volatile Organic Compounds in Drinking Water and Adverse Pregnancy Outcomes.* Interim Report. Division of Health Studies, Atlanta.

Au, W.W., R.G. Lane, M.S. Legator, E.B. Whorton, G.S. Wilkinson, and G.J. Gabehart. 1995. Biomarker monitoring of a population residing near uranium mining activities. *Environ. Health Perspect.* 103:466–470.

Auer, C. 1994. Notice of opportunity to initiate negotiations for TSCA Section 4 enforceable consent agreements. *Fed. Reg.* 59:49934–49938.

Bachrach, K.M. and A.J. Zautra. 1985. Assessing the impact of hazardous waste facilities: Psychology, politics, and environmental impact statements. In: *Advances in Environmental Psychology, Vol. 5*, A. Baum and J.E. Singer, Eds. Lawrence Erlbaum Assoc, Hillsdale, NJ, pp. 185–205.

Baden, B. and D. Coursey. 1997. *The Locality of Waste Sites Within the City of Chicago: A Demographic, Social, and Economic Analysis.* Harris School of Public Policy Studies, University of Chicago, Chicago.

Baker, D.B., S. Greenland, J. Mendlein, and P. Harmon. 1988. A health study of two communities near the Stringfellow waste disposal site. *Arch. Environ. Health* 43:325–334.

Balch, G.I. and S.M. Sutton. 1995. Putting the audience first: Conducting useful evaluation for a risk-related government agency. *Risk Anal.* 15:163–168.

Barry, D. 1996. Personal communication. Agency for Toxic Substances and Disease Registry, Atlanta, GA.

Batstone, R., J.E. Smith, Jr., and D. Wilson. 1989. *The Safe Disposal of Hazardous Wastes: The Special Needs and Problems of Developing Countries, Vol. I.* The World Bank, Washington, DC.

Baum, A., R. Fleming, and J.E. Singer. 1985. Understanding environmental stress: Strategies for conceptual and methodological integration. In: *Advances in Environmental Psychology, Vol. 5*, A. Baum and J.E. Singer, Eds. Lawrence Erlbaum Associates, Hillsdale, NJ, pp. 185–205.

Baum, A., I. Fleming, A. Israel, and M.K. O'Keefe. 1992. Symptoms of chronic stress following a natural disaster and discovery of a human-made hazard. *Environ. Behavior* 24:347–365.

Beaglehole, R., R. Bonita, and T. Kjellström. 1993. *Basic Epidemiology.* World Health Organization, Geneva.

Been, V. 1993. What's fairness got to do with it? Environmental justice and the siting of locally undesirable land uses. *Cornell Law Rev.* 78:1001–1085.

Been, V. 1994. Locally undesirable land uses in minority neighborhoods: Disproportionate siting or market dynamics? *Yale Law J.* 103:1383–1423.

Been, V. 1995. Analyzing evidence of environmental justice. *J. Land Use Environ. Law* 11:1–36.

Bender, M. and R. Preston. 1983. Cytogenetic patterns in persons living near Love Canal, New York. *Morb. Mort. Weekly Report* 32:261–262.

Berry, M. 1994. *Lipari Landfill Birth Weight Study: A 25-Year Trend Analysis.* New Jersey Health Department, Trenton.

Berry, M. and F. Bove. 1997. Birth weight reduction associated with residence near a hazardous waste landfill. *Environ. Health Perspect.* 105:866–861.

Binder, S., D. Forney, W. Kaye, and D. Paschal. 1987. Arsenic exposure in children living near a former copper smelter. *Bull. Environ. Contam. Toxicol.* 39:114–121.

BNA (Bureau of National Affairs). 1992. Senate ratifies Basel Convention on transboundary shipments of waste. *Environ. Reporter*, August 21.

BNA. 1996. Charges filed against two men in Mississippi methyl parathion case. *BNA Chem. Regulat. Daily*, November 26.

BOC. (Bureau of Census). 1992. *Census of Population and Housing, 1990: Summary Tape File 3 on CD-ROM Technical Documentation.* Department of Commerce, Washington, DC.

Bove, F.J., M.C. Fulcomer, J.B. Klotz, J. Esmart, E.M. Dufficy, and J.E. Savrin. 1995. Public drinking water contamination and birth outcomes. *Am. J. Epidemiol.* 141:850–862.

Brown, P. 1995. Race, class, and environmental health: A review and systematization of the literature. *Environ. Res.* 69:15–30.

Browner, C.M. 1995. *Prepared Statement of Carol M. Browner, Administrator, U.S. Environmental Protection Agency, Before the House Committee on Transportation and Infrastructure, Water Resources and Environment Subcommittee.* Washington, DC, U.S. House of Representatives, June 27.

Browner, C.M. 1997. *Prepared Statement of Carol M. Browner, Administrator, U.S. Environmental Protection Agency, Before the Committee on Environment and Public Works.* Washington DC, U.S. Senate, March 5.

Bryant, B. and P. Mohai. 1992. The Michigan conference: A turning point. *EPA J.*, March/April 9–10.

Budnick, L.D., D.C. Sokal, H. Falk, J.N. Logue, and J.M. Fox. 1984. Cancer and birth defects near the Drake Superfund site, Pennsylvania. *Arch. Environ. Health* 39:409–413.

Buffler, P.A., M. Crane, and M.M. Key. 1985. Possibilities of detecting health effects by studies of populations exposed to chemicals from waste disposal sites. *Environ. Health Perspect.* 62:423–456.

Bullard, R.D. 1990. *Dumping in Dixie: Race, Class, and Environmental Quality.* Westview Press, Boulder, CO.

Bullard, R.D. 1992. Use of demographic data to evaluate minority environmental health issues: The role of case studies. In: *National Minority Health Conference: Focus on Environmental Contamination*, B. Johnson, R. Williams, and C. Harris, Eds. Princeton Scientific Publishing Co., Princeton, NJ, pp. 161–171.

Bullard, R.D., Ed. 1994. *Unequal Protection: Environmental Justice and Communities of Color.* Sierra Club, San Francisco.

Bullard, R.D. 1995. *Environmental justice: Strategies for achieving health and sustainable communities.* Paper presented at the International Congress on Hazardous Waste and Public Health and Ecology, Atlanta, GA, June.

Burg, J.R. 1996. Personal communication. Division of Health Studies, Agency for Toxic Substances and Disease Registry, Atlanta.

Burg, J.R. and G.L. Gist. 1998. The national exposure registry: Analysis of health outcomes from the benzene subregistry. *Toxicol. Indust. Health* 14:367–387.

Burg, J.R., G.L. Gist, S.L. Allred, T.M. Radtke, L.L. Pallos, and C.D. Cusack. 1995. The national exposure registry—Morbidity analyses of noncancer outcomes from the trichloroethylene subregistry baseline data. *Int. J. Occupat. Med. Toxicol.* 4:237–257.

Byers, V.S., A.S. Levin, D.M. Ozonoff, and R.W. Baldwin. 1988. Association between clinical symptoms and lymphocyte abnormalities in a population with chronic domestic exposure to industrial solvent-contaminated domestic water supply and a high incidence of leukaemia. *Cancer Immunology Immunotherapy* 27:77–81.

Calabrese, E.J., E.J. Stanek, R.C. James, and S.M. Roberts. 1997. Soil ingestion: A concern for acute toxicity in children. *Environ. Health Perspect.* 105:1354–1358.

Calderon, R.L., C.C. Johnson, G.F. Craun, A.P. Dufour, R.J. Karlin, T. Sinks, and J.L. Valentine. 1993. Health risk from contaminated water: Do class and race matter? *Toxicol. Indust. Health* 9:879–900.

Cantor, K., R. Hoover, P. Hartge, T.J. Mason, D.T. Silverman, R. Altman, D.F. Austin, M.A. Child, C.R. Key, L.D. Marrett, M.H. Myers, A.S. Narayana, L.I. Levin, J.W. Sullivan, G.M. Swanson, D.B. Thomas, and D.W. West. 1987. Bladder cancer, drinking water source, and tap water consumption: A case-control study. *J. Natl. Cancer Inst.* 79:1269–1280.

CBO (Congressional Budget Office). 1994. *The Total Costs of Cleaning up Nonfederal Superfund Sites.* U.S. Government Printing Office, Washington, DC.

CCI (Commission on Chronic Illness). 1957. *Chronic Illness in the United States, Vol. 1.* Commonwealth Fund, Harvard University Press, Cambridge, MA.

CDC. (Centers for Disease Control and Prevention). 1991. *Preventing Lead Poisoning in Young Children.* Atlanta.

CDC. 1997. Update: Blood lead levels—United States, 1991–1994. *Morb. Mort. Weekly Report* 46:141–146.

CDHS (California Department of Health Services). 1985. *Cardiac Defects in Relation to Water Contamination, 1981–1982, Santa Clara County, California; Pregnancy Outcomes in Relation to Water Contamination, 1980–1981, San Jose, California.* Berkeley, CA.

CDOH (Colorado Department of Health). 1989. *Environmental Status Report.* Denver, CO.

CEPA (California Environmental Protection Agency). 1994. *Toward the 21st Century: Planning for the Protection of California's Environment.* California Public Health Foundation, Sacramento.

CEQ (Council on Environmental Quality). 1989. *Risk Analysis: A Guide to Principles and Methods for Analyzing Health and Environmental Risks.* Washington, DC.

Chess, C. and B.J. Hance. 1994. *Communicating with the Public: Ten Questions Environmental Managers Should Ask* (brochure). Center for Environmental Communications, Cook College, Rutgers University, Piscataway, NJ.

Chess, C., B.J. Hance, and P.M. Sandman. 1988. *Improving Dialogue with Communities: A Short Guide to Government Risk Communication.* New Jersey Department of Environmental Protection, Trenton.

Chess, C., K.L. Salomone, and B.J. Hance. 1995. Improving risk communication in government: Research priorities. *Risk Anal.* 15:127–135.

Chou, S. and L. Hutchinson. 1993. Neurotoxic disorders. In: *Priority Health Conditions*, J.A. Lybarger, R.F. Spengler, and C.T. DeRosa, Eds. Agency for Toxic Substances and Disease Registry, Atlanta, pp. 161–193.

Chou, S. and S. Metcalf. 1993. Immune function disorders. In: *Priority Health Conditions*, J.A. Lybarger, R.F. Spengler, and C.T. DeRosa, Eds. Agency for Toxic Substances and Disease Registry, Atlanta, pp. 67–88.

Clinton, W.J. 1994. Federal actions to address environmental justice in minority populations and low-income populations. *Fed. Reg.* 59:7629–7633.

Cohn, P., F. Bove, J. Klotz, M. Berkowitz, and J. Fagliano. 1994. Drinking water contamination and the incidence of leukemia and non-Hodgkin's lymphoma. In: *Hazardous Waste and Public Health*, J. Andrews, H. Frumpkin, B. Johnson, M. Mehlman, C. Xintaras, and J. Bucsela, Eds. Princeton Scientific Publishing, Princeton, NJ, pp. 742–756.

Cole, H. 1996. *Learning from Success: Health Agency Effort to Improve Community Involvement in Communities Affected by Hazardous Waste Sites.* Henry S. Cole & Associates, Washington, DC.

Constan, A.A., R.S. Yang, D.C. Baker, and S.A. Benjamin. 1995. A unique pattern of hepatocyte proliferation in F344 rats following long-term exposures to low levels of a chemical mixture of groundwater contaminants. *Carcinogenesis* 16:303–310.

Constan, A.A., S.A. Benjamin, J.D. Tessari, D.C. Baker, and R.S. Yang. 1996. Increased rate of apoptosis correlates with hepatocellular proliferation in Fisher-344 rats following long-term exposure to a mixture of groundwater contaminants. *Toxicol. Path.* 24:315–322.

Cook, M., W.R. Chappell, R.E. Hoffman, and E.J. Mangione. 1993. Assessment of blood lead levels in children living in a historic mining and smelting community. *Am. J. Epidemiol.* 137:447–455.

Corn, M. and P.N. Breysse. 1983. Human exposure estimates for hazardous waste site risk assessment. In: *Risk Quantitation and Regulatory Policy*, D.G. Hoel, R.A. Merrill, and F.P. Perera, Eds. Cold Spring Harbor Laboratory, pp. 283–291.

COSSMHO (Coalition of Hispanic Health and Human Services Organizations). 1996. Hispanic environmental health: Ambient and indoor air pollution. *Otolaryngol. Head Neck Surg.* 114:256–264.

Covello, V. and F. Allen. 1988. *Seven Cardinal Rules of Risk Communication.* U.S. Environmental Protection Agency, Office of Policy Analysis, Washington, DC.

Covello, V. and R.G. Peters. 1996. The determinants of trust and credibility in environmental risk communication: An empirical study. In: *Scientific Uncertainty and Its Influence on the Public Communication Process*, V.H. Sublet, V.T. Covello, and T.L. Tinker, Eds. Kluwer Academic Publishers, Dordrecht, pp. 33–64.

Covello, V.T., W.G. Flamm, J.V. Rodricks, and R.G. Tardiff. 1983. *The Analysis of Actual Versus Perceived Risks*. Plenum Press, New York.

Covello, V.T., D.B. McCallum, and M.T. Pavlova. 1987. *Effective Risk Communication*. Plenum Press, New York.

CRARM (Commission on Risk Assessment and Risk Management). 1997. *Framework for Environmental Health Risk Management*. U.S. Environmental Protection Agency, Washington, DC.

Croen, L.A., G.M. Shaw, L. Sanbonmatsu, S. Selvin, and P.A. Buffler. 1997. Maternal residential proximity to hazardous waste sites and risk for selected congenital malformations. *Epidemiol.* 8:347–354.

CSJ (Commission on Social Justice). 1987. *Toxic Waste and Race in the United States*. United Church of Christ, New York.

Cutler, J.J., G.S. Parker, S. Rosen, B. Prenney, R. Healey, and G.C. Caldwell. 1986. Childhood leukemia in Woburn, Massachusetts. *Public Health Reports* 101:201–205.

Dawson, B.V., P.D. Johnson, S.J. Goldberg, and J.B. Ulreich. 1990. Cardiac teratogenesis of trichloroethylene and dichloroethylene in a mammalian model. *J. Am. Coll. Cardiol.* 16:1304–1309.

Deane, M., S.H. Swan, J.A. Harris, D.M. Epstein, and R.R. Neutra. 1989. Adverse pregnancy outcomes in relation to water contamination, Santa Clara County, California, 1980–1981. *Am. J. Epidemiol.* 129:894–904.

Deloraine, A., D. Zmirou, C. Tillier, A. Boucharlat, and H. Bouti. 1995. Case-control assessment of the short-term health effects of an industrial toxic waste landfill. *Environ. Res.* 68:124–132.

DEQ (Department of Environmental Quality). 1991a. *Leap to 2000, Louisiana's Environmental Action Plan*. Baton Rouge, LA.

DEQ. 1991b. *Leap to 2000, Louisiana's Environmental Action Plan. Technical Supplement Appendix D: Compilation of Health Effect Assessment Methodology and Issue Analysis Reports*. Baton Rouge, LA.

DeRosa, C.T., Y.-W. Stevens, and B.L. Johnson. 1993. Cancer policy framework for public health assessment of carcinogens in the environment. *Toxicol. Indust. Health* 9:559–575.

Detels, R. and L. Breslow. 1991. Current scope and concerns in public health. In: *Oxford Textbook of Public Health*, 2nd ed., Vol. 1. W.W. Holland, R. Detels, and G. Knox, Eds. Oxford Medical Publications, Oxford University Press, Oxford.

DHHS (U.S. Department of Health and Human Services). 1985. *Risk Assessment and Risk Management of Toxic Substances*. Technical Report from the Committee to Coordinate Environmental Health and Related Programs, Washington, DC.

DHHS. 1991. *Healthy People 2000: National Health Promotion and Disease Prevention Objectives*. Washington, DC.

DHHS. 1994. *Case Studies of Applied Evaluation for Health Risk Communication*. Subcommittee on Risk Communication and Education, Environmental Health Policy Committee, Department of Health and Human Services, Washington, DC, 6.

Díaz-Barriga, F., M. Santos, L. Yanez, J. Cuellar, P. Ostrosky-Wegman, R. Montero, A. Perez, E. Ruiz, A. Garcia, and H. Gomez. 1993. Biological monitoring of workers at a recently opened hazardous waste disposal site. *J. Exposure Anal. Environ. Epidemiol.* 3(Supp 1):63–71.

Dicker, R.C. 1996. Analyzing and Interpreting Data. In: *Field Epidemiology*, M.B. Gregg, Ed. Oxford University Press, New York.

DOJ (U.S. Department of Justice). 1996. *"Superfund" Fact Sheet*. Department of Justice, Washington, DC.

Doll, R. and R. Peto. 1981. The causes of cancer: Quantitative estimates of avoidable risks of cancer in the United States. *J. Nat. Cancer Inst.* 66:1191–1308.

Dunham, F. 1995. *Testimony of the citizens against toxic exposure given during hearings before the Subcommittee on Water Resources and Environment*. Committee on Transportation and Infrastructure, U.S. House of Representatives, Washington, DC, 1432–1440.

Dunne, M.P., P. Burnett, J. Lawton, and B. Raphael. 1990. The health effects of chemical waste in an urban community. *Med. J. Australia* 152:592–597.

Eaton, D.L. and C.D. Klaassen. 1996. Principles of Toxicology. In: *Casarett and Doull's Toxicology*, 5th ed., C.D. Klaassen, Ed. McGraw-Hill, New York.

EDF (Environmental Defense Fund). 1997. *Toxic Ignorance: The Continuing Absence of Basic Health Testing for Top-Selling Chemicals in the United States*. Washington, DC.

ELI (Environmental Law Institute). 1989. *Environmental Law Deskbook*. Environmental Law Institute, Washington, DC, pp. 166,203,221,245.

ELI. 1992. *European Community Deskbook*. Environmental Law Deskbook. Environmental Law Institute, Washington, DC, p. 357.

Elliott, P., M. Hills, J. Beresford, I. Kleinschmidt, D. Jolley, S. Pattenden, L. Rodrigues, A. Westlake, and G. Rose. 1992. Incidence of cancer of the larynx and lung near incinerators of waste solvents and oils in Great Britain. *Lancet* 339:854–858.

Elliott, P., G. Shaddick, I. Kleinschmidt, D. Jolley, P. Walls, J. Beresford, and C. Grundy. 1996. Cancer incidence near municipal solid waste incinerators in Great Britain. *Brit. J. Cancer* 73:702–710.

el-Masri, H.A., K.F. Reardon, and R.S. Yang. 1997. Integrated approaches for the analysis of toxicologic interactions of chemical mixtures. *Crit. Rev. Toxicol.* 27:175–197.

English, P.B., G.M. Shaw, G.C. Windham, and R.R. Neutra. 1989. Illness and absenteeism among California highway patrol officers responding to hazardous material spills. *Arch. Environ. Health* 44:117–119.

EPA (U.S. Environmental Protection Agency). 1986. *Air Quality Criteria for Lead*. EPA Report EPA-600/83/028aF-Df, Environmental Criteria and Assessment Office, Research Triangle Park, NC.

EPA. 1987. *Unfinished Business: A Comparative Assessment of Environmental Problems*. U.S. Environmental Protection Agency, Washington, DC.

EPA. 1989. *Superfund, Emergency Planning, and Community Right-To-Know Programs, Worker Protection*. Office of Solid Waste and Emergency Response, Washington, DC.

EPA. 1992a. *Environmental Equity: Reducing Risk for All Communities*, Vol. 1 and 2. Office of Solid Waste and Emergency Response, Washington, DC, pp. 2, 8, 9.

EPA. 1992b. *Tribes at Risk: The Wisconsin Tribes Comparative Risk Project*. Report 230-R-92-017. Office of Policy, Planning, and Evaluation, Washington, DC.

EPA. 1992c. *Superfund Progress—Aficionado's Version*. Report PB92-963267. Office of Solid Waste and Emergency Response, Washington, DC.

EPA. 1995. *Guidance for Community Advisory Groups at Superfund Sites*. Report EPA 540-k-96-001. Office of Solid Waste and Emergency Response, Washington, DC.

EPA. 1996a. *Comprehensive Environmental Response, Compensation, and Liability Information System (CERCLIS)*. Office of Solid Waste and Emergency Response, Washington, DC, December.

EPA. 1996b. National priorities list for uncontrolled hazardous waste sites: Final rule. *Fed. Reg.* 61:67655–67677.

EPA. 1996c. National priorities list for hazardous waste sites: Proposed rule. *Fed. Reg.* 61:67678–67682.

EPA. 1996d. *EPA Region 4 Press Release*. U.S. Environmental Protection Agency, Region 4, Atlanta.

EPA. 1998a. Personal communication from R. Richardson. Office of Emergency Response and Remediation, Center for Cost Analysis, Washington, DC.

EPA. 1998b. Personal communication from S. Luftig. Office of Emergency Response and Remediation, Washington, DC.

Evans, R.G., K.B. Webb, A.P. Knutsen, S.T. Roodman, D.W. Roberts, J.R. Bagby, W.A. Garret, and J.A. Andrews. 1988. A medical follow-up of the health effects of long term exposure to 2,3,7,8-tetrachlorodibenzo-p-dioxin. *Arch. Environ. Health* 43:273–278.

Fagliano, J., M. Berry, F. Bove, and T. Burke. 1990. Drinking water contamination and the incidence of leukemia: An ecologic study. *Am. J. Public Health* 80:1209–1212.

Favata, E., S. Barnhart, E. Bresnitz, and V. Campbell. 1990. Clinical experiences: Development of a medical surveillance protocol for hazardous waste workers. *J. Occup. Med.* 5:117–126.

Fay, R.M. 1997. Personal communication. Division of Toxicology, Agency for Toxic Substances and Disease Registry, Atlanta.

FDEP (Florida Department of Environmental Protection). 1995. *Comparing Florida's Environmental Risks.* Tallahassee, September, pp. 42, 102, 115–119.

FDHRS (Florida Department of Health and Rehabilitative Services). 1996. *Public Health Assessment of the Escambia Wood-Pensacola Site.* Agency for Toxic Substances and Disease Registry, Atlanta.

Feese, D., R.J. Dhara, and F.L. Stallings. 1996. *Lead and Cadmium Exposure Study, Galena, Kansas.* Final Report. Division of Health Studies, Agency for Toxic Substances and Disease Registry, Atlanta.

Feldman, R., J. Chirico-Post, and S. Proctor. 1988. Blink reflex latency after exposure to trichloroethylene in well water. *Arch. Environ. Health* 43:143–148.

FFPG (Federal Facilities Policy Group). 1995. *Improving Federal Facilities Cleanup.* Council on Environmental Quality, Washington, DC.

Fiedler, N., I. Udasin, M. Gochfeld, G. Buckler, K. Kelly-McNeil, and H. Kipen. 1996. *Medical and Neuropsychological Evaluation of Mercury-Exposed Grant Street Residents, Hoboken, New Jersey.* Unpublished Clinic Report. Environmental and Occupational Health Sciences Institute, Piscataway, NJ.

Fields, T., Jr. 1998. *Prepared Statement of Timothy Fields, Jr, Acting Assistant Administrator, U.S. Environmental Protection Agency, Before the House Subcommittee on Finance and Hazardous Materials.* Washington, DC, U.S. House of Representatives, February 4.

Fischhoff, B. 1995. Ranking risks. *Risk: Health, Safety Environ.* 6:191–202.

Fischhoff, B., S. Lichtenstein, P. Slovic, and D. Keeney. 1981. *Acceptable Risk.* Cambridge University Press, Cambridge, MA.

Fischhoff, B., P. Slovic, S. Lichtenstein, S. Read, and S. Combs. 1978. How safe is safe enough? A psychometric study of attitudes towards technological risks and benefits. *Policy Sci.* 9:127–152.

Fischhoff, B., S. Watson, and C. Hope. 1984. Defining risk. *Policy Sci.* 17:123–139.

Flynn, J., P. Slovic, and C.K. Mertz. 1994. Gender, race, and perception of environmental health risks. *Risk Anal.* 14:1101–1108.

Foreman, C.H., Jr. 1996. A winning hand? The uncertain future of environmental justice. *Brookings Rev.,* Spring, pp. 22–25.

GAO (General Accounting Office). 1983. *Siting of Hazardous Waste Landfills and Their Correspondence with Racial and Economic Status of Surrounding Communities.* General Accounting Office, Washington, DC.

GAO. 1991. *Superfund Public Health Assessments Incomplete and of Questionable Value.* General Accounting Office, Washington, DC.

GAO. 1992. *Superfund: Problems with the Completeness and Consistency of Site Cleanup Plans.* General Accounting Office, Washington, DC.

GAO. 1995a. *Operating and Maintenance Activities Will Require Billions of Dollars*. General Accounting Office, Washington, DC.

GAO. 1995b. *Superfund: Information on Current Health Risks*. General Accounting Office, Washington, DC.

GAO. 1995c. *Demographics of People Living Near Waste Facilities*. General Accounting Office, Washington, DC.

GAO. 1996. *Federal Facilities: Consistent Relative Risk Evaluations Needed for Prioritizing Cleanups*. General Accounting Office, Washington, DC.

GAO. 1997. *Times to Assess and Clean Up Hazardous Waste Sites Exceed Program Goals*. General Accounting Office, Washington, DC.

Germolec, D.R., R.S.H. Yang, M.F. Ackermann, G.J. Rosenthal, G.A. Boorman, P. Blair, and M.I. Luster. 1989. Toxicology studies of a chemical mixture of 25 groundwater contaminants: II. Immunosuppression in B6C3F1 mice. *Fundam. Appl. Toxicol.* 13:377–387.

Geschwind, S.A., J.A.J. Stolwijk, M. Bracken, E. Fitzgerald, A. Stark, C. Olsen, and J. Melius. 1992. Risk of congenital malformations associated with proximity to hazardous waste sites. *Am. J. Epidemiol.* 135:1197–1207.

Gibbs, L.M. 1981. The need for effective governmental response to hazardous waste sites. *J. Public Health Policy* 2:42–48.

Gill, D.A. and J.S. Picou. 1991. The social psychological impacts of a technological accident: Collective stress and perceived health risks. *J. Hazard. Mat.* 27:77–89.

Gist, G.L. and J.R. Burg. 1995. Methodology for selecting substances for the National Exposure Registry. *J. Expos. Anal. Environ. Epidemiol.* 5:197–208.

Gist, G.L., J.R. Burg, and T.M. Radtke. 1994. The site selection process for the National Exposure Registry. *J. Environ. Health* 56:7–12.

Gochfeld, M. 1990. Medical Surveillance of Hazardous Waste Workers. In: *Principles and Problems in Occupational Medicine State of Art Reviews: Hazardous Waste Workers*, M. Gochfeld and E.A. Favata, Eds. Hanley & Belfus, Inc., 5(1):1–8.

Gochfeld, M. 1995. Incineration: Health and environmental consequences. *Mt. Sinai J. Med.* 62:365–374.

Goldberg, M.S., L. Goulet, H. Riberdy, and Y. Bonvalot. 1995a. Low birth weight and preterm births among infants born to women living near a municipal solid waste landfill site in Montreal, Quebéc. *Environ. Res.* 69:37–50.

Goldberg, M.S., N. Al-Homsi, L. Goulet, and H. Riberdy. 1995b. Incidence of cancer among persons living near a municipal solid waste landfill in Montreal, Quebéc. *Arch. Environ. Health* 50:416–424.

Goldberg, S.J., M.B. Lebowitz, E.J. Graver, and S. Hicks. 1990. An association of human congenital cardiac malformations and drinking water contaminants. *J. Am. Coll. Cardiol.* 16:155–164.

Golding, J. 1997. Unnatural constituents of breast milk–medication, lifestyle, pollutants, viruses. *Early Hum. Dev.* 49 (Suppl):S29–43.

Goodman, R.A. and J.V. Peavy. 1996. Describing epidemiologic data. In: *Field Epidemiology*, M.B. Gregg, Ed. Oxford University Press, New York, pp. 60–81.

Gordon, L.J. and D.R. McFarlane. 1996. Public health practitioner incubation plight: Following the money trail. *J. Public Health Policy* 17:59–71, 68.

Gottlieb, K. and J.R. Koehler. 1994. Blood lead levels in children from lower socioeconomic communities in Denver, Colorado. *Arch. Environ. Health* 49:260–266.

Green, B.L., J.D. Lindy, and M.C. Grace. 1994. Psychological Effects of Toxic Contamination. In: *Industrial and Community Responses to Trauma and Disasters: The Structure of Human Chaos*, R.J. Ursano, B.G. McCaughey, and C.S. Fullerton, Eds. Cambridge University Press, New York, pp. 154–176.

Gregg, M.B., R.C. Dicker, and R.A. Goodman, Eds. 1996. *Field Epidemiology*. Oxford University Press, New York.

Griffith, J., R.C. Duncan, W.B. Riggan, and A.C. Pellom. 1989. Cancer mortality in U.S. counties with hazardous waste sites and ground water pollution. *Arch. Environ. Health* 44:69–74.

Grisham, J., Ed. 1986. *Health Aspects of the Disposal of Waste Chemicals*. Pergamon Press, New York.

Gurunathan, S., M. Robson, B. Freeman, B. Buckley, A. Roy, R. Meyer, J. Burkowski, and P.J. Lioy. 1998. Accumulation of chlorpyrifos on residential surfaces and toys accessible to children. *Environ. Health Perspect.* 106:9–16.

Gustavsson, P. 1989. Mortality among workers at a municipal waste incinerator. *Am. J. Indust. Med.* 15:245–253.

Habicht, H. 1991. Plans, problems, and promises for the years ahead. *Hazard. Mat. Control* 60:3–8.

Hall, H.I., V.R. Dhara, W.E. Kaye, and P.A. Price-Green. 1994. Surveillance of hazardous substance releases and related health effects. *Arch. Environ. Health* 49:45–48.

Hall, H.I., V.R. Dhara, P.A. Price-Green, W.E. Kaye, and G.S. Haugh. 1996b. Risk factors for hazardous substance releases that result in injuries and evaluations: Data from 9 states. *Am. J. Public Health* 86:855–857.

Hall, H.I., W.E. Kaye, L.S. Gensburg, and E.G. Marshall. 1996a. Residential proximity to hazardous waste sites and risk of end-state renal disease. *J. Environ. Health* September:17–22.

Hall, H.I., P.A. Price-Green, V.R. Dhara, and W.E. Kaye. 1995. Health effects related to releases of hazardous substances on the Superfund priority list. *Chemosphere* 31:2455–2461.

Hamar, G.B., M.A. McGeehin, and B.L. Phifer. 1996. *Symptom and Disease Prevalence with Biomarkers Health Study, Cornhusker Army Ammunition Plant, Hall County, Nebraska*. Final Report, Division of Health Education, Agency for Toxic Substances and Disease Registry, Atlanta.

Hammad, T.A., M. Sexton, and P. Langenberg. 1996. Relationship between blood lead and dietary iron intake in preschool children. *Ann. Epidemiol.* 6:30–33.

Harris, C. 1994. *Memorandum: Draft final report on similarities and differences between ATSDR public health assessments and USEPA baseline risk assessments under Superfund*. Division of Health Assessment and Consultation, Agency for Toxic Substances and Disease Registry, Atlanta, December 9.

Hartsough, D.M. and J.C. Savitsky. 1984. Three Mile Island: Psychology and environmental policy at a crossroads. *Am. Psychologist* 39:1113–1122.

HDOH (Hawaii Department of Health). 1992. *Environmental Risks to Hawaii's Public Health and Ecosystems*. Honolulu.

Heath, C.W., Jr. 1983. Field epidemiologic studies of populations exposed to waste dumps. *Environ. Health Perspect.* 48:3–7.

Heath, C.W., M.R. Nadel, M.M. Zack, A.T.L. Chen, M.A. Bender, and J. Preston. 1984. Cytogenetic findings in persons living near the Love Canal. *J. Am. Med. Assoc.* 251:1437–1440.

Heitgerd, J.L., J.R. Burg, and H.G. Strickland. 1995. A geographic information systems approach to estimating and assessing National Priorities List site demographics: Racial and Hispanic origin composition. *Internat. J. Occupat. Med. Toxicol.* 4:343–363.

Hensyl, W.R., Ed. 1987. *Stedman's Pocket Medical Dictionary*. Williams & Wilkins, Baltimore, MD.

Hernan, R.E. 1994. A state's right to recover punitive damages in a public nuisance action: The Love Canal case study. *Environ. Law J.* 1:45–110.

Herrman, D., P.G. Rollins, E.D. Rhoades, and K. Cadaret. 1997. *Ottawa County Blood Lead Testing Project*. Oklahoma State Department of Health, Oklahoma City.

Hertzman, C., M. Hayes, J. Singer, and J. Highland. 1987. Upper Ottawa Street landfill site health study. *Environ. Health Perspect.* 75:173–195.

Hicks, H.H., G. Terracciano, and B. Hamar. 1993. Birth defects and reproductive disorders. In: *Priority Health Conditions*, J.A. Lybarger, R.F. Spengler, C.T. DeRosa, Eds. Agency for Toxic Substances and Disease Registry, Atlanta, GA, pp. 24, 13–48.

Hill, A.B. 1965. The environment and disease: Association or causation? *Proc. Royal Soc. Med.* 58:295–300.

Hill, R.H., Jr., S.L. Head, S. Baker, C. Rubin, E. Esteban, S.L. Bailey, D.B. Shealy, and L.L. Needham. 1995. The use of reference range concentrations in environmental health investigations. *Environ. Res.* 71:99–108.

Hird, J. 1993. Environmental policy and equity: The case of Superfund. *J. Policy Anal. Manage.* 12:323–343.

Hodgkinson, H. 1992. *A Demographic Look At Tomorrow*. Institute for Educational Leadership, Washington, DC.

Hodgson, E. and P.E. Levi. 1987. *A Textbook of Modern Toxicology*. Elsevier, New York, pp. 274, 361.

Hoffman, R.E., P.A. Stehr-Green, K.B. Webb, R.G. Evans, A.P. Knutsen, W.F. Schramm, J.L. Staake, B.B. Gibson, and K.K. Steinberg. 1986. Health effects of long-term exposure to 2,3,7,8-tetrachlorodibenzo-*p*-dioxin. *J. Am. Med. Assoc.* 255:2031–2038.

Hoskin, A.F., J.P. Leigh, and T.W. Planek. 1994. Estimated risk of occupational fatalities associated with hazardous waste site remediation. *Risk Anal.* 14:1011–1017.

Hutchinson, L.J. 1992. *Investigation of a Cluster of Pancreatic Cancer Deaths, Livingston and Park County, Montana*. Atlanta.

Ikatsu, H., T. Nakajima, T. Okino, and N. Murayama. 1989. Health care of workers engaged in waste water treatment. *Sangyo Igaku* 31:355–362.

IOM (Institute of Medicine). 1988. *The Future of Public Health*. National Academy Press, Washington, DC, pp. 38–42.

Isacson, P., J. Bean, R. Splinter, D. Olson, and J. Kohler. 1985. Drinking water and cancer incidence in Iowa. *Am. J. Epidemiol.* 121:856–869.

Jacobson, J.L. and S.W. Jacobson. 1996. Intellectual impairment in children exposed to polychlorinated biphenyls in utero. *New Engl. J. Med.* 335:783–789.

Janerich, D.T., W.S. Burnett, G. Feck, M. Hoff, P. Nasca, A.P. Polednak, P. Greenwald, and N. Vianna. 1981. Cancer incidence in the Love Canal area. *Science* 212:1404–1407.

Jansson, B. and L. Voog. 1989. Dioxin from Swedish municipal incinerators and the occurrence of cleft lip and palate malformations. *Intern. J. Environ. Studies* 34:99–104.

Johnson, B.L. 1990. Implementation of Superfund's health-related provisions by the Agency for Toxic Substances and Disease Registry. *Environ. Law Reporter* 20:10277–10282.

Johnson, B.L. 1992. A précis on exposure assessment. *J. Environ. Health* 55:6–9.

Johnson, B.L. 1993. *Testimony to Subcommittee on Superfund, Recycling, and Solid Waste Management*. Senate, U.S. Congress, Washington, DC, May 6.

Johnson, B.L. 1995. *Testimony to the Subcommittee on Commerce, Trade, and Hazardous Materials*. U.S. House of Representatives, Committee on Commerce, Washington, DC, May 23.

Johnson, B.L. 1997. Hazardous waste: Human health effects. *Toxicol. Indust. Health* 13:121–144.

Johnson, B.L., E. Baker, M. El Batawi, R. Gilioli, H. Hänninen, A. Seppälainen, and C. Xintaras, Eds. 1987. *Prevention of Neurotoxic Illness in Working Populations*. John Wiley, New York.

Johnson, B.L., J. Boyd, J.R. Burg, S.T. Lee, C. Xintaras, and B.E. Albright. 1983. Effects on the peripheral nervous system of workers' exposure to carbon disulfide. *Neurotoxicol.* 4:53–66.

Johnson, B.L. and C. DeRosa. 1995. Chemical mixtures released from hazardous waste sites: Implication for health risk assessment. *Toxicol.* 105:145–156.

Johnson, B.L. and C. DeRosa. 1997. The toxicologic hazard of Superfund hazardous-waste sites. *Rev. Environ. Health* 12:235–251.

Johnson, B.L. and D. Weimer. 1995. Legal and International Considerations: One World. In: *Ecotoxicity and Human Health*, F. deSerres and A. Bloom, Eds. Lewis Publishers, Boca Raton, FL, pp. 271–313.

Johnson, B.L., R. Williams, and C. Harris, Eds. 1992. *National Minority Health Conference: Focus on Environmental Contamination*. Princeton Scientific Publishing, Princeton, NJ.

Jones, P.A. and M.A. McGeehin. 1996. *McClellan Air Force Base Cross-Sectional Health Study, Sacramento, Sacramento County, California*. Final Report, Division of Health Education, Agency for Toxic Substances and Disease Registry, Atlanta.

Kardestuncer, T. and H. Frumkin. 1997. Systemic lupus erythematosus in relation to environmental pollution: An investigation in an African-American community in North Georgia. *Arch. Environ. Health* 52:85–90.

Kavanaugh, K. 1996. Cleanup earns official praise: Agencies cooperated to rid homes of dangerous pesticide. *The Plain Dealer*, Cleveland, OH, April 17.

Kawamoto, M. 1992. *Health Hazard Evaluation: The Caldwell Group*. National Institute for Occupational Safety and Health, Centers for Disease Control and Prevention, Cincinnati.

Kiec-Swierczynska, M. 1989. Waste volatile ashes, a new source of hypersensitivity to metals. *Przegl. Dermatol.* 76:410.

Kilburn, K.H. and R.H. Warshaw. 1992. Prevalence of symptoms of systemic lupus erythematosus (SLE) and of fluorescent antinuclear antibodies associated with chronic exposure to trichloroethylene and other chemicals in well water. *Environ. Res.* 57:1–9.

Kilburn, K.H. and R.H. Warshaw. 1993a. Effects on neurobehavioral performance of chronic exposure to chemically contaminated well water. *Toxicol. Indust. Health* 9:391–404.

Kilburn, K.H. and R.H. Warshaw. 1993b. Neurobehavioral testing of subjects exposed residentially to groundwater contaminated from an aluminum die-casting plant and local referents. *J. Toxicol. Environ. Health* 39:483–496.

Kilburn, K.H. and R.H. Warshaw. 1995. Neurotoxic effects from residential exposure to chemicals from an oil reprocessing facility and Superfund site. *Neurotox. Toxicol.* 17:89–102.

Kilburn, K.H., R.H. Warshaw, and B. Hanscom. 1994. Balance measured by head (and trunk) tracking and a force platform in chemically (PCB and TCE) exposed and referent subjects. *Occupat. Environ. Med.* 51:381–385.

Kimbrough, R.D., M. LeVois, and D.R. Webb. 1994. Management of children with slightly elevated blood lead levels. *Pediatrics* 93:188–91.

Klaassen, C.D. 1996. *Casarett and Doull's Toxicology: The Basic Science of Poisons*. McGraw-Hill, New York.

Knox, E.G. and E.A. Gilman. 1997. Hazard proximities of childhood cancers in Great Britain from 1953–1980. *J. Epidemiol. Community Health* 51:151–159.

KNREPC (Kentucky Natural Resources and Environmental Protection Cabinet). 1995. *Kentucky Outlook 2000: Futures Comparative Risk*. Natural Resources and Environmental Protection Cabinet, Frankfort, KY.

Knutsen, A.P. 1984. Immunologic effects of TCDD exposure in humans. *Bull. Environ. Contam. Toxicol.* 33:673–681.

Koopman-Esseboom, C., N. Weisglas-Kuperus, M.A. de Ridder, G.C. Van der Paauw, L.G.M. Tuinstra, and P.J.J. Sauer. 1996. Effects of polychlorinated biphenyl/dioxin exposure and feeding type on infants' mental and psychomotor development. *Pediatrics* 97:700–706.

Krimsky, S. and D. Golding. 1992. *Social Theories of Risk*. Praeger-Greenwood, Westport, CT.

Lagakos, S.W., B.J. Wessen, and M. Zelen. 1986. An analysis of contaminated well water and health effects in Woburn, Massachusetts. *J. Am. Statistical Assoc.* 81:583–596.

Landrigan, P.J. 1983. Epidemiologic approaches to persons with exposures to waste chemicals. *Environ. Health Perspect.* 48:93–97.

Lautenberg, F. 1985. *Superfund Improvement Act of 1985*. 99th Congress, 1st Session. *Cong. Record* 131, Part 17:23946.

Lavelle, M. and M. Coyle. 1992. Unequal protection: The racial divide in environmental law. *National Law J.*, September 21.

Lee, V.L. 1997. *Study of Female Former Workers at a Lead Smelter: An Examination of the Possible Association of Lead Exposure with Decreased Bone Density*. Agency for Toxic Substances and Disease Registry, Atlanta.

Levitin, H. and H. Siegelson. 1996. Hazardous materials: Disaster medical planning and response. *J. Emergency Med. Clinics North Am.* 14:327–348.

Leviton, L.C., G.M. Marsh, E. Talbott, D. Pavlock, and C. Callahan. 1991. Drake Chemical workers' health registry: Coping with community tension over toxic exposures. *Am. J. Public Health* 81:689–693.

Lewit, E.M., L.S. Baker, H. Corman, and P.H. Shiono. 1995. The direct cost of low birth weight. *Future of Children* 5:35–57.

Lichtveld, M.Y. 1997. Personal communication. Division of Health Education and Promotion, Agency for Toxic Substances and Disease Registry, Atlanta.

Lichtveld, M.Y. and J.J. Clinton. 1997. Science put to service: Enhancing environmental health services in communities affected by hazardous substances. *Toxicol. Indust. Health* 13: 267–285.

Lioy, P.J. 1990. Assessing total human exposure to contaminants. *Environ. Sci. Technol.* 24:938–945.

Lioy, P.J., L. Wallace, and E. Pellizzari. 1991. Indoor/outdoor, and personal monitor and breath analysis relationships for selected volatile organic compounds measured at three homes during New Jersey TEAM-1987. *J. Exposure Anal. Environ. Epidemiol.* 1:45–61.

Lipscomb, J. 1989. *The Epidemiology of Symptoms Reported by Persons Living Near Hazardous Waste Sites*. Dissertation. University of California, Berkeley.

Lipscomb, J.A., L.R. Goldman, K.P. Satin, D.F. Smith, W.A. Vance, and R.R. Neutra. 1992. A study of current residents' knowledge of a former environmental health survey of their community. *Arch. Environ. Health* 47:270–273.

Logue, J.N. and J.M. Fox. 1986. Residential health study of families living near the Drake Chemical Superfund site in Lock Haven, Pennsylvania. *Arch. Environ. Health* 41:222–228.

Logue, J.M., R.M. Stroman, D. Reid, C.W. Hayes, and K. Sivarajah. 1985. Investigation of potential health effects associated with well water chemical contamination in Londonberry Township, Pennsylvania, U.S.A. *Arch. Environ. Health* 40:155–160.

Lum, M.R. and T.T. Tinker. 1994. *A Primer on Health Risk Communication Principles and Practices*. Agency for Toxic Substances and Disease Registry, Atlanta.

Lwanga, S.K. and S. Lemeshow. 1991. *Sample Size Determination in Health Studies*. World Health Organization, Geneva.

Lybarger, J.A., R. Lee, D.P. Vogt, R.M. Perhac, R.F. Spengler, and D.R. Brown. 1998. Medical costs and lost productivity from health conditions at volatile organic compound contaminated Superfund sites. *Environ. Res.*, 79:9–19.

Lybarger, J.A., R.F. Spengler, and C.T. DeRosa, Eds. 1993. *Priority Health Conditions*. Agency for Toxic Substances and Disease Registry, Atlanta.

Lynn, F.M. and G.J. Busenberg. 1995. Citizen advisory committees and environmental policy: What we know, what's left to discover. *Risk Anal.* 15:147–162.

Ma, X.F., J.G. Babish, J.M. Scarlett, W.H. Gutenmann, and D.J. Lisk. 1992. Mutagens in urine sampled repetitively from municipal refuse incinerator workers and water treatment workers. *J. Toxicol. Environ. Health* 37:483–494.

Malkin, R., P. Brandt-Rauf, J. Graziano, and M. Parides. 1992. Blood lead levels in incinerator workers. *Environ. Res.* 59:265–270.

Mallin, K. 1990. Investigation of a bladder cancer cluster in northwestern Illinois. *Am. J. Epidemiol.* 132, Suppl No. 1:S96–S106.

Marsh, G.M. and R.J. Caplan. 1987. Evaluating Health Effects of Exposure at Hazardous Waste Sites: A Review of the State-of-the-Art, with Recommendations for Future Research. In: *Health Effects from Hazardous Waste Sites*, J.B. Andelman and D.W. Underhill, Eds. Lewis Publishers, Boca Raton, FL, pp. 3–80.

Marshall, E., L. Gensburg, N. Geary, D. Deres, and M. Cayo. 1995. *Analytic Study to Evaluate Associations Between Hazardous Waste Sites and Birth Defects*. New York State Health Department, Albany.

Marshall, E.G., L.J. Gensburg, D.A. Deres, N.S. Geary, and M.R. Cayo. 1997. Maternal residential exposure to hazardous wastes and risk of central nervous system and musculoskeletal birth defects. *Arch. Environ. Health* 52:416–425.

McGeehin, M.A., J.S. Reif, J.C. Becher, and E.J. Mangione. 1993. Case-control of bladder cancer and water disinfection methods in Colorado. *Am. J. Epidemiol.* 138:492–501.

McLachlan, J.A. 1993. Functional toxicology: A new approach to detect biologically active xenobiotics. *Environ. Health Perspect.* 101:386–387.

McMichael, A., L.P. Heiskanen, and P.W. Callan. 1993. National Health and Medical Research Council Air Quality Program-Current Initiatives and Future Perspectives. In: *International Workshop on Human Health and Environmental Effects of Motor Vehicle Fuels and Their Exhaust Emissions*, R. Manuell, P. Callan, K. Bentley, S. McPhail, and E. Smith, Eds. International Programme on Chemical Safety, World Health Organization, Geneva, pp. 180–181.

MDNR (Michigan Department of Natural Resources). 1992. *Michigan's Environment and Relative Risk*. Lansing.

MDPH (Massachusetts Department of Public Health). 1996. *Woburn Childhood Leukemia Follow-Up Study*. Boston.

Melius, J. 1990. OSHA standard for medical surveillance of hazardous waste workers. *J. Occupat. Med.* 5:143–150.

Melius, J.M. 1995. *Current Worksite Monitoring and Performance Measurement for Environmental Remediation in the United States*. National Environmental Education and Training Center, Indiana, PA.

MEPP (Maine Environmental Priorities Project). 1995. *Summary of the Reports from the Technical Working Groups to the Steering Committee*. Maine State House, Augusta.

MEPP. 1996. *Report from the Steering Committee, Consensus Ranking of Environmental Risks Facing Maine*. Maine State House, Augusta.

Metcalf, S. and W. Williams. 1993. Lung and Respiratory Diseases. In: *Priority Health Conditions*, J.A. Lybarger, R.F. Spengler, and C.T. DeRosa, Eds. Agency for Toxic Substances and Disease Registry, Atlanta, GA, pp. 24, 141–60.

Metzger, R., J.L. Delgado, and R. Herrell. 1995. Environmental health and Hispanic children. *Environ. Health Perspect.* 103(Suppl 6):25–32.

Meyer, C. 1983. Liver dysfunction in residents exposed to leachate from a toxic waste dump. *Environ. Health Perspect.* 48:9–13.

Miller, M.S. and M.A. McGeehin. 1997. Reported Health Outcomes Among Residents Living Adjacent to a Hazardous Waste Site, Harris County, Texas, 1992. In: *Hazardous Waste: Impacts on Human and Ecological Health*, B.L. Johnson, C. Xintaras, J.S. Andrews, Eds. Princeton Scientific Publishing, Princeton, NJ, pp. 390–398.

Miller, M.S., M.A. McGeehin, R. Rao, and Z. Taylor. 1995. *Southbend Subdivision Health Outcomes Study: Harris County, Texas.* Agency for Toxic Substances and Disease Registry, Atlanta.

Minard, R.A., Jr. 1996. Comparative risk and the states. *Resources*, Winter:6–10. Resources for the Future, Washington, DC.

Mitchell, G. 1986. Superfund Amendment and Reauthorization Act-Conference Report, 99th Congress, 2nd Session. *Cong. Rec.* S 14895, Vol. 132, No. 135.

Mohai, P. and B. Bryant. 1992. Environmental racism: Reviewing the evidence. In: *Race and the Incidence of Environmental Hazards: A Time for Discourse*, B. Bryant and P. Mohai, Eds. Westview Press, Bolder, CO, pp. 163–246.

Monster, A. and J. Smolders. 1984. Tetrachloroethylene in exhaled air of persons living near pollution sources. *Inter. Arch. Occupat. Environ. Health* 53:331–336.

Moran, J. 1994. Safety and Health Issues Arising at Superfund Sites. In: *Hazardous Waste and Public Health*, J. Andrews, H. Frumpkin, B. Johnson, M. Mehlman, C. Xintaras, and J. Bucsela, Eds. Princeton Scientific Publishing, Princeton, NJ, pp. 70–77.

Morgan, M.G., B. Fischhoff, L. Lave, and P. Fischbeck. 1996. *Comparing Environmental Risks*, J.C. Davies, Ed. Resources for the Future, Washington DC, pp. 111–147.

Mumtaz, M.M., C.T. DeRosa, and P.R. Durkin. 1994. Approaches and Challenges in Risk Assessments of Chemical Mixtures. In: *Toxicology of Chemical Mixtures*, R.S.H. Yang, Ed. Academic Press, New York, pp. 565–597.

Murgueytio, A.M., R.G. Evans, D. Roberts, and T. Moehr. 1996. Prevalence of childhood lead poisoning in a lead mining area. *J. Environ. Health*: June, 12–17.

NACCHO (National Association of County & City Health Officials). 1995. *Don't Hazard a Guess.* National Association of County & City Health Officials, Washington, DC, pp. 49–54.

NACCHO. 1997a. *Improving Community Collaboration: A Self-Assessment Guide for Local Health Departments.* National Association of County and City Health Officials, Washington, DC.

NACCHO. 1997b. *Partnerships for Environmental Health Education: Performing a Community Needs Assessment at Hazardous Waste Sites.* National Association of County and City Health Officials, Washington, DC.

NACHO (National Association of County Health Officials). 1992. *Current Roles and Future Challenges of Local Health Departments in Environmental Health.* National Association of County Health Officials, Washington, DC.

Najem, G.R., D.B. Louria, M.A. Lavenhar, and M. Feuerman. 1985. Clusters of cancer mortality in New Jersey municipalities; with special reference to chemical toxic waste disposal sites and per capita income. *Internat. J. Epidemiol.* 14:528–537.

Najem, G.R., T. Strunck, and M. Feuerman. 1994. Health effects of a Superfund hazardous chemical waste disposal site. *Am. J. Prev. Med.* 10:151–155.

Najem, G.R., I.S. Thind, M.A. Lavenhar, and D.B. Louria. 1983. Gastrointestinal cancer mortality in New Jersey counties, and the relationship with environmental variables. *Internat. J. Epidemiol.* 12:276–289.

Najem, G.R. and L. Voyce. 1990. Health effects of a thorium waste disposal site. *Am. J. Public Health* 80:478–480.

NCI (National Cancer Institute). 1992. *Making Health Communication Programs Work: A Planner's Guide.* Report 92-1493. Office of Cancer Communications, National Cancer Institute, Washington, DC.

Needham, L.L., R.H. Hill, Jr., D.L. Ashley, J.L. Pirkle, and E.J. Sampson. 1995. The priority toxicant reference range study: interim report. *Environ. Health Perspect.* 103:89–94.

Needleman, H.L. 1994. Childhood lead poisoning. *Curr. Opin. Neurol.* 7:187–190.

Needleman, H.L. 1995. Behavioral toxicology. *Environ. Health Perspect.* 103 (Suppl 6):77–79.

Needleman, H.L., J.A. Riess, M.J. Tobin, G.E. Biesecker, and J.B. Greenhouse. 1996. Bone lead levels and delinquent behavior. *J. Am. Med. Assoc.* 275:363–369.

Neuberger, J.S., M. Mulhall, M.C. Pomatto, J. Sheverbush, and R.S. Hassanein. 1990. Health problems in Galena, Kansas: A heavy metal mining Superfund site. *Sci. Total Environ.* 94:261–272.

Neutra, R., J. Lipscomb, K. Satin, and D. Shusterman. 1991. Hypotheses to explain the higher symptom rates observed around hazardous waste sites. *Environ. Health Perspect.* 94:31–38.

Nez Perce Tribe. 1996. *Our Environment, Our Health.* Environmental Restoration & Waste Management Dept., Lapwai, ID.

Nogueras, M.M., F.J. Bove, and W.E. Kaye. 1995. *A Case-Control Study to Determine Risk Factors for Elevated Blood Lead Levels in Children.* Agency for Toxic Substances and Disease Registry, Atlanta.

Nossiter, A. 1996. Villain is dioxin, relocation is response, but judgment is in dispute. *New York Times*, October 21, pp. 49–50.

NRC (National Research Council). 1983. *Risk Assessment in the Federal Government: Managing the Process.* National Academy Press, Washington, DC, pp. 3, 18.

NRC. 1989a. *Biologic Markers in Reproductive Toxicology.* National Academy Press, Washington, DC, pp. 20, 21.

NRC. 1989b. *Biologic Markers in Pulmonary Toxicology.* National Academy Press, Washington, DC, pp. 20, 21.

NRC. 1989c. *Improving Risk Communication.* National Academy Press, Washington, DC, pp. 20, 21.

NRC. 1991a. *Rethinking the Ozone Problem in Urban and Regional Air Pollution.* National Academy Press, Washington, DC.

NRC. 1991b. *Environmental Epidemiology: Public Health and Hazardous Wastes, Vol. 1.* National Academy Press, Washington, DC, pp. 20, 34, 114.

NRC. 1991c. *Human Exposure to Airborne Pollutants.* National Academy Press, Washington, DC, pp. 3, 43.

NRC. 1991d. *Animals as Sentinels of Environmental Health Hazards.* National Academy Press, Washington, DC.

NRC. 1992a. *Environmental Neurotoxicology.* National Academy Press, Washington, DC.

NRC. 1992b. *Biologic Markers in Immunotoxicology.* National Academy Press, Washington, DC.

NRC. 1994. *Ranking Hazardous-Waste Sites.* National Academy Press, Washington, DC, pp. 29, 37.

NRC. 1995. *Biologic Markers in Urinary Toxicology.* National Academy Press, Washington, DC, pp. 29, 37.

NRC. 1996. *Understanding Risk: Informing Decisions in a Democratic Society.* National Academy Press, Washington, DC.

NRC. 1997. *Environmental Epidemiology, Vol. 2.* National Academy Press, Washington, DC, pp. 12–25.

NYDOH (New York State Department of Health). 1989. *Summary of Forest Glen Exposure Study.* Albany.

Oakes, J., D. Anderton, and A. Anderson. 1996. A longitudinal analysis of environmental equity in communities with hazardous waste facilities. *Social Sci. Res.* 25:125–148.

OEPA (Ohio Environmental Protection Agency). 1995. *Comparing the Risks of Ohio's Environmental Conditions: Executive Summary of Ohio's State of the Environment Report.* Columbus, OH, pp. 22, 309, 317.

Ogunranti, J. 1989. Hematological indices in Nigerians exposed to radioactive waste. *Lancet* 16:667–668.

Ohio University. 1995. *Survey Results for the Athens County Comparative Risk Survey*. Institute for Local Government Administration, Ohio University, Athens.

OTA (Office of Technology Assessment). 1983. *Technologies and Management Strategies for Hazardous Waste Control*. Government Printing Office, Washington, DC.

OTA. 1989. *Coming Clean: Superfund's Problems Can Be Solved*. Washington, DC.

OTA. 1995. *EPA Superfund Actions and ATSDR Public Health Data*. Congress of the United States, Office of Technology Assessment, Washington, DC.

Ozonoff, D., M.E. Colten, A. Cupples, T. Heeren, A. Schatzkin, T. Mangione, M. Dresner, and T. Colton. 1987. Health problems reported by residents of a neighborhood contaminated by a hazardous waste facility. *Am. J. Indust. Med.* 11:581–597.

Ozonoff, D., A. Aschengrau, and P. Coogan. 1994. Cancer in the vicinity of a Department of Defense Superfund Site in Massachusetts. *Toxicol. Environ. Health* 10:119–141.

Paigen, B., L.R. Goldman, J.H. Highland, M.M. Magnant, and A.T. Steegman. 1985. Prevalence of health problems in children living near Love Canal. *Hazardous Waste Hazardous Materials* 2:23–43.

Paigen, B., L. Goldman, M. Magnant, J. Highland, and A. Steegman. 1987. Growth of children living near the hazardous waste site, Love Canal. *Human Biol.* 59:489–508.

Paranzino, G.K., P. Orris, and K. Kirkland. 1996. *Evaluation of Clinical Services at the Del Amo/Montrose Community Health Infestation Project*. Association of Occupational and Environmental Clinics, Washington, DC.

Peavy, J.V. 1996. Surveys and Sampling. In: *Field Epidemiology*, M.B. Gregg, Ed. Oxford University Press, New York, pp. 152–164.

PHS (Public Health Service). 1990. *Healthy People 2000*. U.S. Department of Health and Human Services, Washington, DC.

Picciano, D. 1980. Pilot cytogenetic study of the residents living near Love Canal: A hazardous waste site. *Mammalian Chromosome Newslett.* 21:86–93.

Plotzman, F. 1992. A nation's recycling law puts businesses on the spot. *New York Times*, July 12, pp. K-1, K-5.

Pohl, H.R. and B.F. Hibbs. 1996. Breast-feeding exposure of infants to environmental contaminants–A public health risk assessment viewpoint: Chlorinated dibenzodioxins and chlorinated dibenzofurans. *Toxicol. Indust. Health* 12:593–611.

Pope, A.M. and D.P. Rall (Eds.). 1995. *Environmental Medicine: Integrating a Missing Element into Medical Education*. Institute of Medicine, National Academy Press, Washington, DC.

Pope, A.M., M.A. Snyder, and L.H. Mood, Eds. 1995. *Nursing, Health, & the Environment: Strengthening the Relationship to Improve the Public's Health*. Institute of Medicine, National Academy Press, Washington, DC.

Prausnitz, S. and M.A. McGeehin. 1995. *Pancreatic Cancer Mortality and Residential Proximity to Railroad Refueling Facilities in Montana: A Records-Based Case-Control Pilot Study*. Agency for Toxic Substances and Disease Registry, Atlanta, GA.

Probst, K. 1992. *Testimony to Subcommittee on Oversight*. Ways and Means Committee, House of Representatives, U.S. Congress, Washington, DC, August 12.

Protocol Committee. 1994. *Symposium on Health Research and Needs to Ensure Environmental Justice*. National Institute for Environmental Health Sciences, Research Triangle Park, NC.

Putz-Anderson, V., B.E. Albright, B.J. Taylor, S.T. Lee, B.L. Johnson, D.W. Chrislip, B.J. Taylor, W.S. Brightwell, N. Dickerson, M. Culver, D. Zentmeyer, and P. Smith. 1983. A behavioral examination of workers exposed to carbon disulfide. *Neurotoxicol.* 4:67–78.

Rao, R.A. and M.A. McGeehin. 1996. *Disease and Symptom Prevalence Survey—Tucson International Airport Site*. Division of Health Studies, Agency for Toxic Substances and Disease Registry, Atlanta.

Reilly, W. 1992. Environmental equity: EPA's position. *EPA J.* March/April:18–22.

Reisch, M. 1996. *Congressional Research Report for Congress: Superfund Reauthorization Issues in the 104th Congress*. Committee for the National Institute for the Environment, Washington, DC.

Risher, J.F. and C.T. DeRosa. 1997. The precision, uses, and limitations of public health guidance values. *Human Ecological Risk Assess*. 3: in press.

RiskFocus®. 1990. *Analysis of the Impact of Exposure Assumptions on Risk Assessment of Chemicals in the Environment: Part I*. VERSAR, Inc., Springfield, VA, pp. 123–129.

Robertson, A. and M. Minkler. 1994. New health promotion movement: A critical examination. *Health Educ. Q.* 21:295–312.

Robinson, F.T. 1996. *Prepared testimony of Florence T Robinson*. House Committee on Government Reform and Oversight, Subcommittee on National Economic Growth, Natural Resources and Regulatory Affairs, U.S. Congress, Washington, DC, May 8.

Rodricks, J.V. 1992. *Calculated Risks: Understanding the Toxicity and Human Health Risks of Chemicals in Our Environment*. Cambridge University Press, New York.

Rogan, W. and B. Gladen. 1983. Monitoring breast milk contamination to detect hazards from waste disposal. *Environ. Health Perspect*. 48:87–91.

Roht, L.H., S.W. Vernon, F.W. Weir, S.M. Pier, P. Sullivan, and L.J. Reed. 1985. Community exposure to hazardous waste disposal sites: Assessing reporting bias. *Am. J. Epidemiol*. 122:418–433.

Rom, W.N. 1992. Chlorofluorocarbons and Destruction of the Ozone Layer. In: *Environmental and Occupational Medicine,* 2nd ed, W. Rom, Ed. Little, Brown and Company, Boston, MA.

Roper, W.L. 1993. Health communication takes on new dimensions at CDC. *Public Health Rep*. 108:179–183.

Russell, M., E.M. Colglazier, and M.R. English. 1991. *Hazardous Waste Remediation: The Task Ahead*. University of Tennessee, Knoxville.

Ruttenberg, R., D. Weinstock, and L. Santamaria. 1996. *Labor Market Study of Hazardous Waste Workers and Associated Emergency Responders*. Ruttenberg & Associates, Bethesda, MD.

Sampson, E.J. 1996. Personal communication: blood concentrations (in ppb) of selected volatile organic compounds in a group of nonoccupationally exposed, U.S. adults. National Center for Environmental Health, Centers for Disease Control and Prevention, Atlanta.

Sarasua, S.M., R.F. Vogt, D.C. Middleton, B.A. Slade, M.A. McGeehin, and J.A. Lybarger. 1997. 'CLL-Like' B-Cell Phenotypes Detected in Superfund Studies: Epidemiologic Methods and Findings. In: *Determining the Role of Environmental Exposures as Risk Factors for B-Cell Lymphoproliferative Disorders*, G.E. Marti, R.E. Vogt, and V.E. Zenger, Eds. Agency for Toxic Substances and Disease Registry, Atlanta, pp. 7–19.

Sasso, F., R. Ferraiulolo, G. Garetano, E. Gursky, J. Fagliano, J. Pasqualo, R. Salkie, and J. Rotola. 1996. Mercury exposure among residents of a building formerly used for industrial purposes—New Jersey, 1995. *Morb. Mort. Weekly Report* 45:422–424.

Satchell, M. 1997. A black and green issue moves people. *US News & World Report*, April 21, 1997, pp. 41–42.

Schecter, A., O. Papke, M. Ball, A. Lis, and P. Brandt-Rauf. 1995. Dioxin concentrations in the blood of workers at municipal waste incinerators. *Occup. Environ. Med*. 52:385–387.

Schecter, A., O. Papke, A. Lis, and M. Ball. 1996. Decrease in milk and blood dioxin levels over two years in a mother nursing twins: estimates of decreased maternal and increased infant dioxin body burden from nursing. *Chemosphere* 32:543–549.

Schlaud, M., A. Seidler, A. Salje, and W. Behrendt. 1995. Organochlorine residues in human breast milk: Analysis through a sentinel practice network. *J. Epidemiol. Community Health* 49(Suppl 1):17–21.

Seattle. 1991. *Environmental Risks in Seattle: A Comparative Assessment.* Office for Long-Range Planning, City of Seattle, WA.

Sever, L.E. 1997. Environmental contamination and health effects: What is the evidence? *Toxicol. Indust. Health* 13:145–162.

Sexton, K. and Y. Anderson, Eds. 1993. Equity in environmental health: Research issues and needs. *Toxicol. Indust. Health* 9, No. 5.

Sexton, K., K. Olden, and B.L. Johnson. 1993. Environmental justice: The central role of research in establishing a credible scientific foundation for informed decision making. *Toxicol. Indust. Health* 9:685–727.

Shaw, G.M., J. Schulman, J.D. Frisch, S.K. Cummins, and J.A. Harris. 1992. Congenital malformations and birthweight in areas with potential environmental contamination. *Arch. Environ. Health* 47:147–154.

Shusterman, D., J. Lipscomb, R. Neutra, and K. Satin. 1991. Symptom prevalence and odor-worry interaction near hazardous waste sites. *Environ. Health Perspect.* 94:25–30.

Silverman, J. 1996. Public supports reform of Superfund. *BNA National Environmental Daily*, January 26.

Sim, M.R. and J.J. McNeil. 1992. Monitoring chemical exposure using breast milk: A methodological review. *Am. J. Epidemiol.* 136:1–11.

Simmons, J.E., D.M. DeMarini, and E. Berman. 1988. Lethality and hepatoxicity of complex waste mixtures. *Environ. Res.* 46:74–85.

Skinner, J. 1991. Hazardous waste treatment trends in the U.S. *Waste Manage. Res.* 9:55–63.

Slovic, P. 1987. Perception of risk. *Science* 236:280–285.

Slovic, P. 1992. Perception of Risk: Reflections on the Psychometric Paradigm. In: *Social Theories of Risk*, S. Krimsky and D. Golding, Eds. Praeger, New York, pp. 117–152.

Slovic, P. 1993. Perceived risk, trust, and democracy: A systems perspective. *Risk Anal.* 13:675–682.

Slovic, P., B. Fischhoff, and S. Lichtenstein. 1979. Rating the risks. *Environment* 21:14–39.

Smith, R.P. 1996. Responses of the Blood. In: *Casarett and Doull's Toxicology, 5th Edition*, C.D. Klaassen, Ed. McGraw-Hill, New York.

Smothers, R. 1997. Destruction of a mercury-tainted building in Hoboken, New Jersey. *New York Times*, October 7.

Sosniak, W., W. Kaye, and T. Gomez. 1994. Data linkage to explore the risk of low birthweight associated with maternal proximity to hazardous waste sites from the National Priorities List. *Arch. Environ. Health* 49:251–255.

Stallings, F.L. and P.A. Jones. 1995. *Multisite Lead and Cadmium Exposure Study with Biological Markers Incorporated.* Division of Health Studies, Agency for Toxic Substances and Disease Registry, Atlanta.

Stanek, E.J., E.J. Calabrese, K. Mundt, P. Pekow, and K.B. Yeatts. 1998. Prevalence of soil mouthing/ingestion among healthy children aged 1 to 6. *J. Soil Contam.* 7:227–242.

Stehr-Green, P., V. Burse, and E. Welty. 1988. Human exposure to polychlorinated biphenyls at toxic waste sites: investigations in the United States. *Arch. Environ. Health* 43:420–424.

Straight, M.J., S.G. Folger, B.L. McGeehin, and R.W. Amler. 1993. *Study of Symptom and Disease Prevalence, Caldwell Systems, Inc., Hazardous Waste Incinerator, Caldwell County, North Carolina.* Final Report, Division of Health Education, Agency for Toxic Substances and Disease Registry, Atlanta.

Susten, S. 1997. Personal communication. Agency for Toxic Substances and Disease Registry, Atlanta.

Swan, S.H., G. Shaw, J.A. Harris, and R.R. Neutra. 1989. Congenital cardiac anomalies in relation to water contamination, Santa Clara, California, 1981–1983. *Am. J. Epidemiol.* 129:885–893.

Swanborg, R.H. 1984. Immune Response in Toxicology. In: *Toxicology: Principles and Practice, Vol. 2*, F. Sperling, Ed. pp. 109–119.

Szasz, A. 1994. *EcoPopulism: Toxic Waste and the Movement for Environmental Justice*. University of Minnesota Press, Minneapolis.

Taylor, Z. and O. Faroon. 1993. Kidney Dysfunction. In: *Priority Health Conditions*, J.A. Lybarger, R.F. Spengler, and C.T. DeRosa, Eds. Agency for Toxic Substances and Disease Registry, Atlanta, pp. 89–106.

Teran, S.P., E.R. Blake, L.-J. Schierow, A.L. Casey, and M. Harnly. 1994. Is Biologic Monitoring for Toxic Chemicals Always What Exposed Communities Want? A Practical Ethical Approach to Informed Consent. In: *Hazardous Waste and Public Health*, J. Andrews, H. Frumpkin, B. Johnson, M. Mehlman, C. Xintaras, and J. Bucsela, Eds. Princeton Scientific Publishing, Princeton, NJ, pp. 90–94.

Terris, M. 1992. Concepts of health promotion: dualities in public health theory. *J. Public Health Policy* 13:267–276.

Thacker, S.B. and R.L. Berkelman. 1992. History of Public Health Surveillance. In: *Public Health Surveillance*, W. Halperin, E.L. Baker, and R.R. Monson, Eds. Van Nostrand Reinhold, New York.

Thomas, L.M. 1987. *Preface to Unfinished Business: A Comparative Assessment of Environmental Problems*. Report PB88-127048. Department of Congress, National Technical Information Service, Springfield, VA.

Tinker, T., C. Lewis-Younger, S. Isaacs, L. Neufer, and C. Blair. 1995. Environmental health risk communication: a case study of the Chattanooga Creek site. *J. Tennessee Med. Assoc.* 88:343–349.

Tinker, T., G. Moore, and C. Lewis-Younger. 1996a. Chattanooga Creek: Building partnerships for a healthier environment. *The Tennessee Conservationist* Jan/Feb:14–18.

Tinker, T. 1996b. Recommendations to improve health risk communication: lessons learned from the U.S. Public Health Service. *J. Health Comm.* 1:197–217.

TNRCC (Texas Natural Resource Conservation Commission). 1995. *State of Texas Environmental Priorities Project: Human Health Workgroup Draft Summary Report*. Austin, 3.7.7.8.

Travis, C.C. and S.T. Hester. 1991. Global chemical pollution. *Environ. Sci. Technol.* 25:815–819.

Travis, C.C., P.A. Scofield, and B.P. Blaylock. 1993. Evaluation of remediation worker risk at radioactively contaminated waste sites. *J. Hazardous Materials* 35:387–401.

Tyler, C.W., Jr. and J.M. Last. 1992. Epidemiology. In: *Public Health & Preventive Medicine*, J.M. Last, R.B. Wallace, Eds. Appleton & Lance, Norwalk, pp. 11–39.

UDEQ (Utah Department of Environmental Quality). 1995. *Utah's Environmental Issues Report*. Salt Lake City.

UNEP (United National Environment Program). 1992. *United Nations Conference on Environment and Development: Agenda 21*. United Nations, New York.

University of Hawaii. 1993. *Comparative Risk Analysis for Guam Environmental Protection Agency: Final Report*. Social Science Research Institute, University of Hawaii, Honolulu.

Upton, A.C., T. Kneip, and P. Toniolo. 1989. Public health aspects of toxic chemical disposal sites. *Annu. Rev. Public Health* 10:1–25.

U.S. Congress. 1993. *Government Performance and Results Act of 1993*. Government Printing Office, Washington, DC.

Van Damme, K., L. Casteleyn, E. Heseltine, A. Huici, M. Sorsa, N. van Larebeke, and P. Vineis. 1995. Individual susceptibility and prevention of occupational diseases: Scientific and ethical issues. *J. Occupat. Environ. Med.* 37:91–99.

VANR (Vermont Agency of Natural Resources). 1991. *Environment 1991: Risks to Vermont and Vermonters*. Waterbury.

Vaughan, E. 1995. The significance of socioeconomic and ethnic diversity for the risk communication process. *Risk Anal.* 15:169–179.

Verhaar, H.J.M., J.R. Morroni, K.F. Reardon, S.M. Hays, D.P. Gaver, R.L. Carpenter, and R.S.H. Yang. 1997. A proposed approach to study the toxicology of complex mixtures of petroleum products: The integrated use of QSAR, lumping analysis, and PBPK/PD modeling. *Environ. Health Perspect.* 105 (Suppl 1):179–195.

Vianna, N.J. and A.K. Polan. 1984. Incidence of low birth weight among Love Canal residents. *Science* 226:1217–1219.

Viscusi, W.K. and J.T. Hamilton. 1994. Superfund and real risks. *The American Enterprise* March/April:37–45.

Vleminckx, C., W. Klemans, L. Schriewer, N. Lijsen, M. Ottogali, A. Pays, C. Planard, G. Rigaux, Y. Ros, M. Rivière, J. Vandenvelde, P. De Plaen, and Th. Lakhanisky. 1997. Performance of Cytogenetic Biomarkers on Children Exposed to Environmental Pollutants. In: *Hazardous Waste: Impacts on Human and Ecological Health,* B.L. Johnson, C. Xintaras, and J.S. Andrews, Jr., Eds. Princeton Scientific Publishing, Princeton, NJ, pp. 156–167.

Vogt, R.F., N.K. Meredith, J. Powell, S.F. Ethridge, W. Whitfield, L.O. Henderson, and W.H. Hannon. 1997. Laboratory Results in Eleven Individuals with B-CLL-Like Phenotypes Detected in Environmental Health Studies. In: *Determining the Role of Environmental Exposures as Risk Factors for B-Cell Lymphoproliferative Disorders,* G.E. Marti, R.E. Vogt, and V.E. Zenger, Eds. Agency for Toxic Substances and Disease Registry, Atlanta, pp. 19–29.

Waldo, A.B. and R. Hinds. 1993. *Chemical Hazard Communication Guidebook,* 2nd ed. McGraw-Hill, Inc., New York, pp. 9–32.

Warhit, E.B. 1995. *Labor Demand Projections for the Waste Remediation Market.* National Environmental Education and Training Center, Indiana, PA.

Warren, R.C. 1993. The morbidity/mortality gaps: What is the problem? *Ann. Epidemiol.* 3:127–129.

WCEDM (Western Center for Environmental Decision-Making). 1996a. *Public Involvement in Comparative Risk Projects.* Boulder, CO.

WCEDM. 1996b. *Comparative Risk at the Local Level: Lessons from the Road.* Boulder, CO.

WDOE (Washington Department of Ecology). 1989a. *Environment 2010: The State of the Environment Report.* Olympia.

WDOE. 1989b. Transmittal of Technical Advisory Committee Human Health and Ecological Risk Rankings. Memorandum from Steve Nicholas to Washington Environment 2010 Steering Committee and Public Advisory Committee. Olympia, August 10.

WDOE. 1995. *A Study on Environmental Equity in Washington State.* Report No 95-413, Olympia.

Webb, K., R.G. Evans, P. Stehr, and S.M. Ayres. 1987. Pilot study on health effects of environmental 2,3,7,8-TCDD in Missouri. *Am. J. Indust. Med.* 11:685–692.

Wegman, D.H. 1992. Hazard Surveillance. In: *Public Health Surveillance,* W. Halperin, E.L. Baker, and R.R. Monson, Eds. Van Nostrand Reinhold, New York, pp. 62–73.

Wernette, D. and L. Nieves. 1991. *Minorities and air pollution: A preliminary geo-demographic analysis.* Paper presented at the Socioeconomic Research Analysis Conference, Baltimore, MD, June 27–28.

White, R.F., R.G. Feldman, I.I. Eviator, J.F. Jabre, and C.A. Niles. 1997. Hazardous waste and neurobehavioral effects: a developmental perspective. *Environ. Res.,* in press.

WHO (World Health Organization). 1983. *Guidelines on Studies in Environmental Epidemiology.* Environmental Health Criteria 27. International Programme on Chemical Safety, Geneva.

Williams, M., R. Rao, and R.W. Amler. 1993. Cancer. In: *Priority Health Conditions,* J.A. Lybarger, R.F. Spengler, and C.T. DeRosa, Eds. Agency for Toxic Substances and Disease Registry, Atlanta, pp. 49–65.

Wilson, D. and W. Rathje. 1989. Structure and dynamics of household hazardous wastes. *J. Resource Mgmt. Tech.* 17:200–206.

Wilson, J. and M. Straight. 1993. Liver Dysfunction. In: *Priority Health Conditions*, J.A. Lybarger, R.F. Spengler, and C.T. DeRosa, Eds. Agency for Toxic Substances and Disease Registry, Atlanta, pp. 107–140.

Wolf, A.M.A., L.E. Kettler, J.F. Leahy, and A.H. Spitz. 1997. Surveying household hazardous waste generation and collection. *J. Environ. Health* (March):6–11.

Yang, R.S. 1996. Some current approaches for studying combination toxicology in chemical mixtures. *Food Chem. Toxicol.* 34:1037–1044.

Yang, R.S., H.A. el-Masri, R.S. Thomas, A.A. Constan, and J.D. Tessari. 1995. The application of physiologically based pharmacokinetic/pharmacodynamic (PBPK/PD) modeling for exploring risk assessment approaches of chemical mixtures. *Toxicol. Lett.* 79:193–200.

Zimmerman, R. 1993. Social equity and environmental risk. *Risk Anal.* 13:649–666.

Zmirou, D., A. Deloraine, P. Saviuc, C. Tillier, A. Boucharlat, and N. Maury. 1994. Short-term health effects of an industrial toxic waste landfill: A retrospective follow-up study in Montchanin, France. *Arch. Environ. Health* 49:228–238.

Initialisms

ANOVA	analysis of variance
ARARs	applicable, relevant, appropriate requirements
ATSDR	Agency for Toxic Substances and Disease Registry, DHHS
BRA	baseline risk assessment
CBO	Congressional Budget Office
CDC	Centers for Disease Control and Prevention, DHHS
CEPPS	completed exposure pathways priority substance
CERCLA	Comprehensive Environmental Response, Compensation, and Liability Act
CERCLIS	Comprehensive Environmental Response, Compensation, and Liability Information System
DHHS	U.S. Department of Health and Human Services
DOD	U.S. Department of Defense
DOE	U.S. Department of Energy
DOI	U.S. Department of Interior
DOL	U.S. Department of Labor
EPA	U.S. Environmental Protection Agency
EU	European Union
GAO	General Accounting Office
GIS	geographic information systems
HazDat	hazardous substance data
HazMat	hazardous materials
IARC	International Agency for Research on Cancer, WHO
IOM	Institute of Medicine, NAS
MRL	minimal risk level
NAS	National Academy of Sciences
NIEHS	National Institute of Environmental Health Sciences, NIH
NIH	National Institutes of Health, DHHS
NPL	national priorities list
NRC	National Research Council, NAS
OSHA	Occupational Safety and Health Administration, DOL
OSWER	Office of Solid Waste and Emergency Response, EPA
OTA	Office of Technology Assessment, U.S. Congress
PHA	public health assessment
PHS	Public Health Service, DHHS
ppb	parts per billion
ppm	parts per million
PRPs	potentially responsible parties
QSAR	quantitative structural activity relationships
RCRA	Resource Conservation and Recovery Act

RI   remedial investigation
ROD   record of decision
RQ   reportable quantity
SGOT   serum glutamic-oxaloacetic transaminase (liver enzyme)
SGPT   serum glutamic-pyruvic transaminase (liver enzyme)
SMSA   standard metropolitan statistical area
TCDD   2,3,7,8-Tetrachlorodibenzo-p-dioxin
TEQ   Toxicity equivalent, defined as the product of the concentration of an
                individual "dioxin-like compound" in a complex environmental
                mixture and the corresponding TCDD toxicity equivalency factor for
                that compound.
ToxFAQ   toxicologic frequently asked questions
TRI   toxics release inventory
TSDF   treatment, storage, and disposal facility
VOCs   volatile organic compounds

Glossary

Absorbed dose—The amount of a substance that penetrates across the exchange boundaries of an organism through either physical or biologic processes after contact (exposure) (RiskFocus®, 1990).

Absorption—The process of taking in, incorporation, or reception of gases, liquids, light, or heat (Hensyl, 1987).

Acute—Occurring over a short time, usually a few minutes or hours. An *acute* exposure can result in short-term or long-term health effects. An *acute* effect happens a short time (up to one year) after exposure.

Administered dose—The amount of a substance given to a human or test animal. Administered dose is a measure of exposure because absorption is not considered (RiskFocus®, 1990).

Agent—An entity (chemical, radiologic, mineralogic, or biologic) that may cause effects in an organism exposed to it.

Alloy—A mixture of metals.

Alzheimer's disease—A disease of the nervous system that causes mental deterioration.

Ambient—Surrounding; pertaining to the air, noise, temperature, etc. in which an organism or apparatus functions (Hensyl, 1987)

Analyte—Any substance or chemical constituent that is being analyzed (Hensyl, 1987).

Analytic epidemiologic study—Investigations that evaluate the causal nature of associations between exposure to hazardous substances and disease outcome by testing scientific hypotheses (NRC, 1997)

Anemia—A decreased ability of the blood to transport oxygen; Low numbers of red blood cells or hemoglobin.

Antagonism—A response to a mixture of toxic chemicals that is less than that suggested by the component toxicities.

Anthropogenic pollution—Pollution caused by humans

Apoptosis—Programmed cell death, which involves DNA fragmentation.

Applied dose—The amount of a substance given to a human or test animal, especially through dermal contact. Applied dose becomes a measure of exposure because absorption is not considered (RiskFocus®, 1990).

Assessment—The process of determining the nature and extent of hazards and health problems within a jurisdiction.

Asymptomatic—Without symptoms, or producing no symptoms.

Background level—A typical or average level of a chemical in the environment. *Background* often refers to naturally occurring or uncontaminated levels.

Biologic indicator—A chemical (analyte), its metabolite, or another marker of exposure that can be detected or measured by biomedical testing of human body fluids or tissues to validate human exposure to a hazardous substance.

Biologic monitoring—Measuring chemicals in biologic materials (e.g., blood, urine, breath, hair) to determine whether chemical exposure has occurred in living organisms.

Biologic uptake—The transfer of substances from the environment to living organisms.

Biomedical testing—Biologic testing of persons designed to evaluate any qualitative or quantitative change in physiologic function that may predict a health impairment following exposure to hazardous substance(s).

Blood lead level—The concentration of lead in a sample of blood.

Body burden—The total amount of a chemical in the body. Some chemicals accumulate in the body because they are stored in fat or bone or other tissues.

Carcinogen—A substance that can cause cancer.

Carcinogenicity—Capacity to cause cancer.

Case-control study—A study that compares the exposures of individuals who have specific adverse effects or diseases with the exposures of individuals (controls) who do not have these effects; controls are generally in the same population from which the cases were derived (NRC, 1997).

Case study—The medical or epidemiologic evaluation through interview or biomedical testing of a single person or small number of persons to determine descriptive information about their health status or potential for exposure.

Census block—Small geographic areas enclosed by visible features such as streets, roads, streams, and railroad tracks, or by invisible borders such as city, town, township, and county limits; property lines; or short, imaginary extensions of streets and roads (BOC, 1992).

Census block group—A geographic block group or tabulation block group. The former is a cluster of blocks having the same first digit of their three-digit identifying numbers within a census tract or block numbering area. A tabulation block group is a geographic block group that may be split to present data for every unique combination of county subdivision, place, American Indian and Alaska Native area, urbanized area, voting district, urban/rural and congressional district shown in the data product (BOC, 1992).

Central nervous system—The part of the nervous system that includes the brain and the spinal cord.

CERCLA—The Comprehensive Environmental Response, Compensation, and Liability Act of 1980, also known as Superfund.

Chromosome—The structure (normally 46 in humans) in the cell nucleus that is the bearer of genes.

Chronic—Occurring over a long period of time (e.g., more than 1 year).

Cluster investigation—A review of an unusual number, real or perceived, of health events (for example, reports of cancer) grouped together in time and location. *Cluster investigations* are designed to confirm case reports; determine whether they represent an unusual disease occurrence; and, if possible, explore possible causes and environmental factors (NRC, 1997).

Community—A group or social class having common characteristics.

Community assistance panel (CAP)[1]—A group of community stakeholders, generally 12–15 persons, convened by ATSDR to provide advice from a community impacted by hazardous substances.

Community health investigation—Medical or epidemiologic evaluation of descriptive health information about individual persons or a population of persons to evaluate and determine health concerns and to assess the likelihood that such concerns may be associated with exposure to hazardous substances.

Comparison values—Estimated contaminant concentrations in specific media that are not likely to cause adverse health effects, given a standard daily ingestion rate and standard body weight.

Concentration—The amount of one substance dissolved or contained in a given amount of another.

Confidence interval—An interval of values that has a specified probability of containing a given parameter or characteristic (RiskFocus®, 1990).

Contaminant—Any substance or material that enters a system (e.g., the environment, human body, food) where it is not normally found.

Cytogenetics—The branch of genetics concerned with the structure and function of the cell, especially the chromosomes (Hensyl, 1987).

Demographics—The statistical study of human populations.

Dermal—Referring to the skin. *Dermal* absorption means absorption through the skin.

Descriptive epidemiology—Study of the amount and distribution of disease within a population by person, place, and time (NRC, 1997).

Diagnostic test—A laboratory test used to determine whether a person has a particular health problem.

Disease—Illness; sickness; an interruption, cessation, or disorder of body functions, systems, or organs (Hensyl, 1987).

Disease-and-symptom prevalence study—A study designed to measure the occurrence of self-reported disease that may, in some instances, be validated through medical records or physical examination if available, and to determine those adverse health conditions that may require further investigation because they are considered to have been reported at an excess rate (NRC, 1997).

Disease incidence—The rate of new occurrences of a disease.

Disease registry—A system for collecting and maintaining, in a structured record, information on persons having a common illness or adverse health condition.

Disease surveillance—A data collecting system that monitors the occurrence of specific diseases (e.g., cancer).

Dose—The total amount of radiation or toxicant, drug, or other chemical administered or taken by the organism (adapted from Hodgson and Levi, 1987).

---

[1] Underlined terms are terms specific to the Agency for Toxic Substances and Disease Registry.

Dose-response study—A toxicologic study of the quantitative relationship between the amount of a toxicant administered or taken and the incidence or extent of the adverse effect (Hodgson and Levi, 1987).

Dosimetry—Process of measuring dose.

Edema—An accumulation of excessive watery fluid in cells, tissues, or serous cavities (Hensyl, 1987).

Effluent—Waste material discharged into the environment.

Environmental contamination—The presence of hazardous substances in the environment.

Environmental equity-The proportionate and equitable distribution of environmental benefits and risks among diverse economic and cultural communities (WDOE, 1995).

Environmental health—The area of public health that is concerned with effects on human health of socioeconomic conditions and physical factors in the environment.

Environmental justice—Concern about the disproportionate occurrence of pollution and potential pollution-related health effects affecting low-income, cultural, and ethnic populations and lesser cleanup efforts in their communities (CRARM, 1997).

Environmental medium—Material in the outdoor natural physical environment that surrounds or contacts organisms (e.g., surface water, groundwater, soil, air) and through which substances can move and reach organisms (adapted from RiskFocus®, 1990).

Enzyme—A protein secreted by cells that acts as a catalyst to induce chemical changes in other substances, itself remaining apparently unchanged by the process (Hensyl, 1987).

Epidemiologic surveillance—The ongoing, systematic collection, analysis, and interpretation of health data essential to the planning, implementation, and evaluation of public health practice, closely integrated with the timely dissemination of these data to persons who need to know.

Epidemiology—The study of the occurrence of disease in human populations.

Epithelium—The purely cellular, avascular layer covering all the free surfaces—cutaneous, mucous, and serous—including the glands and other structures derived therefrom (Hensyl, 1987).

Exposure—The amount of a stressor (e.g., a hazardous substance) that living organisms contact over a defined period of time.

Exposure assessment—Determination of the sources, environmental transport and modification, and fate of pollutants and contaminants, including the conditions under which people or other target species could be exposed, and the doses that could result in adverse effects (CRARM, 1997).

Exposure dose reconstruction—The use of computational models and other approximation techniques to estimate cumulative amounts of hazardous substances absorbed by living organisms.

Exposure investigation—The collection and analysis of site-specific information to determine whether human populations have been exposed to hazardous substances. The site-specific information may include environmental sampling, exposure-dose reconstruction, biologic or biomedical testing, and evaluation of medical information.

Exposure pathway—The path by which pollutants travel from sources via air, soil, water, or food to reach living organisms (adapted from CRARM, 1997).

Exposure registry—A system for collecting and maintaining in a structured record, information on persons with documented environmental exposure(s).

Exposure-response relationship—The relationship between exposure level and the incidence of adverse effects.

Exposure route—The way a substance enters an organism after contact (e.g., inhalation, ingestion, dermal absorption).

Fibrosis—Formation of fibrous tissue as a reparative or reactive process (Hensyl, 1987).

Fungicide—A substance that kills molds.

Gavage—Forced feeding by stomach tube.

Gene—The functional unit of heredity that occupies a specific place or locus on a chromosome (Hensyl, 1997).

Genotoxicity—An effect on the genetic material (DNA) of living cells that, upon replication of the cells, is expressed as a mutagenic or a carcinogenic event (Hodgson and Levi, 1987).

Geographic information system (GIS)—A computer hardware and software system designed to collect, manipulate, analyze, and display spatially referenced data for solving complex resource, environmental, and social problems.

Hazard—A stressor with the potency or capacity to cause adverse effects; a measure of the stressor.

Hazard surveillance—A data collecting system that monitors the distribution of specific hazards (e.g., carcinogens).

Hazardous Substances and Health Effects Database (HazDat)—A database that the ATSDR compiled containing data from individual hazardous waste sites on environmental contamination, substance toxicity, human exposure, and human health effects.

Health consultation—An ATSDR response to a specific question or request for information pertaining to a hazardous substance or facility (which includes waste sites). It is a more limited response than a health assessment because it often contains a time-critical element that necessitates a rapid response.

Health education—A program of activities to promote health and provide information and training about reducing exposure, illness, or disease that result from hazardous substances in the environment.

Health investigation—An investigation of a defined population, using epidemiologic methods, that would help determine exposures or possible public health impact by defining health problems which require further investigation through epidemiologic studies, environmental monitoring or sampling, and surveillance.

Health outcome data—Information collected on the health status of an individual or group, including morbidity and mortality data, birth statistics, medical records, tumor and disease registries, surveillance data, and previously conducted health studies.

Health outcomes study—An investigation of exposed persons designed to help identify any adverse health effects. Health studies also define those health problems that require further inquiry by means of, for example, a health surveillance or epidemiologic study.

Health professionals' education—The process of increasing the knowledge, skill, and behavior of health professionals about medical surveillance, screening, and methods of diagnosing, treating, and preventing injury or disease (e.g., that related to exposure to hazardous substances).

Health statistics review—Evaluation of information and relevant health outcome data for an involved population, including reports of injury, disease, or death in the community. Databases can include morbidity and mortality data, tumor and disease registries, birth statistics, and surveillance data.

Health surveillance—The periodic medical screening of a defined population for a specific disease or for biologic markers of disease for which the population is, or is thought to be, at significantly increased risk.

Herbicide—A chemical that kills weeds and other plants.

Hypersensitivity—A greater than normal bodily response to a foreign agent.

Hypochondriasis—Persistent preoccupation with the fear of having a serious disease despite medical reassurance of the absence of disease.

Incidence—The rate of development of disease in a population that can be expressed as either incidence density or cumulative incidence. As Dicker (1996) noted, prevalence refers to existing cases of a health condition in a population, and incidence refers to new cases.

Indeterminate Public Health Hazard—Characterization of sites for which no conclusions about public health hazard can be made because data are lacking.

Ingestion—Taking food or drink into the body. Chemicals can get in or on food, drink, utensils, cigarettes, or hands from which they can be ingested.

Inhalation—Breathing. Exposure can occur from inhaling contaminants because they can be deposited in the lungs, taken into the blood, or both.

Inhibition—The reduction of a substance's toxicity by the addition of another substance that may or may not have a toxic effect of its own (see Antagonism).

Insecticide—An agent that kills insects.

Interaction—An outcome that occurs when exposure to two or more chemicals results in a qualitatively or quantitatively altered biologic response than that predicted from the actions of the components administered separately.

In utero—Within the womb; not yet born.

In vitro—In an artificial environment, as in a test tube or culture medium.

In vivo—In the living body.

Leukemia—Cancer of the blood-forming tissues.

Lipid—"Fat-soluble," denoting substances extracted from animal or vegetable cells by nonpolar or "fat" solvents (Hensyl, 1987).

Lipophilic—Capable of being absorbed in lipids.

Lupus—A term originally used to depict erosion of the skin, now used with modifying terms designating various diseases; for example, systemic lupus erythematosus is an inflammatory connective tissue disease with variable features, frequently including fever, weakness and fatigue, joint pains, diffuse erythematosus skin lesions, and other evidence of autoimmune phenomenon (Hensyl, 1987).

Media—Soil, water, air, plants, animals, or any other parts of the environment that can contain contaminants.

Medical monitoring—The periodic medical testing to screen people who are at significant increased threat of disease.

Metabolism—The sum of chemical changes occurring in tissue. For example, food is *metabolized* (chemically changed) to supply the body with energy. Chemicals can be *metabolized* and made either more or less harmful by the body.

Metabolite—Any product of metabolism.

Microgram (µg)—One one millionth of a gram.

Milligram (mg)—One one thousandth of a gram.

Minimal risk level (MRL)—An estimate of daily human exposure to a substance that is likely to be without an appreciable risk of adverse effects (noncancer) over a specified duration of exposure. MRLs are derived when reliable and sufficient data exist to identify the target organ(s) of effect or the most sensitive health effect(s) for a specific duration via a given route of exposure. MRLs are based on noncancer health effects only. MRLs can be derived for acute, intermediate, and chronic duration exposures by the inhalation and oral routes.

Mitosis—The usual process of cell reproduction that results in the formation of two daughter cells with exactly the same chromosome and DNA content as that of the original cell (Hensyl, 1987).

Mixture—Any set of two or more chemical substances, regardless of their sources, that may jointly contribute to toxicity in the target population.

Morbidity—Illness or disease.

Morbidity rate—The number of illnesses or cases of disease in a population.

Mortality—The condition of being mortal; death.

National exposure registry—An ATSDR exposure registry system for collecting and maintaining in a structured record, information on persons with documented environmental exposure(s).

National Priorities List (NPL)—The Environmental Protection Agency's (EPA) listing of sites that have undergone preliminary assessment and site inspection to determine which locations pose immediate threat to persons living or working near the release.

Necrosis—Pathologic death of one or more cells, or of a portion of tissue or organ, resulting from irreversible damage (Hensyl, 1987).

No Apparent Public Health Hazard—Characterization of sites where human exposure to contaminated media is occurring or has occurred in the past, but the exposure is below the level of a health hazard.

No Public Health Hazard—Characterization of sites for which data indicate no current or past exposure or no potential for exposure and therefore no health hazard.

Peer review—Evaluation of the accuracy or validity of technical data, observations, and interpretation by qualified experts in an organized group process (CRARM, 1997).

Percentile—Any of the values in a series dividing the distribution of the individuals in the series into one hundred groups of equal frequency.

Petitioned public health assessment—A public health assessment conducted at the request of a member of the public.

Phenotype—A category or group in genetic terminology to which an individual may be assigned on the basis of one or more clinically or laboratory observed characteristic that reflects genetic variation or gene-environment interaction (adapted from Hensyl, 1987).

Picogram (pg)—One one trillionth of a gram.

Pilot health study—Any investigation of exposed individuals, using epidemiologic methods, which would help determine exposures or possible public health impacts by defining health problems that require further investigation through epidemiologic studies, environmental monitoring or sampling, surveillance, or registries.

Pleurisy—Pleuritis; inflammation of the pleura, which is the serous membrane enveloping the lungs and lining the walls of the pleural cavity (Hensyl, 1987).

Plume—An area of chemicals in a particular medium, such as air or groundwater, that moves away from its source in a long band or column. For example, a plume can be a column of smoke from a chimney or chemicals moving with groundwater.

Potentially responsible parties—Persons or organizations liable under CERCLA for the costs of remediating NPL sites.

Potentiation—A special case of synergism in which one substance does not have a toxic effect on a certain organ or system, but, when added to another chemical, it makes the latter much more toxic.

Precautionary principle—Decisions about the best ways to manage or reduce risks that reflect a preference for avoiding unnecessary health risks instead of unnecessary economic expenditures when information about potential risks is incomplete (CRARM, 1997).

Prevalence—The proportion of ill persons in a population at a point in time, expressed as a simple percentage. As Dicker noted (1996), prevalence refers to existing cases of a health condition in a population, and incidence refers to new cases.

Primary prevention—The prevention of an adverse health effect in an individual or population through marked reduction or elimination of the hazards known to cause the health effects.

Psychometrics—A method psychologists use to quantitatively measure and depict persons' feelings and attitudes.

Psychophysics—The branch of psychology dealing with the fundamental relations between the mind and physical phenomena.

Public comment—Invited comment from the general public on agency findings or proposed activities.

Public health action—An activity designed to prevent exposures and/or to mitigate or prevent adverse health effects in populations that live near hazardous waste sites or releases.

Public health advisory—A statement by the ATSDR or other health agency that a given release of hazardous substances poses a significant risk to human health. The statement also recommends measures designed to reduce exposure and eliminate or substantially mitigate the substantial hazard to human health.

Public health assessment—An evaluation of data and information on the release of hazardous substances into the environment to assess any current or future impact on public health, develop health advisories or other recommendations, and identify studies or actions needed to evaluate and mitigate or prevent human health effects; also, the document resulting from that evaluation.

Public health hazard—Characterization of sites that pose public health hazards as the result of long-term exposure to hazardous substances.

Public health statement—The first chapter of an ATSDR Toxicological Profile, written in nontechnical language, that summarizes the toxicity and human health effects of the specific hazardous substance being profiled.

Pulmonary—Pertaining to the lungs.

Quantitative structure activity relationships (QSAR)—The relationship between the properties (physical and/or chemical) of substances and their ability to cause particular effects, enter into particular reactions, etc. The goal of QSAR studies in toxicology is to develop procedures whereby the toxicity of a compound can be predicated from its chemical structure by analogy with the known toxic properties of other toxicants of similar structure (adapted from Hodgson and Levi, 1987).

Random samples—Samples selected from a statistical population so that each sample has an equal probability of being selected (RiskFocus*, 1990).

Range—The arithmetic difference between the largest and smallest values in a data set.

Raynaud's Phenomenon—A condition of blanching, numbness, and tingling in the fingers.

Record of decision—An EPA document that discusses the various cleanup techniques that were considered for a site and an explanation of why a particular course of action was selected (EPA, 1992c).

Registry—A system for collecting and maintaining, in a structured record, information on specific persons from a defined population.

Residual risk—The health risk remaining after risk-reduction actions are implemented, such as risks associated with sources of air pollution that remain after maximum achievable control technology has been applied (CRARM, 1997).

Risk—The probability that an exposure will result in harm or loss.

Risk assessment—The use of available information to evaluate and estimate exposure to a substance(s) and its consequent adverse health effects (DHHS, 1985). The four steps are hazard identification, dose-response assessment, exposure assessment, and risk characterization.

Risk communication—An interactive process of exchange of information and opinion among individuals, groups, and institutions (NRC, 1989c).

Route of exposure—The means by which a person may contact a chemical substance. For example, drinking (ingestion) and bathing (skin contact) are two different *routes of exposure* to contaminants that may be found in water.

Screening—A method for identifying asymptomatic individuals as likely, or unlikely, to have a particular health problem.

Screening program—A program of screening for a health problem, diagnostic evaluation of persons who have positive screening-test results, and treatment for persons in whom the health problem is diagnosed.

Secondary prevention—The prevention or slowing of the progression of a health problem attributable to specific hazards through use of education, protective equipment, relocation away from the hazards or other means to avoid contact with the hazard.

Significant health risk—Risk identified in which persons are being or could be exposed to hazardous substances at levels that pose an urgent public health hazard or a public health hazard. Public health advisories are usually issued when urgent public health hazards have been identified.

Sister chromatid exchange (SCE)—A symmetrical exchange of chromosome material that occurs between each of the sister chromatids that together constitute a chromosome. SCE is susceptible to chemical induction and appears to be correlated with the genotoxic potential of chemicals (adapted from Hodgson and Levi, 1987).

Site-specific surveillance—Epidemiologic surveillance activity designed to assess the specific occurrence of one or more defined health conditions among a specific population potentially exposed to hazardous substances in the environment.

Soluble—Dissolves well in liquid.

Solvent—A substance that dissolves another substance.

Stakeholder—An individual or group that has an interest in or will be affected by an action.

Statistical significance—A calculated value that infers the probability whether an observed difference in quantities being measured could be due to variability in the data rather than an actual difference in the quantities themselves.

Stressor—A chemical, material, organism, radiation, noise, temperature change or activity that stresses an organism's health or well-being.

Substance-specific applied research—A program of research designed to fill key data needs for specific priority hazardous substances.

Superfund—Another name for the Comprehensive Environmental Response, Compensation, and Liability Act of 1980 (CERCLA). The term is also used to refer to the Hazard Substance Superfund, the trust fund established by CERCLA.

Superfund Amendments and Reauthorization Act (SARA)—The 1986 legislation that updated CERCLA and broadened the ATSDR's responsibilities in the areas of public health assessments, establishment and maintenance of toxicologic databases, information dissemination, and medical education.

Surveillance—A data-collection system that monitors the occurrence of disease (disease surveillance) or the distribution of hazard (hazard surveillance).

Synergism—A response to a mixture of toxic chemicals that is greater than that suggested by the component toxicities.

Teratogen—An agent that causes abnormal development.

Teratology—The branch of embryology concerned with the production, development, anatomy, and classification of malformed fetuses (Hensyl, 1987).

Tonne—1 tonne equals 1,000 kilograms.

Toxicant—A substance not produced by a living organism that causes a harmful effect when administered to a living organism. See Toxin.

Toxicity—The property of chemicals that causes adverse effects on living organisms.

Toxicokinetics—Toxicodynamics; the study of the quantitative relationship between absorption, distribution, and excretion of toxicants and their metabolites (Hodgson and Levi, 1987).

Toxicological profile—A document prepared by the ATSDR that describes the salient toxicologic properties and human health effects of priority hazardous substances.

Toxicology—The science that deals with poisons (toxicants) and their effects (Hodgson and Levi, 1987).

Toxics Release Inventory—The EPA-maintained database of substances reported as released into the environment by industrial operations.

Toxin—A toxicant produced by a living organism (Hodgson and Levi, 1987).

Tumor—An abnormal mass of tissue.

Urgent Public Health Hazard—Characterization of sites that pose serious risks to the public health as a result of short-term exposures to hazardous substances.

Volatile organic compounds (VOCs)—Substances containing carbon and different proportions of other elements such as hydrogen, oxygen, fluorine, chlorine, bromine, sulfur, or nitrogen. These substances easily become vapors or gases. Many *VOCs* are commonly used as solvents (paint thinners, lacquer thinner, degreasers, and dry cleaning fluids).

Weight of evidence—A systemic method for applying biomedical judgment to empirical observations and mechanistic considerations to qualitatively assess the potential toxicity of a substance, singly or in a chemical mixture, for a given target organ or system.

Index